Measure Theory and Integration

Measure Theory and Integration

Michael E. Taylor

Graduate Studies in Mathematics

Volume 76

American Mathematical Society
Providence, Rhode Island

2000 *Mathematics Subject Classification.* Primary 28–01.

For additional information and updates on this book, visit
www.ams.org/bookpages/gsm-76

Library of Congress Cataloging-in-Publication Data

Taylor, Michael Eugene, 1946–
Measure theory and integration / Michael E. Taylor.
p. cm. — (Graduate studies in mathematics, ISSN 1065-7339 ; v. 76)
Includes bibliographical references.
ISBN-13: 978-0-8218-4180-8
1. Measure theory. 2. Riemann integrals. 3. Convergence. 4. Probabilities. I. Title. II. Series.

QA312.T387 2006
515′.42—dc22 2006045635

∞ The paper used in this book is acid-free and falls within the guidelines
established to ensure permanence and durability.
Visit the AMS home page at http://www.ams.org/

10 9 8 7 6 5 4 3 2 1 11 10 09 08 07 06

Contents

Introduction

This monograph is based on a one-semester course on measure theory, which I have taught several times. The prerequisite for the course is an introductory analysis course, covering such matters as metric spaces, uniform convergence of functions, the contraction mapping principle, and aspects of multi-variable calculus, including the inverse function theorem. For the convenience of the reader, some of this material is briefly treated in some appendices.

The core topic for the course treated here is the theory of measure and integration, associated especially with the work of H. Lebesgue, though of course many other mathematicians have contributed to this central subject. We mention particularly E. Borel, M. Riesz, J. Radon, M. Frechet, G. Fubini, C. Caratheodory, F. Hausdorff, and A. Besicovitch, among the classical founders.

We begin with an introductory chapter on the Riemann integral, for functions defined on an interval $[a, b]$ in $\mathbb{R}$. We develop some of the properties of the Riemann integral, including a proof of the Fundamental Theorem of Calculus. We see that while continuous functions are Riemann integrable, some very reasonable-looking functions are not.

In Chapter 2 we construct Lebesgue measure on $\mathbb{R}$. We emphasize that the key difference between Lebesgue measure of $S \subset \mathbb{R}$ and the "content" of S, arising from the Riemann integral, is that the content is approximated by taking finite coverings of S by intervals, while the Lebesgue measure is approximated by taking *countable* (perhaps infinite) coverings of S by intervals.

In Chapter 3 we define the Lebesgue integral and establish some basic properties, such as the Monotone Convergence Theorem and the Lebesgue

Dominated Convergence Theorem. We integrate measurable functions defined on general measure spaces. Though at this point we have only constructed Lebesgue measure on $\mathbb{R}$, the basic theory of integration is not more complicated on general measure spaces, and pursuing it helps clarify what one should do to construct more general measures. In Chapter 4 we introduce L^p spaces, consisting of measurable functions f such that $|f|^p$ is integrable or, more precisely, of equivalence classes of such functions, where we say $f_1 \sim f_2$ provided these functions differ only on a set of measure zero. If μ is a measure on a space X, we study $L^p(X, \mu)$ as a Banach space, for $1 \le p \le \infty$, and in particular we study $L^2(X, \mu)$ as a Hilbert space. We develop some Hilbert space theory and apply it to establish the Radon-Nikodym Theorem, comparing two measures μ and ν when ν is "absolutely continuous" with respect to μ.

Constructing measures other than Lebesgue measure on $\mathbb{R}$ is an important part of measure theory, and we begin this task in Chapter 5, giving some useful general methods, especially due to Caratheodory, for making such constructions, establishing that they are measures, and identifying certain types of sets as measurable. The first concrete application of this is made in Chapter 6, in the construction of the "product measure" on $X \times Y$, when X has a measure μ and Y has the measure ν. The integral with respect to the product measure $\mu \times \nu$ is compared to "iterated integrals," in theorems of Fubini and Tonelli.

In Chapter 7 we construct Lebesgue measure on $\mathbb{R}^n$, for $n > 1$, as a product measure. We study how the Lebesgue integral on $\mathbb{R}^n$ transforms under an invertible linear transformation on $\mathbb{R}^n$ and, more generally, under a C^1 diffeomorphism. We go a bit further, considering transformation via a Lipschitz homeomorphism, and we establish a result under the hypothesis that the transformation is differentiable almost everywhere, a property that will be studied further in Chapter 11. We extend the scope of the n-dimensional integral in another direction, constructing surface measure on an n-dimensional surface M in $\mathbb{R}^m$. This is done in terms of the Riemann metric tensor induced on M. We go further and discuss integration on more general Riemannian manifolds. This central chapter contains a larger number of exercises than the others, divided into several sets of exercises. After the first set, of a nature parallel to exercise sets for other chapters, there is a set relating the Riemann and Lebesgue integrals on $\mathbb{R}^n$, extending the previous discussion of the relationship between the material of Chapter 1 and that of Chapters 2–3, an exercise set on determinants, and an exercise set on row reduction on matrix products, providing linear algebra background for the proof of the change of variable formula. There is also an exercise set on the connectivity of $Gl(n, \mathbb{R})$ and a set on integration on certain matrix groups.

In Chapter 8 we discuss "signed measures," which differ from the measures considered up to that point only in that they can take negative as well as positive values. The key result established there is the Hahn decomposition, $\nu = \nu^+ - \nu^-$, for a signed measure ν, on X, where ν^+ and ν^- are positive measures, with disjoint supports. This allows us to extend the Radon-Nikodym Theorem to the case of a signed measure ν, absolutely continuous with respect to a (positive) measure μ. We consider a further extension, to complex measures, though this is completely routine.

In Chapter 9 we again take up the study of L^p spaces and pursue it a bit further. We identify the *dual* of the Banach space $L^p(X, \mu)$ with $L^q(X, \mu)$, where $1 \le p < \infty$ and $1/p + 1/q = 1$, making use of the Radon-Nikodym Theorem for signed measures as a tool in the demonstration. We study some integral operators, including convolution operators, amongst others, and derive operator bounds on L^p spaces. We consider the Fourier transform $\mathcal{F}$ and prove that $\mathcal{F}$ is an invertible norm-preserving operator on $L^2(\mathbb{R}^n)$. Some mention of Fourier series was made in Chapter 4, particularly in the exercises.

Chapter 10 discusses Sobolev spaces, $H^{k,p}(\mathbb{R}^n)$, consisting of functions whose derivatives of order $\le k$, defined in a suitable weak sense, belong to $L^p(\mathbb{R}^n)$. Certain Sobolev spaces are shown to consist entirely of bounded continuous functions, even Hölder continuous, on $\mathbb{R}^n$. This subject is of great use in the study of partial differential equations, though such applications are not made here. The most significant application we make of Sobolev space theory in these notes appears in the following chapter. We mention that in Chapter 10 certain results on the weak derivative depend on the Fundamental Theorem of Calculus for the Riemann integral; it is for this reason that we included a proof of this result in Chapter 1.

Chapter 11 deals with various results in the area of almost-everywhere convergence. The basic result, called Lebesgue's differentiation theorem, is that a function $f \in L^1(\mathbb{R}^n)$ is equal for almost all $x \in \mathbb{R}^n$ to the limit of its averages over balls of radius r, centered at x, as $r \to 0$. Under a slightly stronger condition on x, we say x is a Lebesgue point for f; more generally there is the notion of an L^p-Lebesgue point, and one shows that if $f \in L^p(\mathbb{R}^n)$, then almost every $x \in \mathbb{R}^n$ is an L^p-Lebesgue point for f, provided $1 \le p < \infty$. We use the Hardy-Littlewood maximal function as a tool to establish these results. This area also requires a "covering lemma," allowing one to select from a collection of sets covering S a subcollection with desirable properties.

Another important result established in Chapter 11 is Rademacher's Theorem that a Lipschitz function on $\mathbb{R}^n$ is differentiable almost everywhere. We get this as a corollary of the stronger result that if $f \in H^{1,p}(\mathbb{R}^n)$ and $p > n$ (or $p = n = 1$), then f is differentiable almost everywhere; in fact, f is differentiable at every L^p-Lebesgue point of the weak derivative ∇f. With this result, we can complete the demonstration of the result from Chapter 7 that the change of variable formula for the integral extends to Lipschitz homeomorphisms. Making use of Rademacher's Theorem, we show that a Lipschitz function can be altered off a small set to yield a C^1 function. These results are important in the study of Lipschitz surfaces in $\mathbb{R}^n$.

The covering lemma we use for the results of Chapter 11 mentioned above is Weiner's Covering Lemma. We also discuss covering lemmas of Vitali and of Besicovitch and show how Besicovitch's result leads to extensions of the Lebesgue differentiation theorem.

In Chapter 12 we construct r-dimensional Hausdorff measure $\mathcal{H}^r$, on a separable metric space X, for any $r \in [0, \infty)$. In case S is a Borel set in $\mathbb{R}^n$, one has $\mathcal{H}^n(S) = \mathcal{L}^n(S)$, Lebesgue measure; this is not a straightforward consequence of the definition (unless $n = 1$), and its proof requires some effort. In particular, the proof requires a covering lemma. We extend this analysis to show that n-dimensional Hausdorff measure on an n-dimensional manifold with a continuous metric tensor coincides with the volume measure constructed in Chapter 7. This applies to C^1 surfaces in Euclidean space. We go further and study Hausdorff measure on Lipschitz surfaces in Euclidean space. Results on Lipschitz functions established in Chapter 11 are invaluable here. These provide basic results in "geometric measure theory." There is a great deal more to geometric measure theory, which has been developed as a tool in the study of minimal surfaces, amongst other applications. A good overview can be found in [**Mor**], a reading of which should prepare one for the treatise [**Fed**].

While the bulk of Chapter 12 deals with $\mathcal{H}^r$ when r is an integer, there are wild sets, some of incredible beauty, called "fractals," for which $\mathcal{H}^r$ is germane for a value $r \notin \mathbb{Z}$. We touch this only briefly; one can consult [**Ed**], [**Mdb**], [**Fal**], and [**PR**] for more material on fractals.

In Chapter 13 we show how a positive linear functional on $C(X)$, the space of continuous functions on X, gives rise to a (positive) measure on X, when X is a compact metric space, and how a bounded linear functional on $C(X)$ gives rise to a signed measure on X. Out of this come compactness results for bounded sets of measures. These results extend to the case where X is a general compact Hausdorff space; treatments of this can be found in several places, including [**Fol**] and [**Ru**]. The argument is somewhat simpler when X is metrizable; in particular, we can appeal to results from Chapter 5 for a lot of the technical work.

Chapters 14–17 explore connections between measure theory and probability theory. The basic connection is that a probability measure is a positive measure of total mass 1. Chapter 14 treats ergodic theory, which deals with statistical properties of iterates of a measure-preserving map φ on a probability space $(X, \mathfrak{F}, \mu)$. In particular, one studies the map $Tf(x) = f(\varphi(x))$ on L^p spaces and the means $A_k f = (1/k) \sum_{j=0}^{k-1} T^j f(x)$. Mean ergodic theorems and Birkhoff's Ergodic Theorem treat L^p-norm behavior and pointwise a.e. behavior of $A_k f(x)$, tending to a limit $Pf(x)$ as $k \to \infty$. Ergodic transformations are those for which such limits are constant. Knowing that certain transformations are ergodic can provide valuable information, as we will see.

Chapter 15 discusses some of the fundamental results of probability theory, dealing with random variables (i.e., measurable functions) on a probability space. These results include weak and strong laws of large numbers, whose basic message is that means of a large number of independent random variables of the same type (i.e., with the same probability distribution) tend to constant limits, with probability one. We approach the strong law via Birkhoff's Ergodic Theorem. We also treat the Central Limit Theorem, giving conditions under which such means have approximately Gaussian probability distributions.

In Chapter 16, we construct Wiener measure on the set of continuous paths in $\mathbb{R}^n$, describing the probabilistic behavior of a particle undergoing Brownian motion. We begin with a probability measure W on a countable product $\mathfrak{P}$ of compactifications of $\mathbb{R}^n$, first defining a positive linear functional on $C(\mathfrak{P})$ and then getting the measure via the results of Chapter 13. The index set for the countable product is $\mathbb{Q}^+$, the set of rational numbers ≥ 0. The space of continuous paths is naturally identified with a subset $\mathfrak{P}_0$, shown to be a Borel subset of $\mathfrak{P}$, of W-measure one. To illustrate how fuzzy Brownian paths are, we show that when $n \geq 2$, almost all paths have Hausdorff dimension 2.

There is a further structure associated with Wiener measure on path space, namely a filtered family of σ-algebras. Certain families of functions $\mathfrak{f}_t$ on $\mathfrak{P}_0$ are "martingales," i.e., $\mathfrak{f}_t$ is obtained from $\mathfrak{f}_s$ by taking the "conditional expectation," when $t < s$. This is discussed in Chapter 17. We define martingales more generally, prove the Martingale Maximal Inequality, and apply this to a number of convergence results for martingales, obtaining both a variant of the Lebesgue differentiation theorem and another proof of the strong law of large numbers. We also produce several martingales associated with Brownian motion and apply the martingale maximal inequality to these.

This book has several appendices, some providing background material, as mentioned above, and others providing supplementary material. Appendix A contains some basic material on metric spaces, topological spaces, and compactness. In particular, we prove the Stone-Weierstrass Theorem, which gives a very useful sufficient condition for a set $\mathcal{A}$ of continuous functions on a compact space X to be dense in the space of all continuous functions $C(X)$. This result is used in a number of places, including Chapters 4, 6, and 16. In Appendix B we give some basic results in multi-variable differential calculus, including the Inverse Function Theorem, the nature of a diffeomorphism, and the concept of a manifold. This material is useful for appreciating change of variable formulas for integrals and also the results on integration on surfaces and more general manifolds in Chapters 7 and 12.

Appendix C is devoted to the Whitney Extension Theorem, needed for the approximation theorem for Lipschitz functions in Chapter 11. Appendix D treats the Marcinkiewicz Interpolation Theorem, giving L^p estimates on an operator satisfying "weak type" estimates on L^q and L^r, with $q < p < r$. Appendix E discusses Sard's Theorem, that the set of critical values of a C^1 map $F : \mathbb{R}^n \to \mathbb{R}^n$ has measure zero. This result is applied in Appendix F, to help prove a change of variable theorem for a C^1 map of $\mathbb{R}^n$ not assumed to be a diffeomorphism.

In Appendix G we discuss the elements of the theory of differential forms and their integration. These results have many applications to problems in differential equations, differential geometry, and topology. A key result is a general Gauss-Green-Stokes formula. To illustrate the power of this formula, we show how it leads to a short proof of the famous Brouwer Fixed-Point Theorem. In Appendix H we apply the differential form results to obtain another approach to the change of variable formula, a modification of an approach put forward by P. Lax [**La**]. Finally in Appendix I we extend the Gauss-Green-Stokes formula from the setting of smoothly bounded domains considered in Appendix G to the setting of Lipschitz domains.

There are several ways to use this monograph in a course. Chapters 1–4 provide a quick introduction to the basics of Lebesgue integration. I have used this material at the end of analysis courses that precede a full-blown course in measure theory. For a one quarter course on measure theory, Chapters 1–9 would provide a solid background in the subject. For a semester course, one could deepen this background with a selection of material from Chapters 10–17 and the appendices.

Each of the seventeen chapters in this monograph ends with a set of exercises. These form an integral part of our presentation, and thinking about them should sharpen the reader's understanding of the material. On occasion, the results of some of the exercises are used in the development of subsequent material.

ACKNOWLEDGMENTS. Thanks to my friends, at the University of North Carolina and elsewhere, for stimulation and encouragement during the preparation of this monograph. Thanks particularly to Jane Hawkins for enlightenment in ergodic theory, to Mark Pinsky for helpful comments on the material in Chapters 15–17, and to an anonymous reviewer for numerous suggestions for improvements. During the course of writing this book, my research has been supported by NSF grants, including most recently NSF Grant #0456861.

Michael E. Taylor

Chapter 1

The Riemann Integral

The Riemann integral is a fundamental part of calculus and an essential precursor to the Lebesgue integral. In this chapter we define the Riemann integral of a bounded function on an interval $I = [a, b]$ on the real line. To do this, we partition I into smaller intervals. A *partition* $\mathcal{P}$ of I is a finite collection of subintervals $\{J_k : 0 \leq k \leq N\}$, disjoint except for their endpoints, whose union is I. We can order the J_k so that $J_k = [x_k, x_{k+1}]$, where

$$x_0 < x_1 < \cdots < x_N < x_{N+1}, \quad x_0 = a, \ x_{N+1} = b. \tag{1.1}$$

We call the points x_k the *endpoints* of $\mathcal{P}$. We set

$$\begin{aligned} \ell(J_k) = x_{k+1} - x_k, \quad \operatorname{maxsize}(\mathcal{P}) &= \max_{0 \leq k \leq N} \ell(J_k), \\ \operatorname{minsize}(\mathcal{P}) &= \min_{0 \leq k \leq N} \ell(J_k). \end{aligned} \tag{1.2}$$

We then set

$$\begin{aligned} \overline{I}_{\mathcal{P}}(f) &= \sum_k \sup_{J_k} f(x)\, \ell(J_k), \\ \underline{I}_{\mathcal{P}}(f) &= \sum_k \inf_{J_k} f(x)\, \ell(J_k). \end{aligned} \tag{1.3}$$

Note that $\underline{I}_{\mathcal{P}}(f) \leq \overline{I}_{\mathcal{P}}(f)$. These quantities should approximate the Riemann integral of f, if the partition $\mathcal{P}$ is sufficiently "fine."

To be more precise, if $\mathcal{P}$ and $\mathcal{Q}$ are two partitions of I, we say $\mathcal{P}$ *refines* $\mathcal{Q}$, and we write $\mathcal{P} \succ \mathcal{Q}$, if $\mathcal{P}$ is formed by partitioning each interval in $\mathcal{Q}$. Equivalently, $\mathcal{P} \succ \mathcal{Q}$ if and only if all the endpoints of $\mathcal{Q}$ are also endpoints of $\mathcal{P}$. It is easy to see that any two partitions have a common refinement; just take the union of their endpoints to form a new partition. Note also that

$$\mathcal{P} \succ \mathcal{Q} \Longrightarrow \overline{I}_{\mathcal{P}}(f) \le \overline{I}_{\mathcal{Q}}(f) \text{ and } \underline{I}_{\mathcal{P}}(f) \ge \underline{I}_{\mathcal{Q}}(f). \tag{1.4}$$

Consequently, if $\mathcal{P}_j$ are any two partitions and $\mathcal{Q}$ is a common refinement, we have

$$\underline{I}_{\mathcal{P}_1}(f) \le \underline{I}_{\mathcal{Q}}(f) \le \overline{I}_{\mathcal{Q}}(f) \le \overline{I}_{\mathcal{P}_2}(f). \tag{1.5}$$

Now, whenever $f : I \to \mathbb{R}$ is bounded, the following quantities are well defined:

$$\overline{I}(f) = \inf_{\mathcal{P}\in\Pi(I)} \overline{I}_{\mathcal{P}}(f), \quad \underline{I}(f) = \sup_{\mathcal{P}\in\Pi(I)} \underline{I}_{\mathcal{P}}(f), \tag{1.6}$$

where $\Pi(I)$ is the set of all partitions of I. Clearly, by (1.5), $\underline{I}(f) \le \overline{I}(f)$. We then say that f is *Riemann integrable* provided $\overline{I}(f) = \underline{I}(f)$, and in such a case, we set

$$\int_I f(x)\,dx = \overline{I}(f) = \underline{I}(f). \tag{1.7}$$

We will denote the set of Riemann integrable functions on I by $\mathcal{R}(I)$.

We derive some basic properties of the Riemann integral.

Proposition 1.1. *If $f,\ g \in \mathcal{R}(I)$, then $f+g \in \mathcal{R}(I)$, and*

$$\int_I (f+g)\,dx = \int_I f\,dx + \int_I g\,dx. \tag{1.8}$$

Proof. If J_k is any subinterval of I, then

$$\sup_{J_k} (f+g) \le \sup_{J_k} f + \sup_{J_k} g,$$

so, for any partition $\mathcal{P}$, we have $\overline{I}_{\mathcal{P}}(f+g) \le \overline{I}_{\mathcal{P}}(f) + \overline{I}_{\mathcal{P}}(g)$. Also, using common refinements, we can *simultaneously* approximate $\overline{I}(f)$ and $\overline{I}(g)$ by $\overline{I}_{\mathcal{P}}(f)$ and $\overline{I}_{\mathcal{P}}(g)$. Thus the characterization (1.6) implies $\overline{I}(f+g) \le \overline{I}(f) + \overline{I}(g)$. A parallel argument implies $\underline{I}(f+g) \ge \underline{I}(f) + \underline{I}(g)$, and the proposition follows.

Next, there is a fair supply of Riemann integrable functions.

Proposition 1.2. *If f is continuous on I, then f is Riemann integrable.*

Proof. Any continuous function on a compact interval is uniformly continuous; let $\omega(\delta)$ be a modulus of continuity for f, so

$$|x - y| \le \delta \Longrightarrow |f(x) - f(y)| \le \omega(\delta), \quad \omega(\delta) \to 0 \text{ as } \delta \to 0. \tag{1.9}$$

Then

$$\text{maxsize}\,(\mathcal{P}) \le \delta \Longrightarrow \overline{I}_{\mathcal{P}}(f) - \underline{I}_{\mathcal{P}}(f) \le \omega(\delta) \cdot \ell(I), \tag{1.10}$$

which yields the proposition.

This argument, showing that every continuous function is Riemann integrable, also provides a criterion on a partition $\mathcal{P}$ guaranteeing that $\overline{I}_{\mathcal{P}}(f)$ and $\underline{I}_{\mathcal{P}}(f)$ are close to $\int_I f\,dx$, when f is continuous. The following is a useful extension. Let $f \in \mathcal{R}(I)$, take $\varepsilon > 0$, and let $\mathcal{P}_0$ be a partition such that

$$\overline{I}_{\mathcal{P}_0}(f) - \varepsilon \le \int_I f\,dx \le \underline{I}_{\mathcal{P}_0}(f) + \varepsilon. \tag{1.11}$$

Let

$$M = \sup_I |f(x)|, \quad \delta = \text{minsize}(\mathcal{P}_0). \tag{1.12}$$

Proposition 1.3. *Under the hypotheses above, if $\mathcal{P}$ is any partition of I satisfying*

$$\text{maxsize}\,(\mathcal{P}) \le \frac{\delta}{k}, \tag{1.13}$$

then

$$\overline{I}_{\mathcal{P}}(f) - \varepsilon_1 \le \int_I f\,dx \le \underline{I}_{\mathcal{P}}(f) + \varepsilon_1, \quad \textit{with } \varepsilon_1 = \varepsilon + \frac{2M}{k}\ell(I). \tag{1.14}$$

Proof. Consider on the one hand those intervals in $\mathcal{P}$ that are contained in intervals in $\mathcal{P}_0$, and on the other hand those intervals in $\mathcal{P}$ that are *not* contained in intervals in $\mathcal{P}_0$ (whose lengths sum to $\le \ell(I)/k$). Let $\mathcal{P}_1$ be the minimal common refinement of $\mathcal{P}$ and $\mathcal{P}_0$. We obtain

$$\overline{I}_{\mathcal{P}}(f) \le \overline{I}_{\mathcal{P}_1}(f) + \frac{2M}{k}\ell(I), \quad \underline{I}_{\mathcal{P}}(f) \ge \underline{I}_{\mathcal{P}_1}(f) - \frac{2M}{k}\ell(I).$$

Since also $\overline{I}_{\mathcal{P}_1}(f) \le \overline{I}_{\mathcal{P}_0}(f)$ and $\underline{I}_{\mathcal{P}_1}(f) \ge \underline{I}_{\mathcal{P}_0}(f)$, this implies (1.14).

The following corollary is sometimes called Darboux's Theorem.

Corollary 1.4. *Let $\mathcal{P}_\nu$ be any sequence of partitions of I into ν intervals $J_{\nu k}$, $1 \le k \le \nu$, such that*

$$\operatorname{maxsize}(\mathcal{P}_\nu) = \delta_\nu \to 0,$$

and let $\xi_{\nu k}$ be any choice of one point in each interval $J_{\nu k}$ of the partition $\mathcal{P}_\nu$. Then, whenever $f \in \mathcal{R}(I)$,

$$\int_I f(x)\,dx = \lim_{\nu\to\infty} \sum_{k=1}^{\nu} f(\xi_{\nu k})\,\ell(J_{\nu k}). \tag{1.15}$$

The sum on the right side of (1.15) is called a Riemann sum. One should be warned that, once such a specific choice of $\mathcal{P}_\nu$ and $\xi_{\nu k}$ has been made, the limit on the right side of (1.15) might exist for a bounded function f that is *not* Riemann integrable. This and other phenomena are illustrated by the following example of a function which is not Riemann integrable. For $x \in I$, set

$$\vartheta(x) = 1 \ \text{ if } \ x \in \mathbb{Q}, \quad \vartheta(x) = 0 \ \text{ if } \ x \notin \mathbb{Q}, \tag{1.16}$$

where $\mathbb{Q}$ is the set of *rational* numbers. Now every interval $J \subset I$ of positive length contains points in $\mathbb{Q}$ and points not in $\mathbb{Q}$, so for any partition $\mathcal{P}$ of I we have $\overline{I}_{\mathcal{P}}(\vartheta) = \ell(I)$ and $\underline{I}_{\mathcal{P}}(\vartheta) = 0$, and hence

$$\overline{I}(\vartheta) = \ell(I), \quad \underline{I}(\vartheta) = 0. \tag{1.17}$$

Note that, if $\mathcal{P}_\nu$ is a partition of I into ν equal subintervals, then we could pick each $\xi_{\nu k}$ to be rational, in which case the limit on the right side of (1.15) would be $\ell(I)$, or we could pick each $\xi_{\nu k}$ to be irrational, in which case this limit would be zero. Alternatively, we could pick half of them to be rational and half to be irrational, and the limit would be $\ell(I)/2$.

Let $f_k \in \mathcal{R}(I)$ be a uniformly bounded, monotonically increasing sequence of functions. Then there is a bounded function f on I such that, as $k \to \infty$,

$$f_k(x) \nearrow f(x), \quad \forall x \in I. \tag{1.18}$$

It would be desirable to conclude that f is integrable and

$$\int_I f_k(x)\,dx \to \int_I f(x)\,dx. \tag{1.19}$$

A shortcoming of the Riemann integral is that such a limit might not belong to $\mathcal{R}(I)$. For example, since $I \cap \mathbb{Q}$ is countable, let $I \cap \mathbb{Q} = \{c_1, c_2, c_3, \dots\}$, and let

$$\vartheta_k(x) = 1 \;\text{ if }\; x \in \{c_1, \dots, c_k\}, \quad 0 \text{ otherwise.} \tag{1.20}$$

It is easy to see that $\underline{I}(\vartheta_k) = \overline{I}(\vartheta_k) = 0$, so each $\vartheta_k \in \mathcal{R}(I)$. But, as $k \to \infty$,

$$\vartheta_k(x) \nearrow \vartheta(x), \tag{1.21}$$

defined by (1.16), which is not in $\mathcal{R}(I)$. The Lebesgue theory of integration remedies this defect. If f_k are Lebesgue integrable, and if one uses in (1.19) the Lebesgue integral (which coincides with the Riemann integral for functions in $\mathcal{R}(I)$), then (1.18) $\Rightarrow$ (1.19). This is known as the Monotone Convergence Theorem, and it will be seen to be a central result in the Lebesgue theory.

Associated to the Riemann integral is a notion of size of a set S, called *content*. If S is a subset of I, define the "characteristic function"

$$\chi_S(x) = 1 \;\text{ if } x \in S, \;\; 0 \;\text{ if }\; x \notin S. \tag{1.22}$$

We define "upper content" cont^+ and "lower content" cont^- by

$$\text{cont}^+(S) = \overline{I}(\chi_S), \quad \text{cont}^-(S) = \underline{I}(\chi_S). \tag{1.23}$$

We say S "has content," or "is contented" if these quantities are equal, which happens if and only if $\chi_S \in \mathcal{R}(I)$, in which case the common value of $\text{cont}^+(S)$ and $\text{cont}^-(S)$ is

$$m(S) = \int_I \chi_S(x)\, dx. \tag{1.24}$$

It is easy to see that

$$\text{cont}^+(S) = \inf\Bigl\{\sum_{k=1}^{N} \ell(J_k) : S \subset J_1 \cup \cdots \cup J_N\Bigr\}, \tag{1.25}$$

where J_k are intervals. Here, we require S to be in the union of a *finite* collection of intervals.

The key to the construction of Lebesgue measure is to cover a set S by a *countable* (either finite or *infinite*) set of intervals. The *outer measure* of $S \subset I$ will be defined by

$$m^*(S) = \inf\Bigl\{\sum_{k\ge 1} \ell(J_k) : S \subset \bigcup_{k\ge 1} J_k\Bigr\}. \tag{1.26}$$

Here $\{J_k\}$ is a finite or countably infinite collection of intervals. Clearly

$$m^*(S) \le \operatorname{cont}^+(S). \tag{1.27}$$

Note that, if $S = I \cap \mathbb{Q}$, then $\chi_S = \vartheta$, defined by (1.16). In this case it is easy to see that $\operatorname{cont}^+(S) = \ell(I)$, but $m^*(S) = 0$. Zero is the "right" measure of this set.

We develop a few more properties of the Riemann integral. It is useful to note that $\int_I f\,dx$ is additive in I, in the following sense.

Proposition 1.5. *If $a < b < c$, $f : [a, c] \to \mathbb{R}$, $f_1 = f\big|_{[a,b]}$, $f_2 = f\big|_{[b,c]}$, then*

$$f \in \mathcal{R}\big([a, c]\big) \Longleftrightarrow f_1 \in \mathcal{R}\big([a, b]\big) \ \text{ and } \ f_2 \in \mathcal{R}\big([b, c]\big), \tag{1.28}$$

and, if this holds,

$$\int_a^c f\,dx = \int_a^b f_1\,dx + \int_b^c f_2\,dx. \tag{1.29}$$

Proof. Since any partition of $[a, c]$ has a refinement for which b is an endpoint, we may as well consider a partition $\mathcal{P} = \mathcal{P}_1 \cup \mathcal{P}_2$, where $\mathcal{P}_1$ is a partition of $[a, b]$ and $\mathcal{P}_2$ is a partition of $[b, c]$. Then

$$\overline{I}_{\mathcal{P}}(f) = \overline{I}_{\mathcal{P}_1}(f_1) + \overline{I}_{\mathcal{P}_2}(f_2), \quad \underline{I}_{\mathcal{P}}(f) = \underline{I}_{\mathcal{P}_1}(f_1) + \underline{I}_{\mathcal{P}_2}(f_2), \tag{1.30}$$

so

$$\overline{I}_{\mathcal{P}}(f) - \underline{I}_{\mathcal{P}}(f) = \big\{\overline{I}_{\mathcal{P}_1}(f_1) - \underline{I}_{\mathcal{P}_1}(f_1)\big\} + \big\{\overline{I}_{\mathcal{P}_2}(f_2) - \underline{I}_{\mathcal{P}_2}(f_2)\big\}. \tag{1.31}$$

Since both terms in braces in (1.31) are ≥ 0, we have equivalence in (1.28). Then (1.29) follows from (1.30) upon taking sufficiently fine partitions.

Let $I = [a, b]$. If $f \in \mathcal{R}(I)$, then $f \in \mathcal{R}([a, x])$ for all $x \in [a, b]$, and we can consider the function

$$g(x) = \int_a^x f(t)\,dt. \tag{1.32}$$

If $a \le x_0 \le x_1 \le b$, then

$$g(x_1) - g(x_0) = \int_{x_0}^{x_1} f(t)\,dt, \tag{1.33}$$

so, if $|f| \le M$,

$$|g(x_1) - g(x_0)| \le M|x_1 - x_0|. \tag{1.34}$$

In other words, if $f \in \mathcal{R}(I)$, then g is Lipschitz continuous on I.

To finish this section, we want to relate the integral and the derivative. Recall from elementary calculus that a function $g : (a, b) \to \mathbb{R}$ is said to be differentiable at $x \in (a, b)$ provided there exists the limit

$$\lim_{h \to 0} \frac{1}{h}\big[g(x+h) - g(x)\big] = g'(x). \tag{1.35}$$

When such a limit exists, $g'(x)$, also denoted dg/dx, is called the derivative of g at x. Clearly g is continuous wherever it is differentiable.

The next result is part of the Fundamental Theorem of Calculus.

Theorem 1.6. *If $f \in C([a, b])$, then the function g, defined by (1.32), is differentiable at each point $x \in (a, b)$, and*

$$g'(x) = f(x). \tag{1.36}$$

Proof. Parallel to (1.33), we have, for $h > 0$,

$$\frac{1}{h}\big[g(x+h) - g(x)\big] = \frac{1}{h}\int_x^{x+h} f(t)\, dt. \tag{1.37}$$

If f is continuous at x, then, for any $\varepsilon > 0$, there exists $\delta > 0$ such that $|f(t) - f(x)| \le \varepsilon$ whenever $|t - x| \le \delta$. Thus the right side of (1.37) is within ε of $f(x)$ whenever $h \in (0, \delta]$. Thus the desired limit exists as $h \searrow 0$. A similar argument treats $h \nearrow 0$.

The next result is the rest of the Fundamental Theorem of Calculus.

Theorem 1.7. *If G is differentiable and $G'(x)$ is continuous on $[a, b]$, then*

$$\int_a^b G'(t)\ dt = G(b) - G(a). \tag{1.38}$$

Proof. Consider the function

$$g(x) = \int_a^x G'(t)\, dt. \tag{1.39}$$

We have $g \in C([a, b])$, $g(a) = 0$, and, by Theorem 1.6,

$$g'(x) = G'(x), \quad \forall\, x \in (a, b).$$

Thus $f(x) = g(x) - G(x)$ is continuous on $[a, b]$, and

$$f'(x) = 0, \quad \forall\, x \in (a, b). \tag{1.40}$$

We claim that (1.40) implies f is *constant* on $[a, b]$. Granted this, since $f(a) = g(a) - G(a) = -G(a)$, we have $f(x) = -G(a)$ for all $x \in [a, b]$, so the integral (1.39) is equal to $G(x) - G(a)$ for all $x \in [a, b]$. Taking $x = b$ yields (1.38).

The fact that (1.40) implies f is constant on $[a, b]$ is a consequence of the following result, the Mean Value Theorem.

Theorem 1.8. *Let $f : [a, \beta] \to \mathbb{R}$ be continuous, and assume f is differentiable on (a, β). Then $\exists\, \xi \in (a, \beta)$ such that*

$$f'(\xi) = \frac{f(\beta) - f(a)}{\beta - a}. \tag{1.41}$$

Proof. Replacing $f(x)$ by $\tilde{f}(x) = f(x) - \kappa(x - a)$, where κ is the right side of (1.41), we can assume without loss of generality that $f(a) = f(\beta)$. Then we claim that $f'(\xi) = 0$ for some $\xi \in (a, \beta)$. Indeed, since $[a, \beta]$ is compact, f must assume a maximum and a minimum on $[a, \beta]$. If $f(a) = f(\beta)$, *one* of these must be assumed at an interior point, ξ, at which f' clearly vanishes.

We now show that (1.40) implies f is constant on $[a, b]$. If not, $\exists\, \beta \in (a, b]$ such that $f(\beta) \neq f(a)$. Then just apply Theorem 1.8 to f on $[a, \beta]$. This completes the proof of Theorem 1.7.

We mention some useful notation. If a function G is differentiable on (a, b) and G' is continuous on (a, b), we say G is a C^1 function, and we write $G \in C^1\big((a, b)\big)$. Inductively, we say $G \in C^k\big((a, b)\big)$ provided $G' \in C^{k-1}\big((a, b)\big)$. Similarly define $C^k\big([a, b]\big)$. Note that the hypothesis of Theorem 1.7 is that $G \in C^1\big([a, b]\big)$.

Finally, we mention that there are more general versions of the Fundamental Theorem of Calculus involving the Riemann integal; see for example [**BS**]. Since the Riemann integral is not our main focus here, we have been content to present the simpler results above. (See, however, Exercise 14 below.) Our main motivation for taking the space to present these results (which the reader might reasonably be presumed to have seen before) is provided by their role in the study of weak derivatives, in Chapter 10. Furthermore, we present extensions of Theorems 1.6–1.7, involving the Lebesgue integral, in Chapter 11, particularly in Propositions 11.11–11.12.

Exercises

1. Show that, if $f_k \in \mathcal{R}(I)$ and $f_k \to f$ *uniformly* on I, i.e.,

$$\sup_I |f_k(x) - f(x)| \to 0 \text{ as } k \to \infty,$$

then $f \in \mathcal{R}(I)$, and (1.19) holds.

2. Establish the following "monotone convergence theorem" for the Riemann integral. Assume f_k and f are continuous on I. Then (1.18) $\Rightarrow$ (1.19).
Hint. Show that $f_k \to f$ uniformly on I, under these hypotheses. (This result is known as Dini's Theorem.)

3. If $I \cap \mathbb{Q} = \{c_1, c_2, c_3, \dots\}$ as in the construction of (1.20), set

$$\Theta(x) = \begin{aligned} &\frac{1}{k} \ \text{ if } \ x = c_k, \\ &0 \ \text{ if } \ x \in I \setminus \mathbb{Q}. \end{aligned}$$

Show that $\Theta \in \mathcal{R}(I)$ and compute $\int_I \Theta(x)\,dx$.

4. Assume f is bounded on $I = [a, b]$ and *continuous* on $\overset{\circ}{I} = (a, b)$. Show that $f \in \mathcal{R}(I)$ and that $\int_I f(x)\,dx$ is independent of the values of f at $x = a$ and $x = b$.

5. Let $\mathcal{P}_\nu$ be a sequence of partitions of I satisfying the hypotheses of Corollary 1.4. Show that, when $f : I \to \mathbb{R}$ is bounded,

$$\lim_{\nu \to \infty} \overline{I}_{\mathcal{P}_\nu}(f) = \overline{I}(f) \ \text{ and } \ \lim_{\nu \to \infty} \underline{I}_{\mathcal{P}_\nu}(f) = \underline{I}(f). \tag{1.42}$$

6. Let $C(I)$ denote the set of continuous real-valued functions on I. Show that, for any bounded function $f : I \to \mathbb{R}$,

$$\overline{I}(f) = \inf \Bigl\{ \int_I g\,dx : g \ge f,\ g \in C(I) \Bigr\}. \tag{1.43}$$

Similarly characterize $\underline{I}(f)$.

7. Let $c > 0$ and let $f : [ac, bc] \to \mathbb{R}$ be Riemann integrable. Working directly with the definition of integral, show that

$$\int_a^b f(cx)\,dx = \frac{1}{c}\int_{ac}^{bc} f(x)\,dx.$$

More generally, show that

$$\int_{a-d/c}^{b-d/c} f(cx+d)\,dx = \frac{1}{c}\int_{ac}^{bc} f(x)\,dx.$$

8. Let $f : I \times S \to \mathbb{R}$ be continuous, where $I = [a, b]$ and $S \subset \mathbb{R}^n$. Take $\varphi(y) = \int_I f(x, y)\,dx$. Show that φ is continuous on S.
Hint. If $f_j : I \to \mathbb{R}$ are continuous and $|f_1(x) - f_2(x)| \le \delta$ on I, then

$$\Bigl|\int_I f_1\,dx - \int_I f_2\,dx\Bigr| \le \ell(I)\delta.$$

9. With f as in Exercise 8, suppose $g_j : S \to \mathbb{R}$ are continuous and $a \le g_0(y) < g_1(y) \le b$. Take $\varphi(y) = \int_{g_0(y)}^{g_1(y)} f(x, y)\,dx$. Show that φ is continuous on S.
Hint. Make a change of variables, linear in y, to reduce this to Exercise 8.

10. Let $\varphi : [a, b] \to [A, B]$ be C^1 on a neighborhood J of $[a, b]$, with $\varphi'(x) > 0$ for all $x \in [a, b]$. Assume $\varphi(a) = A$, $\varphi(b) = B$. Show that the identity

$$\int_A^B f(y)\,dy = \int_a^b f\bigl(\varphi(t)\bigr)\varphi'(t)\,dt, \tag{1.44}$$

for any $f \in C(J)$, follows from the chain rule and the Fundamental Theorem of Calculus.
Hint. Replace b by x, B by $\varphi(x)$, and differentiate.
Note that this result contains that of Exercise 7.
Try to establish (1.44) directly by working with the definition of the integral as a limit of Riemann sums.

11. Show that, if f and g are C^1 on a neighborhood of $[a, b]$, then

$$\int_a^b f(s)g'(s)\,ds = -\int_a^b f'(s)g(s)\,ds + \bigl[f(b)g(b) - f(a)g(a)\bigr]. \tag{1.45}$$

This transformation of integrals is called "integration by parts."

12. Let $f:(a,b)\to\mathbb{R}$ be a C^{k+1}-function, and take $y\in(a,b)$. Show that for $x\in(a,b)$

$$\begin{aligned} f(x) = f(y) + f'(y)(x-y) + \frac{f''(y)}{2}(x-y)^2 \\ + \cdots + \frac{f^{(k)}(y)}{k!}(x-y)^k + R_k(x,y), \end{aligned} \tag{1.46}$$

where

$$R_k(x,y) = \frac{1}{k!}\int_y^x (x-s)^k f^{(k+1)}(s)\,ds. \tag{1.47}$$

This is Taylor's formula with remainder.
Hint. Apply $\partial/\partial y$ to both sides of (1.46). The left side becomes 0 and there is considerable cancellation on the right side, yielding

$$\frac{\partial}{\partial y}R_k(x,y) = -\frac{f^{(k+1)}(y)}{k!}(x-y)^k, \quad R_k(x,x)=0. \tag{1.48}$$

Integrating then gives (1.47).
Note that a reformulation of (1.47) is

$$R_k(x,y) = \frac{(x-y)^{k+1}}{(k+1)!}\int_0^1 f^{(k+1)}\big(y+(1-t^{1/(k+1)})(x-y)\big)\,dt. \tag{1.49}$$

13. Suppose $[a,b]$ is covered by open intervals J_k, $1\le k\le N$, of length $\ell(J_k)$. Show that

$$b-a \le \sum_{k=1}^N \ell(J_k).$$

Hint. Show that $\sum_k \chi_{J_k}(x) \ge \chi_{[a,b]}(x)$ and deduce consequences for the Riemann integrals of these functions.
Alternative. Try an induction on N.

14. Extend Theorem 1.7 to the case where G is differentiable and G' is Riemann integrable on $[a,b]$.
Hint. Use

$$\begin{aligned} G(b)-G(a) &= \sum_{k=0}^{n-1}\Big[G\Big(a+(b-a)\frac{k+1}{n}\Big) - G\Big(a+(b-a)\frac{k}{n}\Big)\Big] \\ &= \frac{1}{n}\sum_{k=0}^{n-1} G'(\xi_{kn}). \end{aligned}$$

Chapter 2

Lebesgue Measure on the Line

In this chapter we discuss the concept of Lebesgue measure of subsets of the real line $\mathbb{R}$. It is convenient to begin with a discussion of the measure of subsets of a bounded interval.

If S is a subset of an interval $I = [a, b]$, then, as indicated in Chapter 1, we define the *outer measure* of S by

$$m^*(S) = \inf\Bigl\{\sum_{k\geq 0} \ell(J_k) : S \subset \bigcup_{k\geq 0} J_k\Bigr\}, \quad J_k \text{ intervals.} \tag{2.1}$$

It is easy to see that the result is not affected if one insists that all the intervals J_k be open in I or that they all be closed (or half-open, etc.). We can let J_k be intervals in $\mathbb{R}$, or we can require $J_k \subset I$. In particular, if $\mathcal{O} \subset (a, b)$ is open, then $\mathcal{O}$ is a disjoint union of a countable collection of open intervals $\mathcal{O}_k$, and (see Exercise 1 at the end of this chapter)

$$m^*(\mathcal{O}) = \sum \ell(\mathcal{O}_k). \tag{2.2}$$

Furthermore, for any $S \subset (a, b)$,

$$m^*(S) = \inf\bigl\{m^*(\mathcal{O}) : \mathcal{O} \supset S,\ \mathcal{O} \text{ open}\bigr\}. \tag{2.3}$$

Replacing I by a slightly larger interval, we see that (2.3) holds for any $S \subset I$.

Obviously the outer measure of a single point $p \in I$ is zero. Under the most liberal allowance for intervals in (2.1), p itself is an interval, of length zero. If we insist on open intervals, then let J_ε be an interval of length ε centered at p. More generally, if $C = \{c_1, c_2, c_3, \dots\}$ is a *countable* subset of I, write $C \subset \bigcup J_k(\varepsilon)$, where $J_k(\varepsilon)$ is an open interval of length $2^{-k}\varepsilon$, centered at c_k. Thus $m^*(C) \le \sum 2^{-k}\varepsilon = \varepsilon$, so

$$C \subset I \text{ countable} \Longrightarrow m^*(C) = 0. \tag{2.4}$$

Note that, if $\{J_{1k} : k \ge 0\}$ covers S_1 and $\{J_{2k} : k \ge 2\}$ covers S_2, then $\{J_{1k}, J_{2k} : k \ge 0\}$ is a cover of $S_1 \cup S_2$, so

$$m^*(S_1 \cup S_2) \le m^*(S_1) + m^*(S_2). \tag{2.5}$$

This subadditivity property is shared by upper content, defined in Chapter 1, but outer measure is distinguished from upper content by also having the property of *countable subadditivity*:

Proposition 2.1. *If $\{S_j : j \ge 0\}$ is a countable family of subsets of I, then*

$$m^*\Bigl(\bigcup_j S_j\Bigr) \le \sum_j m^*(S_j). \tag{2.6}$$

Proof. Pick $\varepsilon > 0$. Each S_j has a countable cover $\{J_{jk} : k \ge 0\}$, by intervals, such that $m^*(S_j) \ge \sum_k \ell(J_{jk}) - 2^{-j}\varepsilon$. Then $\{J_{jk} : j, k \ge 0\}$ is a countable cover of $\bigcup_j S_j$ by intervals, so $m^*\bigl(\bigcup S_j\bigr) \le \sum m^*(S_j) + 2\varepsilon$, for all $\varepsilon > 0$. Letting $\varepsilon \searrow 0$, we get (2.6).

Our main goal in this section is to produce a large class (call it $\mathfrak{L}$) of subsets of I with the property that m^* is "countably additive" on $\mathfrak{L}$, in the sense that if $S_j \in \mathfrak{L}$ is a countable collection of mutually disjoint sets (indexed by $j \in \mathbb{Z}^+$), then

$$m^*\Bigl(\bigcup_{j\ge1} S_j\Bigr) = \sum_{j\ge1} m^*(S_j).$$

We will pursue this in stages, showing first that such an identity holds when $S_1 = K$ is compact and $S_2 = I \setminus K$. In preparation for this, we take a closer look at the outer measure of a compact set $K \subset I$. Since any open cover of K has a finite subcover, we can say that

$$K \text{ compact} \Longrightarrow m^*(K) = \inf\Bigl\{\sum_{k=1}^N \ell(J_k) : K \subset \bigcup_{k=1}^N J_k\Bigr\}, \tag{2.7}$$

where J_k are open intervals (which we are free to close up). This coincides with the definition of upper content given in Chapter 1. It implies that, given $\varepsilon > 0$, one can pick a finite collection of disjoint open intervals $\{J_k : 1 \le k \le N\}$ such that $\mathcal{O} = \bigcup_k J_k \supset K$ and such that we have

$$m^*(K) \le m^*(\mathcal{O}) = \sum_{k=1}^{N} \ell(J_k) \le m^*(K) + \varepsilon. \tag{2.8}$$

The following result is sharper than (2.8).

Lemma 2.2. *Given $\varepsilon > 0$, we can construct $\mathcal{O} = \bigcup_{k=1}^N J_k \supset K$ such that*

$$m^*(\mathcal{O} \setminus K) \le \varepsilon. \tag{2.9}$$

Proof. Start with the $\mathcal{O}$ described above. Then $\mathcal{O} \setminus K = \mathcal{A}$ is open, so write $\mathcal{A} = \bigcup_{k \ge 1} \mathcal{A}_k$, a countable disjoint union of open intervals. To achieve (2.9), we need to arrange that $\sum_{k \ge 1} \ell(\mathcal{A}_k) \le \varepsilon$, at least after possibly shrinking $\mathcal{O}$.

To do this, pick M large enough that

$$\sum_{k > M} \ell(\mathcal{A}_k) \le \frac{\varepsilon}{2}.$$

We want to replace $\mathcal{O}$ by $\mathcal{O} \setminus \bigcup_{k=1}^M \mathcal{A}_k^\#$, which should still be a cover of K by a finite number of open intervals. It would be tempting to take $\mathcal{A}_k^\# = \overline{\mathcal{A}}_k$, but note that the endpoints of $\mathcal{A}_k$ might belong to K. Instead, let $\mathcal{A}_k^\# \subset \mathcal{A}_k$ be a closed interval with the same center as $\mathcal{A}_k$, such that $\ell(\mathcal{A}_k^\#) \ge \ell(\mathcal{A}_k) - \varepsilon/2M$. With the new $\mathcal{O}$ we have the lemma.

Using Lemma 2.2, we can establish the following important result, advertised above.

Proposition 2.3. *If $K \subset I$ is compact, then*

$$m^*(K) + m^*(I \setminus K) = \ell(I). \tag{2.10}$$

Proof. To begin, we note that if $\mathcal{O} = \bigcup_{k=1}^N J_k$ is a cover of K satisfying the conditions of Lemma 2.2, then $I \setminus \mathcal{O}$ is a finite disjoint union of intervals, say $I \setminus \mathcal{O} = \bigcup_{j=1}^{\nu} J_j'$, and clearly $m^*(I \setminus \mathcal{O}) = \sum_{j=1}^{\nu} \ell(J_j')$, so

$$m^*(\mathcal{O}) + m^*(I \setminus \mathcal{O}) = \ell(I). \tag{2.11}$$

Furthermore, $I \setminus K = (I \setminus \mathcal{O}) \cup (\mathcal{O} \setminus K)$, so by (2.5) and (2.9),

$$m^*(I \setminus K) \le m^*(I \setminus \mathcal{O}) + \varepsilon. \tag{2.12}$$

It follows that

$$m^*(K) + m^*(I \setminus K) \leq m^*(\mathcal{O}) + m^*(I \setminus \mathcal{O}) + \varepsilon = \ell(I) + \varepsilon, \tag{2.13}$$

for all $\varepsilon > 0$, which implies that the left side of (2.10) is $\leq \ell(I)$. The reverse inequality is automatic from the subadditivity property (2.5).

The *additivity* property, that $m^*(S_1 \cup S_2) = m^*(S_1) + m^*(S_2)$ when S_1 and S_2 are disjoint, just verified for $S_1 = K$ compact and $S_2 = I \setminus K$, does not hold for all disjoint sets S_j, though it holds for the "measurable" ones, as we will see below. At this point we record three easy cases of additivity.

Lemma 2.4. *If $\mathcal{O}_1$ and $\mathcal{O}_2$ are two disjoint open sets, then*

$$m^*(\mathcal{O}_1 \cup \mathcal{O}_2) = m^*(\mathcal{O}_1) + m^*(\mathcal{O}_2). \tag{2.14}$$

If $S_1, S_2 \subset I$ and

$$\rho = \rho(S_1, S_2) = \inf \{|x_1 - x_2| : x_j \in S_j\} > 0, \tag{2.15}$$

then

$$m^*(S_1 \cup S_2) = m^*(S_1) + m^*(S_2). \tag{2.16}$$

Furthermore, if K_j, $1 \leq j \leq N$, is a finite collection of mutually disjoint compact subsets of I, then

$$m^*\Big(\bigcup_{j=1}^N K_j\Big) = \sum_{j=1}^N m^*(K_j). \tag{2.17}$$

Proof. The identity (2.14) is immediate from (2.2). To establish (2.16), given $\delta > 0$, pick an open set $\mathcal{O} \supset S_1 \cup S_2$ such that $m^*(\mathcal{O}) \leq m^*(S_1 \cup S_2) + \delta$. Note that each open set

$$\mathcal{O}_j = \mathcal{O} \cap \Big\{x : \operatorname{dist}(x, S_j) < \frac{\rho}{2}\Big\}$$

contains S_j. We see that $\mathcal{O}_1 \cap \mathcal{O}_2 = \emptyset$, so (2.14) applies. Thus

$$m^*(S_1) + m^*(S_2) \leq m^*(\mathcal{O}_1) + m^*(\mathcal{O}_2) = m^*(\mathcal{O}_1 \cup \mathcal{O}_2) \leq m^*(S_1 \cup S_2) + \delta,$$

for all $\delta > 0$. Thus $m^*(S_1) + m^*(S_2) \leq m^*(S_1 \cup S_2)$, when (2.15) holds. The reverse inequality follows from (2.5), so we have (2.16).

Finally, to establish (2.17), it suffices to treat the case $N = 2$, but two disjoint compact sets $S_j = K_j$ necessarily satisfy (2.15), so the lemma is proved.

Granted this result, we can establish the important property of *countable additivity* of m^* on disjoint compact sets.

Proposition 2.5. *If K_j, $j \geq 1$, is a countable collection of mutually disjoint compact sets in I, then*

$$m^*\Big(\bigcup_{j\geq 1} K_j\Big) = \sum_{j\geq 1} m^*(K_j). \tag{2.18}$$

Proof. The left side of (2.17) is $\leq$ the left side of (2.18) for all N, while the right side of (2.17) converges to the right side of (2.18). Hence the left side of (2.18) is $\geq$ the right side. The reverse inequality follows from (2.6).

We now define *inner measure*: if $S \subset I$, we set

$$m_*(S) = \ell(I) - m^*(I \setminus S). \tag{2.19}$$

By (2.5), $m_*(S) \leq m^*(S)$ for all S. Using the characterization (2.3) of outer measure together with Proposition 2.3, we have

$$m_*(S) = \sup\,\big\{m^*(K) : K \subset S, K \text{ compact}\big\}. \tag{2.20}$$

Definition. *$S \subset I$ is measurable if and only if $m^*(S) = m_*(S)$.*

If S is measurable, we set

$$m(S) = m^*(S) = m_*(S). \tag{2.21}$$

Clearly, by (2.19), $S \subset I$ is measurable if and only if $I \setminus S$ is measurable. In view of (2.20), we see that any compact $K \subset I$ is measurable. Hence any open $\mathcal{O} \subset I$ is measurable.

One useful measurability result follows easily from Proposition 2.5:

Proposition 2.6. *If K_j, $j \geq 1$, is a countable collection of mutually disjoint compact subsets of I, then $L = \bigcup_{j\geq 1} K_j$ is measurable, and*

$$m(L) = \sum_{j\geq 1} m(K_j). \tag{2.22}$$

Proof. Since $\bigcup_{j=1}^N K_j = L_N \subset L$ for all $N < \infty$, we have

$$m_*(L) \geq m\Big(\bigcup_{j=1}^N K_j\Big) = \sum_{j=1}^N m(K_j).$$

Hence $m_*(L) \geq \sum_{j\geq 1} m(K_j)$. In view of (2.18) and the inequality $m_*(L) \leq m^*(L)$, we have the proposition.

In general, in counterpoint to countable subadditivity of outer measure, we have countable *superadditivity* of inner measure:

Proposition 2.7. *Let $S_k \subset I$ be a countable family of mutually disjoint sets. Then*

$$m_*\Big(\bigcup_{k\geq 1} S_k\Big) \geq \sum_{k\geq 1} m_*(S_k). \tag{2.23}$$

Proof. Pick $\varepsilon > 0$; then pick $K_k \subset S_k$, compact, such that $m(K_k) \geq m_*(S_k) - 2^{-k}\varepsilon$. Then $\bigcup_{k\geq 1} K_k = L \subset \bigcup_{k\geq 1} S_k$, and, by Proposition 2.6,

$$m_*(L) = m^*(L) = \sum_{k\geq 1} m(K_k).$$

Since $m_*(L) \leq m_*(S)$, where $S = \bigcup S_k$, we have

$$m_*(S) \geq \sum_{k\geq 1} m(K_k) \geq \sum_{k\geq 1} m_*(S_k) - \varepsilon,$$

for all $\varepsilon > 0$, yielding (2.23).

We have the following important conclusion, extending Proposition 2.6.

Theorem 2.8. *If S_k, $k \geq 1$, is a countable family of mutually disjoint measurable subsets of I, then $S = \bigcup_{k\geq 1} S_k$ is measurable, and*

$$m(S) = \sum_{k\geq 1} m(S_k). \tag{2.24}$$

Proof. Using Proposition 2.1 and Proposition 2.7, we have

$$m^*(S) \leq \sum_{k\geq 1} m^*(S_k) = \sum_{k\geq 1} m_*(S_k) \leq m_*(S),$$

and since $m_*(S) \leq m^*(S)$, we have $m^*(S) = m_*(S)$ and the identity (2.24).

The identity (2.24) asserts *countable additivity* of m and is at the heart of Lebesgue measure theory.

We will derive some further sufficient conditions for a set to be measurable. The following criterion for measurability is useful.

Lemma 2.9. *A set $S \subset I$ is measurable if and only if, for each $\delta > 0$, there exist a compact K and an open $\mathcal{O}$ such that $K \subset S \subset \mathcal{O}$ and $m(\mathcal{O} \setminus K) < \delta$.*

Since both K (compact) and $\mathcal{O} \setminus K$ (open) are measurable, we have $m(\mathcal{O}) - m(K) = m(\mathcal{O} \setminus K)$ by Theorem 2.8, so the proof of Lemma 2.9 is straightforward. Here is an application.

Proposition 2.10. *If S_1 and S_2 are measurable subsets of I, the following are also measurable:*

$$S_1 \cup S_2, \quad S_1 \cap S_2, \quad S_1 \setminus S_2. \tag{2.25}$$

Proof. Given $\delta > 0$, take K_j compact, $\mathcal{O}_j$ open, so that $K_j \subset S_j \subset \mathcal{O}_j$ and $m(\mathcal{O}_j \setminus K_j) < \delta/2$. Then $K = K_1 \cup K_2$ is compact, $\mathcal{O} = \mathcal{O}_1 \cup \mathcal{O}_2$ is open, $K \subset S_1 \cup S_2 \subset \mathcal{O}$, and $\mathcal{O} \setminus K \subset (\mathcal{O}_1 \setminus K_1) \cup (\mathcal{O}_2 \setminus K_2)$, so subadditivity implies $m(\mathcal{O} \setminus K) < \delta$. This shows that $S_1 \cup S_2$ is measurable.

We know that S_j is measurable $\Leftrightarrow I \setminus S_j$ is measurable, so, by the last argument, $(I \setminus S_1) \cup (I \setminus S_2) = I \setminus (S_1 \cap S_2)$ is measurable; hence $S_1 \cap S_2$ is measurable.

Finally, $S_1 \setminus S_2 = S_1 \cap (I \setminus S_2)$, so this is measurable.

There is a more incisive countable counterpart.

Theorem 2.11. *If S_j, $j \geq 1$, is a countable collection of measurable subsets of I, the following are measurable:*

$$\bigcup_{j \geq 1} S_j, \quad \bigcap_{j \geq 1} S_j. \tag{2.26}$$

Proof. To treat $U = \bigcup_{j \geq 1} S_j$, let $S'_j = \bigcup_{\ell=1}^{j} S_j$, which is measurable by Proposition 2.10. Then $S'_1 \subset S'_2 \subset \cdots$ and $U = \bigcup_{j \geq 1} S'_j$. Now let $T_1 = S'_1 = S_1$ and $T_j = S'_j \setminus S'_{j-1}$ for $j \geq 2$, also measurable by Proposition 2.10. Thus $U = \bigcup_{j \geq 1} T_j$. Since the T_j are mutually *disjoint*, we can apply Theorem 2.8, to conclude that U is measurable. Finally, $\bigcap_{j \geq 1} S_j$ is the complement in I of $\bigcup_{j \geq 1} (I \setminus S_j)$, so the theorem is proved.

A nonempty family $\mathfrak{F}$ of subsets of some set X is called a σ-algebra if it is closed under the formation of countable unions, countable intersections, and complements. Thus we see that the family $\mathfrak{L}$ of measurable subsets of I is a σ-algebra. The *smallest* σ-algebra of subsets of a topological space X containing all the closed sets is called the algebra of Borel sets. (Cf. Exercises 11–12 for the existence of such a smallest σ-algebra.) We see that each Borel set in I is measurable.

We record a criterion for measurability that, while first perhaps appearing curious, will actually serve as a convenient *definition* of measurability, in the more general construction of measures to be discussed in Chapter 5.

Proposition 2.12. *A set $S \subset I$ is measurable if and only if, for all $Y \subset I$ (measurable or not),*

$$m^*(Y) = m^*(Y \cap S) + m^*(Y \setminus S). \tag{2.27}$$

Proof. Since the validity of (2.27) for $Y = I$ was the definition of measurability, via (2.19), it remains only to show that (2.27) holds whenever S is measurable. Of course, we always have $m^*(Y) \le m^*(Y \cap S) + m^*(Y \setminus S)$. Now take $\varepsilon > 0$ and pick an open set $\mathcal{O} \supset Y$ such that $m(\mathcal{O}) \le m^*(Y) + \varepsilon$. We know that $\mathcal{O} \cap S$ and $\mathcal{O} \setminus S$ are measurable if S is, and thus

$$m(\mathcal{O}) = m(\mathcal{O} \cap S) + m(\mathcal{O} \setminus S) \ge m^*(Y \cap S) + m^*(Y \setminus S).$$

Hence $m^*(Y \cap S) + m^*(Y \setminus S) \le m^*(Y) + \varepsilon$ for all $\varepsilon > 0$, and (2.27) follows.

We now discuss Lebesgue measure on the line, which is very much like that on a bounded interval, except that now some sets can have measure $+\infty$. Let us partition $\mathbb{R}$ into a countable set of bounded intervals (of positive length):

$$\mathbb{R} = \bigcup_k I_k, \tag{2.28}$$

so two different intervals can intersect at most at one point. We say a subset $S \subset \mathbb{R}$ is measurable if and only if $S \cap I_k$ is measurable, for all k. If this holds, we then set

$$m(S) = \sum_k m(S \cap I_k). \tag{2.29}$$

Note that possibly $m(S) = +\infty$; for example, $\mathbb{R}$ is measurable and $m(\mathbb{R}) = +\infty$. It is easy to show that this characterization is independent of the choice of partition and to extend the results established above, to show that the family of measurable subsets of $\mathbb{R}$ is a σ-algebra and Lebesgue measure is countably additive, as in Theorem 2.8. We leave the job as an exercise.

We mention that (2.1) could still be used to define outer measure, and one has, for all $S \subset \mathbb{R}$,

$$m^*(S) = \sum_k m^*(S \cap I_k). \tag{2.30}$$

Also, (2.20) could be used to define inner measure. Furthermore, *provided* $m_*(S) < \infty$, then $S \subset \mathbb{R}$ is measurable $\Leftrightarrow m_*(S) = m^*(S)$. However, if $m_*(S) = \infty$, then certainly $m^*(S) = \infty$, but S might possibly not be measurable.

There is also a natural Lebesgue measure on the circle $\mathbb{T}$, which can be identified with $\mathbb{R}/\mathbb{Z}$. In fact, we have a natural 1-1 correspondence between $\mathbb{T}$ and $I = [0,1]$, with 0 and 1 identified, and Lebesgue measure on $\mathbb{T}$ is simply the same as that on I. We note that $\mathbb{T}$ is a group, and it acts on itself in a measure-preserving fashion. That is, if $S \subset \mathbb{T}$ and $\alpha \in \mathbb{T}$, we can form $S_\alpha = S + \alpha = \{x + \alpha : x \in S\}$, where "+" denotes addition mod 1, and

$$m^*(S_\alpha) = m^*(S_\beta), \quad m_*(S_\alpha) = m_*(S_\beta), \tag{2.31}$$

for all $\alpha, \beta \in \mathbb{T}$.

We end this section with a standard example of a subset S of $I = [0,1]$ that is not measurable. Actually, it is convenient to construct S as a subset of $\mathbb{T}$. As noted above, $\mathbb{T}$ is a group. It has a subgroup $\mathfrak{Q} = \mathbb{Q}/\mathbb{Z}$, acting on $\mathbb{T}$ as a countable family of measure-preserving transformations. We form S by picking *one* element from each orbit of $\mathfrak{Q}$ in $\mathbb{T}$. Doing this requires the "axiom of choice." Now, for each $\alpha \in \mathfrak{Q}$, consider $S_\alpha = S + \alpha = \{x + \alpha : x \in S\}$, as in the previous paragraph. Note that

$$\bigcup_{\alpha \in \mathfrak{Q}} S_\alpha = \mathbb{T}, \quad \alpha \neq \beta \in \mathfrak{Q} \Rightarrow S_\alpha \cap S_\beta = \emptyset. \tag{2.32}$$

Also (2.31) holds, for all $\alpha, \beta \in \mathfrak{Q}$. Now respective applications of countable subadditivity of m^* and countable superadditivity of m_* to $\mathbb{T} = \bigcup_\alpha S_\alpha$ yield $1 \le \sum_\alpha m^*(S_\alpha)$ and $1 \ge \sum\limits_\alpha m_*(S_\alpha)$. Hence

$$m^*(S_\alpha) > 0 \quad \text{and} \quad m_*(S_\alpha) = 0, \quad \forall\, \alpha \in \mathfrak{Q}. \tag{2.33}$$

Thus none of the sets S_α are measurable.

Exercises

1. In order to establish the identity (2.2), start by demonstrating the following:

Claim. If $\mathfrak{I}$ is a bounded interval and $\{J_k : k \ge 1\}$ a countable cover of $\mathfrak{I}$ by intervals,

$$\sum_{k \ge 1} \ell(J_k) \ge \ell(\mathfrak{I}). \tag{2.34}$$

Hint. Reduce to the case where $\mathfrak{I}$ is *closed* and J_k are *open*. Consult Exercise 13 of Chapter 1.
From here finish the proof of (2.2).
Hint. The definition (2.1) clearly gives $m^*(\mathcal{O}) \le \sum \ell(\mathcal{O}_k)$. For the converse inequality, fix $\varepsilon > 0$ and pick a finite number of closed intervals $J_k \subset \mathcal{O}_k$, $1 \le k \le N$, such that $\sum_1^N \ell(J_k) \ge \sum \ell(\mathcal{O}_k) - \varepsilon$. Then show that $m^*(\mathcal{O}) \ge m^*(\bigcup_k J_k) = \sum_1^N \ell(J_k)$.

2. If S_j is an *increasing* sequence of measurable subsets of I, i.e., $S_1 \subset S_2 \subset S_3 \subset \cdots$, and $S = \bigcup_{j \ge 1} S_j$, show that

$$m(S_j) \nearrow m(S), \text{ as } j \to \infty. \tag{2.35}$$

Hint. Examine the proof of Theorem 2.11.

3. If S_j is a *decreasing* sequence of measurable subsets of I, i.e., $S_1 \supset S_2 \supset S_3 \supset \cdots$, and $S = \bigcap_{j \ge 1} S_j$, show that

$$m(S_j) \searrow m(S), \text{ as } j \to \infty. \tag{2.36}$$

4. If S_j is an increasing sequence of measurable subsets of $\mathbb{R}$, with union S, show that (2.35) continues to hold. If S_j is a decreasing sequence of measurable subsets of $\mathbb{R}$, with intersection S, show that (2.36) holds, *provided* $m(S_j) < \infty$ for some j. Give a counterexample to (2.36) when this provision does not hold.

5. Show that a nonempty family $\mathfrak{F}$ of subsets of X which is closed under countable unions and complements is automatically closed under countable intersections and hence is a σ-algebra.

6. A nonempty family $\mathfrak{F}$ of subsets of X is called a σ-ring if it is closed under countable unions and under *differences* (i.e., $S_j \in \mathfrak{F} \Rightarrow S_1 \setminus S_2 \in \mathfrak{F}$). Show that a σ-ring is also closed under countable intersections. Show that such a σ-ring $\mathfrak{F}$ is a σ-algebra if and only if $X \in \mathfrak{F}$.

7. Show that every measurable set $X \subset \mathbb{R}$ with *positive* measure contains a non measurable subset.
Hint. If $X \subset I$, consider $X \cap S_\alpha$, with S_α as in (2.31)–(2.33). Show that $m_*(X \cap S_\alpha) = 0$ for all α, while $m^*(X \cap S_\alpha) > 0$ for some α.

8. Form the Cantor middle third set as follows. Let $K_0 = [0, 1]$. Form K_1 by removing the open interval in the middle of K_0, of length $1/3$. Then K_1 consists of two intervals, each of length $1/3$. Next, remove from

each of the intervals making up K_1 the open interval in the middle, of length $1/3^2$, to get K_2. Continue this process. Thus K_ν is a union of 2^ν disjoint closed intervals, each of length $3^{-\nu}$. Then

$$K_0 \supset K_2 \supset K_2 \supset K_3 \supset \cdots \searrow K,$$

and $K = \bigcap_\nu K_\nu$ is the Cantor middle third set. Show that

$$m(K) = 0.$$

9. Modify the construction in Exercise 8 as follows. Let $L_0 = [0,1]$. Form L_1 by removing the open interval in the middle of L_0 of length $1/5$. Next, remove from each of the intervals making up L_1 the open intervals in the middle, of length $1/5^2$, to get L_2. Continue this process. Thus L_ν is obtained from $[0,1]$ by omitting one open interval of length $1/5$, two of length $1/5^2$, and so on, up to omitting $2^{\nu-1}$ open intervals of length $5^{-\nu}$. Then $L_0 \supset L_1 \supset L_2 \supset \cdots \searrow L$, where $L = \bigcap_\nu L_\nu$ is a compact set. Show that
$$m(L) > 0.$$

10. With K and L as in Exercises 8–9, show that there is a homeomorphism $F : [0,1] \to [0,1]$ such that
$$F(L) = K.$$

In Exercises 11–12, let X be a nonempty set and $\mathcal{C}$ a nonempty collection of subsets of X.

11. Show that the collection of σ-algebras of subsets of X that contain $\mathcal{C}$ is nonempty.
Hint. Consider the set of all subsets of X.

12. If $\{\mathcal{F}_\alpha : \alpha \in \mathcal{A}\}$ is the collection of all σ-algebras of subsets of X that contain $\mathcal{C}$, show that
$$\bigcap_{\alpha \in \mathcal{A}} \mathcal{F}_\alpha = \mathcal{F}$$
is a σ-algebra of subsets of X, containing $\mathcal{C}$, and is in fact the *smallest* such σ-algebra. One says $\mathcal{F}$ is the σ-algebra generated by $\mathcal{C}$ and writes $\mathcal{F} = \sigma(\mathcal{C})$.

Chapter 3

Integration on Measure Spaces

In this chapter we introduce the general notion of a measure on a space X, define the class of measurable functions, and define the integral, first on a class of positive "simple functions," then on general positive, measurable functions, and finally on measurable, integrable functions. We establish several key results about integrals of limits of sequences of functions.

It is useful to select the crucial features of Lebesgue measure, as developed in Chapter 2, and apply them to more general situations; this leads to the notion of a "measure space." To define this, we start with a set X and a σ-algebra $\mathfrak{F}$ of subsets. As in Chapter 2, this means $\mathfrak{F}$ is a nonempty family of subsets of X, closed under countable unions and under complements, and hence also under countable intersections. We call $(X, \mathfrak{F})$ a *measurable space.* Note that, if $S \in \mathfrak{F}$, then $X \setminus S \in \mathfrak{F}$, so $X = S \cup (X \setminus S) \in \mathfrak{F}$, and hence $X^c = \emptyset \in \mathfrak{F}$.

A *measure* μ on $(X, \mathfrak{F})$ is a function $\mu : \mathfrak{F} \to [0, \infty]$ satisfying the following two conditions:

$$\mu(\emptyset) = 0, \tag{3.1}$$

$$S_j \in \mathfrak{F} \text{ disjoint, countable} \Longrightarrow \mu\Big(\bigcup_j S_j\Big) = \sum_j \mu(S_j). \tag{3.2}$$

We call $(X, \mathfrak{F}, \mu)$ a *measure space.* It is easy to verify that μ automatically has the following four additional properties:

$$S_j \in \mathfrak{F},\ S_1 \subset S_2 \Longrightarrow \mu(S_1) \le \mu(S_2), \tag{3.3}$$

$$S_j \in \mathfrak{F} \text{ countable } \Longrightarrow \mu\Bigl(\bigcup_j S_j\Bigr) \le \sum_j \mu(S_j), \tag{3.4}$$

$$S_j \in \mathfrak{F},\ S_1 \subset S_2 \subset S_3 \subset \cdots \Longrightarrow \mu(S_j) \nearrow \mu(\bigcup S_j), \tag{3.5}$$

and, *provided* $\mu(S_j) < \infty$ for some j,

$$S_j \in \mathfrak{F},\ S_1 \supset S_2 \supset S_3 \supset \cdots \Longrightarrow \mu(S_j) \searrow \mu(\bigcap S_j). \tag{3.6}$$

Property (3.3) is called *monotonicity*, and property (3.4) is called countable subadditivity; the contrast with *countable additivity* in (3.2) is that in (3.4) the sets S_j need not be disjoint.

To prove (3.3), note that, if $S_1 \subset S_2$, then $Y = S_2 \setminus S_1 = S_2 \cap (X \setminus S_1) \in \mathfrak{F}$, and $S_2 = S_1 \cup Y$ is a disjoint union; hence, by (3.2), $\mu(S_2) = \mu(S_1) + \mu(Y)$, a sum of terms that are ≥ 0. We leave the proofs of (3.4)–(3.6) as exercises. Compare Exercises 2 and 3 in Chapter 2.

So far we have as examples of measure spaces $X = [a, b]$ and $X = \mathbb{R}$, with $\mathfrak{F}$ either the class of Lebesgue measurable subsets of X or the class of Borel subsets of X, and $\mu =$ Lebesgue measure. More generally, we could let X be any Lebesgue measurable subset of $\mathbb{R}$. We will study further constructions of measures in Chapters 5–7 and Chapters 12–14.

Now we look at the issue of integrating a function $f : X \to \mathbb{R}$, when $(X, \mathfrak{F}, \mu)$ is a measure space. We first define what it means for f to be *measurable.* The most liberal-sounding characterization is that $f^{-1}(J) \subset X$ belongs to $\mathfrak{F}$ for every open interval $J \subset \mathbb{R}$. If this property holds, it is an automatic consequence that $f^{-1}(S) \in \mathfrak{F}$ for every *Borel* set $S \subset \mathbb{R}$. (Proof: The collection of subsets $S \subset \mathbb{R}$ such that $f^{-1}(S) \in \mathfrak{F}$ is a σ-algebra. If it contains all open intervals, then it contains all Borel subsets of $\mathbb{R}$.) We regard this latter characterization as the definition of measurability. To emphasize the role of $\mathfrak{F}$, we sometimes say f is $\mathfrak{F}$-measurable.

Note that, if X is a topological space and $\mathfrak{B}$ is the σ-algebra of Borel sets in X, i.e., the smallest σ-algebra containing the closed subsets of X, then any continuous $f : X \to \mathbb{R}$ is $\mathfrak{B}$-measurable.

By the definition, $f : \mathbb{R} \to \mathbb{R}$ is Lebesgue measurable provided $f^{-1}(S) \in \mathfrak{L}$ for every $S \in \mathfrak{B}$, where $\mathfrak{B}$ is the class of Borel sets in $\mathbb{R}$ and $\mathfrak{L}$ is the class of Lebesgue measurable subsets of $\mathbb{R}$. Recall that $\mathfrak{B} \subset \mathfrak{L}$. We note, however, that it does *not* follow that $f^{-1}(S) \in \mathfrak{L}$ for all $S \in \mathfrak{L}$. Indeed, there exist *continuous* $f : \mathbb{R} \to \mathbb{R}$ and $S \in \mathfrak{L}$ such that $f^{-1}(S) \notin \mathfrak{L}$. (See Exercise 21.)

We can also define $f : X \to [-\infty, \infty]$ to be $\mathfrak{F}$-measurable provided $f^{-1}(S) \in \mathfrak{F}$ for all Borel sets S in $[-\infty, \infty]$.

It is valuable to know that certain basic operations preserve measurability. For example, we have the following.

Proposition 3.1. *If $f_j : X \to \mathbb{R}$ are $\mathfrak{F}$-measurable, so are $f_1 + f_2$ and $f_1 f_2$.*

Proof. Let J be an open interval in $\mathbb{R}$. If $g = f_1 + f_2$, then

$$g^{-1}(J) = \{x \in X : \big(f_1(x), f_2(x)\big) \in \mathcal{O}\}, \tag{3.7}$$

where $\mathcal{O} = \{(x, y) \in \mathbb{R}^2 : x + y \in J\}$. Now we can write $\mathcal{O}$ as a countable disjoint union of rectangles, of the form $\mathcal{O}_k = I_k \times I'_k$ (where the intervals I_k, I'_k are open). Then

$$\begin{aligned} g^{-1}(J) &= \bigcup_k \{x \in X : f_1(x) \in I_k,\ f_2(x) \in I'_k\} \\ &= \bigcup_k [f_1^{-1}(I_k) \cap f_2^{-1}(I'_k)] \end{aligned} \tag{3.8}$$

is clearly an element of $\mathfrak{F}$, provided f_1 and f_2 are $\mathfrak{F}$-measurable. The measurability of $f_1 f_2$ is proved similarly.

Remark 3.1. The argument used above to prove Proposition 3.1 readily extends to the following more general setting. Assume $f_j : X \to \mathbb{R}$ are $\mathfrak{F}$-measurable, for $j = 1, \dots, n$, and that $\Phi : \mathbb{R}^n \to \mathbb{R}$ is continuous, and define $g : X \to \mathbb{R}$ by

$$g(x) = \Phi(f_1(x), \dots, f_n(x)).$$

Then g is $\mathfrak{F}$-measurable.

The following result on measurability of sups, infs, etc., will be very valuable.

Proposition 3.2. *If f_j are $\mathfrak{F}$-measurable, for $j \in \mathbb{Z}^+$, then*

$$\begin{aligned} g_1(x) &= \sup_j f_j(x), \quad & g_2(x) &= \inf_j f_j(x), \\ g_3(x) &= \limsup_j f_j(x), \quad & g_4(x) &= \liminf_j f_j(x) \end{aligned} \tag{3.9}$$

are all $\mathfrak{F}$-measurable functions, from X to $[-\infty, +\infty]$. If $g(x) = \lim f_j(x)$ exists for all $x \in X$, then g is $\mathfrak{F}$-measurable.

Proof. Here the σ-algebra of Borel sets in $[-\infty, +\infty]$ is generated by intervals, either of the form $(a, +\infty]$ or of the form $[-\infty, a)$. Note that

$$g_1^{-1}\big((a, \infty]\big) = \bigcup_{j \geq 1} f_j^{-1}\big((a, \infty]\big), \tag{3.10}$$

so the measurability of g_1 follows. That of g_2 is similar. Now

$$g_3(x) = \inf_{k\ge 1} h_k(x), \quad \text{where} \quad h_k(x) = \sup_{j\ge k} f_j(x), \tag{3.11}$$

so measurability of g_3 follows, and measurability of g_4 is similar. Finally, if $g(x) = \lim f_j(x)$ exists everywhere, then $g_3 = g_4 = g$, so we have measurability of g.

If $(X, \mathfrak{F}, \mu)$ is a measure space and $S \subset \mathfrak{F}$, then the characteristic function χ_S, defined as in (1.22) by

$$\chi_S(x) = 1 \ \text{ if } \ x \in S, \quad 0 \ \text{ if } \ x \in X \setminus S, \tag{3.12}$$

is $\mathfrak{F}$-measurable, as is any finite linear combination

$$\varphi(x) = \sum_{j=1}^{N} a_j \chi_{S_j}(x), \quad a_j \in \mathbb{R}, \ S_j \in \mathfrak{F}. \tag{3.13}$$

Such a function is called a simple function.

The sets S_j in (3.13) can overlap. It is convenient to rewrite (3.13) using characteristic functions of disjoint sets. We can do this as follows. For each subset $\sigma \subset \{1, \dots, N\}$, let

$$\widetilde{S}_\sigma = \Bigl(\bigcap_{j\in\sigma} S_j\Bigr) \cap \Bigl(\bigcap_{j\notin\sigma} S_j^c\Bigr), \quad S_j^c = X \setminus S_j. \tag{3.14}$$

Then the sets $\widetilde{S}_\sigma$ are mutually disjoint, and for each $j \in \{1, \dots, N\}$, we have

$$S_j = \bigcup_{\sigma\ni j} \widetilde{S}_\sigma. \tag{3.15}$$

Furthermore, if φ is given by (3.13), then

$$\varphi(x) = \sum_{j=1}^{N} \sum_{\sigma\ni j} a_j \chi_{\widetilde{S}_\sigma}(x) = \sum_{\sigma} \tilde{a}_\sigma \chi_{\widetilde{S}_\sigma}(x), \tag{3.16}$$

where $\tilde{a}_\sigma = \sum_{j\in\sigma} a_j$.

Let us denote by $\mathfrak{S}^+(X)$ the class of simple functions that are ≥ 0 on X. It follows from (3.16) that such functions can be put in the form (3.13) with $a_j \ge 0$. It also follows from (3.16) that $\mathfrak{S}^+(X)$ is the class of $\mathfrak{F}$-measurable functions with range in a finite subset of $[0, \infty)$.

Given $\varphi \in \mathfrak{S}^+(X)$, we define the integral

$$\int \varphi\, d\mu = \sum_{j=1}^{N} a_j \mu(S_j). \tag{3.17}$$

This quantity is possibly $+\infty$, if $\mu(S_j) = \infty$ for some S_j. However, if $a_j = 0$, then by convention we take $a_j \mu(S_j) = 0$ whether or not $\mu(S_j)$ is finite.

Note that by (3.16)

$$\sum_j a_j \mu(S_j) = \sum_j \sum_{\sigma \ni j} a_j \mu(\widetilde{S}_\sigma) = \sum_\sigma \tilde{a}_\sigma \mu(\widetilde{S}_\sigma). \tag{3.18}$$

Note also that if the set of nonzero values taken on by φ is $\{b_1, \dots, b_M\}$, then

$$\varphi(x) = \sum_{k=1}^{M} b_k \chi_{T_k}(x), \quad T_k = \{x \in X : \varphi(x) = b_k\}. \tag{3.19}$$

Each T_k is a union of sets $\widetilde{S}_\sigma$ in (3.16), and each corresponding coefficient $\tilde{a}_\sigma$ is equal to b_k. It follows that

$$\int \varphi\, d\mu = \sum_\sigma \tilde{a}_\sigma \mu(\widetilde{S}_\sigma) = \sum_k b_k \mu(T_k), \tag{3.20}$$

the first identity by (3.18) and the second by additivity of μ. This identity shows that $\int \varphi\, d\mu$ is well defined, independently of the particular representation of φ in (3.13).

If $\varphi, \psi \in \mathfrak{S}^+(X)$, one also verifies immediately that

$$\int c\varphi\, d\mu = c \int \varphi\, d\mu, \quad \int (\varphi + \psi)\, d\mu = \int \varphi\, d\mu + \int \psi\, d\mu, \tag{3.21}$$

and

$$\varphi \le \psi \Longrightarrow \int \varphi\, d\mu \le \int \psi\, d\mu. \tag{3.22}$$

The following is also easy, though worthy of the status of a lemma.

Lemma 3.3. *If $\varphi \in \mathfrak{S}^+(X)$, then*

$$\lambda(A) = \int_A \varphi\, d\mu = \int \chi_A\, \varphi\, d\mu \tag{3.23}$$

is a measure on $\mathfrak{F}$.

Here, the last identity is a definition of $\int_A \varphi \, d\mu$. Note that if φ is given by (3.13), then

$$\lambda(A) = \sum_{j=1}^{N} a_j \mu(A \cap S_j). \tag{3.24}$$

The proof of Lemma 3.3 is an exercise.

Now let $\mathcal{M}^+(X)$ denote the set of all measurable functions $f : X \to [0, \infty]$. For $f \in \mathcal{M}^+(X)$, we define the integral

$$\int f \, d\mu = \sup \Bigl\{ \int \varphi \, d\mu : 0 \le \varphi \le f, \ \varphi \in \mathfrak{S}^+(X) \Bigr\}. \tag{3.25}$$

This is somewhat analogous to the definition of the lower Riemann integral $\underline{I}(f)$. One difference is that the quantity defined by (3.25) might be ∞. Our hypothesis that f is measurable makes it unnecessary to introduce also an analogue of the upper integral. That (3.25) agrees with (3.17) for $f = \varphi \in \mathfrak{S}^+(X)$ follows from (3.22).

Note that, for any $f \in \mathcal{M}^+(X)$, we can find a sequence

$$\varphi_j \in \mathfrak{S}^+(X), \quad \varphi_j \nearrow f. \tag{3.26}$$

For example, set $\varphi_j(x) = k/2^j$ on $\{x \in X : k 2^{-j} \le f(x) < (k+1)2^{-j}\}$ if $k < j \cdot 2^j$, and set $\varphi_j(x) = j$ on $\{x \in X : j \le f(x)\}$. If $K < \infty$ and $0 \le f(x) \le K$ on $S \subset X$, then $\varphi_j \to f$ uniformly on S, as $j \to \infty$.

When (3.26) holds, one should expect $\int \varphi_j \, d\mu$ to converge to $\int f \, d\mu$. That this is true is a special case of the following central result, known as the Monotone Convergence Theorem.

Theorem 3.4. *If* $f_j \in \mathcal{M}^+(X)$ *and* $f_j \nearrow$, *then* $f_j \nearrow f \in \mathcal{M}^+(X)$, *and*

$$\int f \, d\mu = \lim_{n \to \infty} \int f_n \, d\mu. \tag{3.27}$$

Proof. Given that $f_j(x) \nearrow$, the limit certainly exists for each x, though it might be $+\infty$. The measurability of the limit f follows from Proposition 3.2. Clearly $\int f_n \, d\mu \le \int f \, d\mu$ for all n, and $\int f_n \, d\mu$ is increasing, so $\Phi = \lim \int f_n \, d\mu$ exists, and $\Phi \le \int f \, d\mu$. We need the reverse inequality.

To get it, take $\alpha \in (0,1)$. Let $\varphi \in \mathfrak{S}^+(X)$, $\varphi \le f$. Let $E_n = \{x \in X : f_n(x) \ge \alpha \varphi(x)\}$. Then $\{E_n\}$ is an increasing sequence of measurable sets whose union is X, and

$$\int f_n \, d\mu \ge \int_{E_n} f_n \, d\mu \ge \alpha \int_{E_n} \varphi \, d\mu = \alpha \lambda(E_n), \tag{3.28}$$

where $\lambda(E_n)$ is as in (3.23). By Lemma 3.3, together with the property (3.5), we have

$$\lim \int_{E_n} \varphi \, d\mu = \int \varphi \, d\mu. \tag{3.29}$$

Hence (3.28) implies $\Phi \geq \alpha \int \varphi \, d\mu$, for all $\alpha \in (0,1)$. Letting $\alpha \to 1$, we obtain $\Phi \geq \int \varphi \, d\mu$, for all $\varphi \in \mathfrak{S}^+(X)$ such that $\varphi \leq f$. In view of the definition (3.25), this implies $\Phi \geq \int f \, d\mu$, and the proof is done.

We next establish an additivity result:

Proposition 3.5. *If $f_n \in \mathcal{M}^+(X)$ is a countable sequence and $f = \sum_{n\geq 1} f_n$, then*

$$\int f \, d\mu = \sum_{n\geq 1} \int f_n \, d\mu. \tag{3.30}$$

Proof. First consider the case $f = f_1 + f_2$. As in (3.26), take $\varphi_{nj} \in \mathfrak{S}^+(X)$ such that $\varphi_{nj} \nearrow f_n$ as $j \to \infty$. Then $\varphi_{1j} + \varphi_{2j} \nearrow f_1 + f_2$, so, by the Monotone Convergence Theorem together with (3.21), we have

$$\begin{aligned} \int (f_1 + f_2) \, d\mu &= \lim \int (\varphi_{1j} + \varphi_{2j}) \, d\mu \\ &= \lim \int \varphi_{1j} \, d\mu + \lim \int \varphi_{2j} \, d\mu \\ &= \int f_1 \, d\mu + \int f_2 \, d\mu. \end{aligned} \tag{3.31}$$

By induction,

$$\int \Big(\sum_{n=1}^{N} f_n \Big) \, d\mu = \sum_{n=1}^{N} \int f_n \, d\mu, \tag{3.32}$$

given $f_n \in \mathcal{M}^+(X)$, for any $N < \infty$, and applying the Monotone Convergence Theorem again yields (3.30).

The next result is known as Fatou's Lemma.

Proposition 3.6. *If $f_n \in \mathcal{M}^+(X)$, then*

$$\int \big(\liminf f_n\big) \, d\mu \leq \liminf \int f_n \, d\mu. \tag{3.33}$$

Proof. For each $k \in \mathbb{Z}^+$, $j \geq k \Rightarrow \inf_{n \geq k} f_n \leq f_j$, so

$$\int \inf_{n \geq k} f_n \, d\mu \leq \inf_{j \geq k} \int f_j \, d\mu. \tag{3.34}$$

Of course, $\inf_{n \geq k} f_n \nearrow \liminf f_n$ as $k \to \infty$. Thus, letting $k \to \infty$ in (3.25) and applying the Monotone Convergence Theorem, we have

$$\int (\liminf f_n) \, d\mu = \lim_{k \to \infty} \int \big(\inf_{n \geq k} f_n\big) \, d\mu \leq \liminf \int f_j \, d\mu, \tag{3.35}$$

which gives (3.33).

We now tackle the definition of $\int f \, d\mu$ when $f(x)$ is not necessarily ≥ 0 everywhere. For a general measurable function $f : X \to \mathbb{R}$, we can write $f = f^+ - f^-$, where $f^+(x) = f(x)$ when $f(x) \geq 0$, $f^+(x) = 0$ otherwise, and $f^-(x) = -f(x)$ where $f(x) \leq 0$, $f^-(x) = 0$ otherwise. Thus $f^+, f^- \in \mathcal{M}^+(X)$. We define the integral

$$\int f \, d\mu = \int f^+ \, d\mu - \int f^- \, d\mu, \tag{3.36}$$

provided at least one of the terms on the right is finite. If they are both finite or, equivalently, if

$$\int |f| \, d\mu < \infty, \tag{3.37}$$

we say f is *integrable* and write

$$f \in \mathcal{L}^1(X, \mu). \tag{3.38}$$

For short, we might write $f \in \mathcal{L}^1(X)$ or simply $f \in \mathcal{L}^1$.

The following basic result is parallel in statement, though not in proof, to Proposition 1.1.

Proposition 3.7. *The set $\mathcal{L}^1(X, \mu)$ is a linear space, and the map*

$$\mathfrak{I} : \mathcal{L}^1(X, \mu) \longrightarrow \mathbb{R}, \quad \mathfrak{I}(f) = \int f \, d\mu \tag{3.39}$$

is a linear map.

Proof. From $|f_1 + f_2| \le |f_1| + |f_2|$, it easily follows that $f_j \in \mathcal{L}^1(X, \mu) \Rightarrow f_1 + f_2 \in \mathcal{L}^1(X, \mu)$. Also, clearly $a_j \in \mathbb{R} \Rightarrow a_j f_j \in \mathcal{L}^1(X, \mu)$ in this case. Let us establish that, if $f_j \in \mathcal{L}^1(X, \mu)$, then

$$\int (f_1 + f_2)\, d\mu = \int f_1\, d\mu + \int f_2\, d\mu. \tag{3.40}$$

Indeed, if $g = f_1 + f_2$, then $g^+ - g^- = f_1^+ - f_1^- + f_2^+ - f_2^-$, so $g^+ + f_1^- + f_2^- = g^- + f_1^+ + f_2^+$. Hence, by Proposition 3.5,

$$\begin{aligned} \int (g^+ + f_1^- + f_2^-)\, d\mu &= \int g^+\, d\mu + \int f_1^-\, d\mu + \int f_2^-\, d\mu \\ = \int (g^- + f_1^+ + f_2^+)\, d\mu &= \int g^-\, d\mu + \int f_1^+\, d\mu + \int f_2^+\, d\mu. \end{aligned} \tag{3.41}$$

Equating the right sides, we get (3.40). The other ingredient in linearity,

$$\int af\, d\mu = a \int f\, d\mu, \quad \text{for } f \in \mathcal{L}^1(X, \mu),\ a \in \mathbb{R}, \tag{3.42}$$

is easy.

We can also integrate a *complex-valued* function. If $f : X \to \mathbb{C}$, write $f = f_1 + i f_2$ where f_j are real valued. We say f is measurable if f_1 and f_2 are both measurable. If $f_1, f_2 \in \mathcal{L}^1(X, \mu)$, we set

$$\int f\, d\mu = \int f_1\, d\mu + i \int f_2\, d\mu. \tag{3.43}$$

Often, we use the same notation, $\mathcal{L}^1(X, \mu)$, to denote the set of such complex-valued integrable functions. The rest of the results of this section will deal specifically with real-valued functions, but the reader should be able to formulate extensions to complex-valued functions.

The next result is known as the Dominated Convergence Theorem.

Theorem 3.8. *Let $f_n \in \mathcal{L}^1(X, \mu)$, for $n \in \mathbb{Z}^+$. Suppose $f : X \to \mathbb{R}$ and*

$$f_n(x) \to f(x) \tag{3.44}$$

for all $x \in X$, and suppose that there is a function g satisfying

$$g \in \mathcal{L}^1(X, \mu), \quad |f_n| \le g, \ \text{for all } n. \tag{3.45}$$

Then

$$f \in \mathcal{L}^1(X, \mu) \ \text{ and } \int f\, d\mu = \lim_{n \to \infty} \int f_n\, d\mu. \tag{3.46}$$

Proof. We know f is measurable; of course then $|f| \le g \Rightarrow f \in \mathcal{L}^1$. Now if g satisfies (3.45), then $g + f_n \ge 0$. Thus, by Fatou's Lemma and (3.40),

$$\begin{aligned} \int g\,d\mu + \int f\,d\mu &= \int (g+f)\,d\mu \\ &\le \liminf \int (g+f_n)\,d\mu \\ &= \int g\,d\mu + \liminf \int f_n\,d\mu, \end{aligned} \tag{3.47}$$

and hence

$$\int f\,d\mu \le \liminf \int f_n\,d\mu, \tag{3.48}$$

whenever the hypotheses of Theorem 3.8 hold. Then the same argument applies with f_n replaced by $-f_n$, so $\int(-f)\,d\mu \le \liminf \int(-f_n)\,d\mu$ or, equivalently,

$$\int f\,d\mu \ge \limsup \int f_n\,d\mu. \tag{3.49}$$

The inequalities (3.48)–(3.49) yield (3.46).

The next result is known as Egoroff's Theorem. It has many uses, including an alternative approach to the proof of the Dominated Convergence Theorem (explored in some of the exercises below).

Proposition 3.9. *Let $(X, \mathfrak{F}, \mu)$ be a measure space such that $\mu(X) < \infty$. Assume f_j are $\mathfrak{F}$-measurable and $f_j(x) \to f(x)$ for all $x \in X$. Then, given any $\delta > 0$, there exists $B \in \mathfrak{F}$ such that*

$$\mu(B) \le \delta \text{ and } f_j \to f \text{ uniformly on } X \setminus B. \tag{3.50}$$

Proof. For $\varepsilon > 0$, $n \in \mathbb{Z}^+$, set

$$F_{n\varepsilon} = \{x \in X : \text{ for some } j \ge n,\ |f_j(x) - f(x)| \ge \varepsilon\}. \tag{3.51}$$

Note that each set $F_{n\varepsilon}$ belongs to $\mathfrak{F}$. The hypothesis implies $\bigcap_{n\ge 1} F_{n\varepsilon} = \emptyset$, for each $\varepsilon > 0$, so $\mu(F_{n\varepsilon}) \to 0$ as $n \to \infty$.

In particular, given any $\delta > 0$, if we set $\varepsilon = 2^{-k}$, there exists $n = n(k)$ such that

$$\mu\big(F_{n(k),2^{-k}}\big) \le 2^{-k}\delta. \tag{3.52}$$

Let $B = \bigcup_{k\geq 1} F_{n(k),2^{-k}}$, $B \in \mathfrak{F}$. We see that $\mu(B) \leq \delta$ and

$$x \notin B,\ j \geq n(k) \Longrightarrow |f_j(x) - f(x)| < 2^{-k}. \tag{3.53}$$

This establishes (3.50).

We mention a point which is quite simple but important for applications. Namely, in all the convergence theorems established above, the hypothesis that a sequence of measurable functions f_j converges for all $x \in X$ can be weakened to the hypothesis that f_j converges *almost everywhere*, i.e., f_j converges for all $x \in X_1$, where $X_1 \subset X$ is a measurable subset such that $\mu(X \setminus X_1) = 0$. The proof is easy; just restrict attention to the measure space $(X_1, \mathfrak{F}_1, \mu_1)$, where $\mathfrak{F}_1 = \{S \in \mathfrak{F} : S \subset X_1\} = \{S \cap X_1 : S \in \mathfrak{F}\}$ and μ_1 is μ restricted to $\mathfrak{F}_1$.

To close this chapter, we relate the Riemann integral, considered in Chapter 1, to the Lebesgue integral, which is the integral defined in this chapter when μ is Lebesgue measure. The theory of the Lebesgue integral allows us to specify precisely which bounded functions on an interval $I = [a,b]$ are Riemann integrable. Indeed, let $\varphi : I \to \mathbb{R}$ be a bounded function, and recall the definition (1.3)–(1.6) of $\overline{I}(\varphi)$ and $\underline{I}(\varphi)$. Let $\mathcal{P}_j$ be the sequence of partitions of I such that $\mathcal{P}_0 = \{I\}$ and $\mathcal{P}_{j+1}$ is obtained by dividing each interval of $\mathcal{P}_j$ in half. Define f_j to be $\sup_J \varphi(x)$ on each interval $J \in \mathcal{P}_j$ and define g_j to be $\inf_J \varphi(x)$ on each such interval J. (Here, omit the right endpoint of J, unless it is the point b.) Then $f_j \searrow$ while $g_j \nearrow$ and, by (1.42),

$$\int_I f_j\, dx \searrow \overline{I}(\varphi), \quad \int_I g_j\, dx \nearrow \underline{I}(\varphi). \tag{3.54}$$

On the other hand, we see that

$$f_j \searrow f \geq \varphi, \quad g_j \nearrow g \leq \varphi, \tag{3.55}$$

where f and g are bounded and *measurable*, and, by the Monotone Convergence Theorem, together with (3.54), we have

$$\overline{I}(\varphi) = \int_I f\, dx, \quad \underline{I}(\varphi) = \int_I g\, dx. \tag{3.56}$$

Consequently φ is Riemann integrable if and only if $\int (f - g)\, dx = 0$, hence if and only if $f(x) = g(x)$ for almost all $x \in I$. (See Exercise 4 below.) On the other hand, if E is the collection of endpoints of all the intervals in the partitions $\mathcal{P}_j$ (a countable set, hence of Lebesgue measure zero), we have

$$\forall\, x \notin E, \quad f(x) = g(x) \Longleftrightarrow \varphi \text{ is continuous at } x. \tag{3.57}$$

Thus we have the following proposition.

Proposition 3.10. *If $\varphi : I \to \mathbb{R}$ is bounded, then φ is Riemann integrable if and only if the set of points $x \in I$ at which φ is discontinuous has Lebesgue measure zero. If this condition holds, then the Riemann integral and the Lebesgue integral of φ coincide.*

Exercises

1. Suppose $\mathfrak{F}$ is an algebra of subsets of X. Show that $\mathfrak{F}$ is a σ-algebra, provided it is closed under countable *disjoint* unions.

2. Establish the properties (3.3)–(3.6) of a measure.

3. Prove Lemma 3.3.
Hint. Show that, if $\mu_1, \dots, \mu_N$ are measures on X and $a_j \in [0, \infty)$, then
$$\lambda = a_1\mu_1 + \cdots + a_N\mu_N$$
is a measure. Also show that, if μ is a measure on X and $E_j \subset X$ are measurable, then
$$\mu_j(S) = \mu(S \cap E_j) = \int \chi_{E_j}\, d\mu \tag{3.58}$$
are measures.

4. Let $(X, \mathfrak{F}, \mu)$ be a measure space and $f \in \mathcal{M}^+(X)$. Show that
$$f = 0 \ \mu\text{-a.e.} \Longleftrightarrow \int f\, d\mu = 0. \tag{3.59}$$
To say that $f = 0$ μ-a.e. is to say that, for $\mathcal{N} = \{x \in X : f(x) \neq 0\}$, $\mu(\mathcal{N}) = 0$.

5. A measure μ on $(X, \mathcal{M})$ is said to be *complete* provided
$$A \in \mathcal{M},\ \mu(A) = 0,\ S \subset A \Longrightarrow S \in \mathcal{M},\ \text{and}\ \mu(S) = 0.$$
Suppose μ is a measure on $(X, \mathfrak{F})$ that is not complete. Show that
$$\overline{\mathfrak{F}} = \{E \cup S : E \in \mathfrak{F},\ S \subset F \in \mathfrak{F},\ \mu(F) = 0\} \tag{3.60}$$
is a σ-algebra, that there is a unique measure $\overline{\mu}$ on $(X, \overline{\mathfrak{F}})$ such that $\overline{\mu} = \mu$ on $\mathfrak{F}$, and that $\overline{\mu}$ is complete. We call $\overline{\mu}$ the *completion* of μ.

6. Let μ be a measure on $(X, \mathfrak{F})$, as above, and $\overline{\mu}$ its completion, on $(X, \overline{\mathfrak{F}})$. If $f : X \to \mathbb{R}$ is $\overline{\mathfrak{F}}$-measurable, show that there exists f_0 that is $\mathfrak{F}$-measurable, such that $f_0 = f$ outside a set of $\overline{\mu}$-measure zero.
Hint. First treat the case of simple functions, then positive functions, via the approximation (3.26).

7. Let $f \in \mathcal{M}^+(X)$ and set, for $A \in \mathfrak{F}$,

$$\lambda(A) = \int_A f \, d\mu = \int \chi_A f \, d\mu. \tag{3.61}$$

Show that λ is a measure.
Hint. Use the Monotone Convergence Theorem.
Note that this result is much stronger than Lemma 3.3.

8. Let $f \in \mathcal{L}^1(X, \mu)$ be given. Show that, for every $\varepsilon > 0$, there is a $\delta > 0$ such that

$$S \in \mathfrak{F}, \ \mu(S) < \delta \Longrightarrow \int_S |f| \, d\mu < \varepsilon. \tag{3.62}$$

Hint. Pick $\varphi \in \mathfrak{S}^+(X)$ such that $0 \le \varphi \le |f|$ and $\int \varphi \, d\mu \ge \int |f| \, d\mu - \varepsilon/2$. Then $A = \sup \ \varphi < \infty$, so pick $\delta < \varepsilon/2A$.

9. Give an alternative proof of the Dominated Convergence Theorem, deducing it from Egoroff's Theorem and the result of Exercise 8, rather than from Fatou's Lemma.
Hint. Given $|f_n| \le g \in \mathcal{L}^1(X, \mu), \ f_n \to f, \ \varepsilon > 0$, pick a measurable set $Y \subset X$ such that

$$\mu(Y) < \infty, \qquad \int_{X \setminus Y} g \, d\mu < \varepsilon, \tag{3.63}$$

and pick a measurable $S \subset Y$, such that $\mu(S) < \varepsilon$ and $f_n \to f$ uniformly on $Y \setminus S$.

10. Let f_j be measurable functions on $(X, \mathfrak{F}, \mu)$. We say $f_j \to f$ *in measure* provided $\mu(E_{j\varepsilon}) \to 0$ as $j \to \infty$, for each $\varepsilon > 0$, where

$$E_{j\varepsilon} = \{x \in X : |f_j(x) - f(x)| > \varepsilon\}.$$

Show that, if $f_j \to f$ in measure, then some subsequence $f_{j_\nu} \to f$ almost everywhere.

Hint. Arange that $\mu(E_{j_\nu, 2^{-\nu}}) \le 2^{-\nu}$.

11. Show that the Dominated Convergence Theorem continues to hold if the hypothesis (3.44) that $f_n \to f$ pointwise is changed to $f_n \to f$ in measure.

12. Let $(X, \mathfrak{F}, \mu)$ be a measure space and assume $\mu(X) < \infty$. Let $\mathcal{M}(X, \mathfrak{F})$ denote the space of measurable functions and let $M(X, \mathfrak{F}, \mu)$ denote the set of equivalence classes of measurable functions, where

$$f \sim g \Longleftrightarrow f(x) = g(x), \quad \mu\text{-a.e.}$$

For measurable f and g, set

$$d(f, g) = \int_X \frac{|f - g|}{1 + |f - g|} \, d\mu. \tag{3.64}$$

Show that this is a distance function, making $M(X, \mathfrak{F}, \mu)$ a metric space, and that $f_n \to f$ in this metric if and only if $f_n \to f$ in measure, as defined in Exercise 10.

13. Show that, if $\mu(X) < \infty$, then $f_n \to f$, μ-a.e. implies $f_n \to f$ in measure. Give a counterexample for $X = \mathbb{Z}$, with counting measure.

In Exercises 14–18, we deal with Lebesgue measure on the line $\mathbb{R}$ or on an interval $I = [a, b]$. A set $S \subset \mathbb{R}$ is called an F_σ set if $S = \bigcup_{j \ge 1} F_j$, F_j closed. S is called a G_δ set if $S = \bigcap_{j \ge 1} G_j$, G_j open.

14. If $\mathcal{O} \subset I$ is open, show that there exist $f_j \in C(I)$ such that $0 \le f_j(x) \nearrow \chi_{\mathcal{O}}(x)$ for all $x \in I$. If $K \subset I$ is closed, show that there exist $f_j \in C(I)$ such that $1 \ge f_j(x) \searrow \chi_K(x)$ for all $x \in I$.

15. If S is an F_σ set or a G_δ set, show that there exist $f_j \in C(I)$ such that $0 \le f_j \le 1$ and $f_j(x) \to \chi_S(x)$ for almost all $x \in I$.
Hint. Tackle the question of convergence in measure.

16. Let $S \subset I$ be Lebesgue measurable. Show that there exists an F_σ set S_0 and a G_δ set S_1 such that $S_0 \subset S \subset S_1$ and $m(S_1 \setminus S_0) = 0$. Deduce that there exist $f_j \in C(I)$ such that $0 \le f_j \le 1$ and $f_j \to \chi_S$ almost everywhere.

17. Let $f \in \mathcal{L}^1(I, dx)$. Show that there exist $f_j \in C(I)$ such that $f_j \to f$ almost everywhere.
Hint. First approximate f by simple functions, as in (3.26).

18. Given $f \in \mathcal{L}^1(I, dx)$ and $\delta > 0$, show that there is a Lebesgue measurable $S \subset I$ such that $m(S) < \delta$ and $f\big|_{I \setminus S}$ is *continuous.*
Hint. Apply Egoroff's Theorem to the results of Exercise 17.
This result is known as Lusin's Theorem.

19. If $(X, \mathfrak{F}, \mu)$ is a measure space and $f : X \to [0, \infty]$ is measurable, show that there exist $A_j \in \mathfrak{F}$ such that

$$f = \sum_{j \geq 1} \frac{1}{j} \, \chi_{A_j}. \tag{3.65}$$

Hint. Set $A_1 = \{x \in X : f(x) \geq 1\}$ and, inductively, for $k \geq 2$,

$$A_k = \Big\{ x \in X : f(x) \geq \frac{1}{k} + \sum_{j=1}^{k-1} \frac{1}{j} \chi_{A_j}(x) \Big\}.$$

20. Let $(X, \mathfrak{F}, \mu)$ be a measure space. Given $f \in \mathcal{M}^+(X)$, $t \in (0, \infty)$, set

$$S_f(t) = \{x \in X : f(x) > t\}, \quad \Psi_f(t) = \mu\big(S_f(t)\big). \tag{3.66}$$

Show that

$$\int_X f \, d\mu = \int_0^\infty \Psi_f(t) \, dt. \tag{3.67}$$

Hint. First verify this for simple functions. Then use (3.26) and the Monotone Convergence Theorem. Show that, if $\varphi_j \nearrow f$, then, for each $t \in (0, \infty)$,

$$S_{\varphi_1}(t) \subset S_{\varphi_2}(t) \subset \cdots \nearrow S_f(t)$$

to deduce that $\int \Psi_{\varphi_j}(t) \, dt \nearrow \int \Psi_f(t) \, dt$.

21. Let $F : [0, 1] \to [0, 1]$ be the homeomorphism described in Exercise 10 of Chapter 2, satisfying

$$F(L) = K, \quad m(L) > 0, \quad m(K) = 0.$$

Show that there is a subset $S \subset K$ such that $F^{-1}(S)$ is not Lebesgue measurable.

Hint. Apply Exercise 7 of Chapter 2 to $X = L$.

22. Let u_n be measurable functions on $(X, \mathfrak{F}, \mu)$. Assume

$$u_n \to u \text{ pointwise}, \quad u_n \geq 0, \quad \int u_n \, d\mu \to \int u \, d\mu < \infty.$$

Prove that

$$\int |u - u_n| \, d\mu \to 0.$$

Hint. Write

$$|u - u_n| = 2(u - u_n)^+ - (u - u_n)$$

and apply the Dominated Convergence Theorem to the first term on the right side.

23. Below we display the "guts" (though not the skeleton or the flesh) of the derivation of Fatou's Lemma and the Dominated Convergence Theorem (DCT) from the Monotone Convergence Theorem (MCT), as done in the proofs of Proposition 3.6 and Theorem 3.8. Familiarize yourself with the details of these arguments, to the point where you can look at the material displayed below and see the entire argument.

$$\textbf{MCT} \Longrightarrow \textbf{Fatou} \Longrightarrow \textbf{DCT.}$$

(Fatou)

$$\int \liminf f_n \, d\mu = \underset{\substack{\uparrow \\ \text{MCT}}}{\lim_{k\to\infty}} \int \Big(\inf_{n \geq k} f_n\Big) \, d\mu \leq \liminf \int f_j \, d\mu,$$

(DCT)

$$\int (g + f) \, d\mu \underset{\substack{\uparrow \\ \text{Fatou}}}{\leq} \liminf \int (g + f_n) \, d\mu = \int g \, d\mu + \liminf \int f_n \, d\mu$$

$$\Big(\text{analogue for } f_n \text{ replaced by } - f_n\Big).$$

Chapter 4

L^p Spaces

Let (X, μ) be a measure space. As in Chapter 3, we say a measurable function f belongs to $\mathcal{L}^1(X, \mu)$ provided

$$\|f\|_{L^1} = \int_X |f(x)|\, d\mu(x) < \infty. \tag{4.1}$$

Elements of $L^1(X, \mu)$ consist of equivalence classes of elements of $\mathcal{L}^1(X, \mu)$, where we say

$$f \sim \tilde{f} \Leftrightarrow f(x) = \tilde{f}(x) \text{ for } \mu\text{-almost every } x. \tag{4.2}$$

With a slight abuse of notation, we denote by f both a measurable function in $\mathcal{L}^1(X, \mu)$ and its equivalence class in $L^1(X, \mu)$. Also we say f, defined only almost everywhere on X, belongs to $L^1(X, \mu)$, if there exists $\tilde{f} \in \mathcal{L}^1(X, \mu)$, equal μ-almost everywhere to f. The quantity $\|f\|_{L^1}$ defined by (4.1) is called the L^1 *norm* of f.

In general, a normed linear space is a vector space equipped with a positive function $\|v\|$ having the properties

$$\begin{aligned} \|av\| &= |a| \cdot \|v\|, \text{ for } v \in V,\ a \in \mathbb{C} \text{ (or } \mathbb{R}), \\ \|v + w\| &\le \|v\| + \|w\|, \\ \|v\| &> 0, \text{ unless } v = 0. \end{aligned} \tag{4.3}$$

The second of these conditions is called the *triangle inequality*. Given a norm on V, setting $d(u, v) = \|u - v\|$ defines a distance function on V, making it a metric space.

It is easy to see that $\mathcal{L}^1(X, \mu)$ is a vector space and that $\|f\|_{L^1}$ satisfies the first two conditions in (4.3). However, $\|f\|_{L^1} = 0$ if and only if $f = 0$

almost everywhere. (Recall Exercise 4 of Chapter 3.) That is the reason we define $L^1(X,\mu)$ to consist of equivalence classes defined by (4.2), so $L^1(X,\mu)$ becomes a normed linear space.

Generally speaking, a sequence (v_j) in a normed linear space is said to be a Cauchy sequence if $\|v_j - v_k\| \to 0$ as $j,k \to \infty$. If every Cauchy sequence has a limit in V, then V is said to be complete; a complete normed linear space is called a Banach space.

Theorem 4.1. *$L^1(X,\mu)$ is a Banach space.*

The proof of completeness of $L^1(X,\mu)$ makes use of the following two lemmas, which are essentially restatements of the Monotone Convergence Theorem and the Dominated Convergence Theorem, respectively.

Lemma 4.2. *If $f_j \in \mathcal{L}^1(X,\mu)$, $0 \le f_1(x) \le f_2(x) \le \cdots$, and $\|f_j\|_{L^1} \le C < \infty$, then $\lim\limits_{j\to\infty} f_j(x) = f(x)$, with $f \in L^1(X,\mu)$ and $\|f_j - f\|_{L^1} \to 0$ as $j \to \infty$.*

Proof. We know that $f \in \mathcal{M}^+(X)$. The Monotone Convergence Theorem implies $\int f_j\,d\mu \nearrow \int f\,d\mu$. Thus $\int f\,d\mu \le C$. Since $\|f_j - f\|_{L^1} = \int f\,d\mu - \int f_j\,d\mu$ in this case, the lemma follows.

Lemma 4.3. *If $f_j \in \mathcal{L}^1(X,\mu), \lim f_j(x) = f(x)$ μ-a.e., and if there is an $F \in \mathcal{L}^1(X,\mu)$ such that $|f_j(x)| \le F(x)$ μ-a.e., for all j, then $f \in L^1(X,\mu)$, and $\|f_j - f\|_{L^1} \to 0$.*

Proof. Apply the Dominated Convergence Theorem to $g_j = |f_j - f| \to 0$ a.e. Note that $|g_j| \le 2F$.

To show $L^1(X,\mu)$ is complete, suppose (f_n) is Cauchy in L^1. Passing to a subsequence, we can assume $\|f_{n+1} - f_n\|_{L^1} \le 2^{-n}$. Consider the infinite series

$$f_1(x) + \sum_{n=1}^{\infty} \big[f_{n+1}(x) - f_n(x)\big]. \tag{4.4}$$

Now the partial sums are dominated by

$$|f_1(x)| + \sum_{n=1}^{m} |f_{n+1}(x) - f_n(x)| = |f_1(x)| + G_m(x), \tag{4.5}$$

and since $0 \le G_1 \le G_2 \le \cdots$ and $\|G_m\|_{L^1} \le \sum 2^{-n} \le 1$, we deduce from Lemma 4.2 that $G_m \nearrow G$ μ-a.e. and in L^1-norm. Hence the infinite series

(4.4) is convergent a.e., to a limit $f(x)$, and via Lemma 4.3 we deduce that $f_n \to f$ in L^1-norm. This proves completeness.

Continuing with a description of L^p spaces, we define $\mathcal{L}^\infty(X,\mu)$ to consist of bounded measurable functions, $L^\infty(X,\mu)$ to consist of equivalence classes of such functions, via (4.2), and we define $\|f\|_{L^\infty}$ to be the smallest sup of $\tilde{f} \sim f$. It is easy to show that $L^\infty(X,\mu)$ is a Banach space.

For $p \in (1,\infty)$, we define $\mathcal{L}^p(X,\mu)$ to consist of measurable functions f such that

$$\left[\int_X |f(x)|^p \, d\mu(x)\right]^{1/p} \tag{4.6}$$

is finite. $L^p(X,\mu)$ consists of equivalence classes, via (4.2), and the L^p-norm $\|f\|_{L^p}$ is given by (4.6). This time it takes a little work to verify the triangle inequality. That this holds is the content of *Minkowski's inequality*:

$$\|f+g\|_{L^p} \le \|f\|_{L^p} + \|g\|_{L^p}. \tag{4.7}$$

One neat way to establish this is by the following characterization of the L^p-norm. Suppose p and q are related by

$$\frac{1}{p} + \frac{1}{q} = 1. \tag{4.8}$$

We claim that, if $f \in L^p(X,\mu)$,

$$\|f\|_{L^p} = \sup\{\|fh\|_{L^1} : h \in L^q(X,\mu),\ \|h\|_{L^q} = 1\}. \tag{4.9}$$

We can apply (4.9) to $f+g$, which belongs to $L^p(X,\mu)$ if f and g do, since $|f+g|^p \le 2^p(|f|^p + |g|^p)$. Given this, (4.7) follows easily from the inequality $\|(f+g)h\|_{L^1} \le \|fh\|_{L^1} + \|gh\|_{L^1}$.

The identity (4.9) can be regarded as two inequalities. The "$\le$" part can be proved by choosing $h(x)$ to be an appropriate multiple $C|f(x)|^{p-1}$. We leave this as an exercise. The converse inequality, "$\ge$," is a consequence of *Hölder's inequality*:

$$\int |f(x)g(x)| \, d\mu(x) \le \|f\|_{L^p}\|g\|_{L^q}, \quad \frac{1}{p} + \frac{1}{q} = 1. \tag{4.10}$$

Hölder's inequality can be proved via the following inequality for positive numbers:

$$ab \le \frac{a^p}{p} + \frac{b^q}{q} \quad \text{for } a, b > 0, \tag{4.11}$$

assuming that $p \in (1, \infty)$ and (4.8) holds. In fact, we claim that, given $a, b > 0$, $1/p + 1/q = 1$,

$$\varphi(t) = \frac{a^p}{p} t^p + \frac{b^q}{q} t^{-q} \Longrightarrow \inf_{t>0} \varphi(t) = ab, \tag{4.12}$$

which implies (4.11) since the right side of (4.11) is $\varphi(1)$. As for (4.12), note that $\varphi(t) \to \infty$ as $t \searrow 0$ and as $t \nearrow \infty$, and the unique critical point occurs for $a^p t^p = b^q t^{-q}$, i.e., for $t = b^{1/p}/a^{1/q}$, giving the desired conclusion.

Applying (4.11) to the integrand in (4.10) gives

$$\int |f(x)g(x)|\, d\mu(x) \le \frac{1}{p} \|f\|_{L^p}^p + \frac{1}{q} \|g\|_{L^q}^q. \tag{4.13}$$

This looks weaker than (4.10), but now replace f by tf and g by $t^{-1}g$, so that the left side of (4.13) is dominated by

$$\frac{t^p}{p} \|f\|_{L^p}^p + \frac{1}{qt^q} \|g\|_{L^q}^q, \tag{4.14}$$

for all $t > 0$. Another application of (4.12) then gives Hölder's inequality. Consequently (4.6) defines a norm on $L^p(X, \mu)$. Completeness follows as in the $p = 1$ case discussed above.

In detail, given (f_n) Cauchy in $L^p(X, \mu)$, we can pass to the case $\|f_{n+1} - f_n\|_{L^p} \le 2^{-n}$ and define G_m as in (4.5). We have $\|G_m\|_{L^p} \le 1$, and hence (via the Monotone Convergence Theorem) deduce that $G_m \nearrow G$, μ-a.e., and in L^p-norm. Hence the series (4.4) converges, μ-a.e., to a limit $f(x)$. Since $|f - f_{m+1}| \le G - G_m$, we have by the Dominated Convergence Theorem that

$$\int_X |f - f_{m+1}|^p \, d\mu \le \int_X (G - G_m)^p \, d\mu \to 0,$$

as $m \to \infty$. Hence $L^p(X, \mu)$ is complete. To summarize, we have

Theorem 4.4. *For $p \in [1, \infty)$, $L^p(X, \mu)$, with norm given by (4.6), is a Banach space.*

It is frequently useful to show that a certain linear subspace L of a Banach space V is *dense*. We give an important case of this here; $C(X)$ denotes the space of continuous functions on X.

Proposition 4.5. *If μ is a finite Borel measure on a compact metric space X, then $C(X)$ is dense in $L^p(X, \mu)$ for each $p \in [1, \infty)$.*

Proof. First, let K be any compact subset of X. The functions

$$f_{K,n}(x) = \big[1 + n \operatorname{dist}(x, K)\big]^{-1} \in C(X) \tag{4.15}$$

are all ≤ 1 and decrease monotonically to the characteristic function χ_K equal to 1 on K, 0 on $X \setminus K$. The Monotone Convergence Theorem gives $f_{K,n} \to \chi_K$ in $L^p(X, \mu)$ for $1 \leq p < \infty$. Now let $A \subset X$ be any measurable set. Any Borel measure on a compact metric space is *regular*, i.e.,

$$\mu(A) = \sup\{\mu(K) : K \subset A,\ K \text{ compact}\}. \tag{4.16}$$

In case $X = I = [a, b]$ and $\mu = m$ is Lebesgue measure, this follows from (2.20) together with the consequence of Theorem 2.11, that all Borel sets in I are Lebesgue measurable. The general case follows from results that will be established in the next chapter; see (5.60).

Thus there exists an increasing sequence K_j of compact subsets of A such that $\mu(A \setminus \bigcup_j K_j) = 0$. Again, the Monotone Convergence Theorem implies $\chi_{K_j} \to \chi_A$ in $L^p(X, \mu)$ for $1 \leq p < \infty$. Thus all *simple* functions on X are in the closure of $C(X)$ in $L^p(X, \mu)$ for $p \in [1, \infty)$. Construction of $L^p(X, \mu)$ directly shows that each $f \in L^p(X, \mu)$ is a norm limit of simple functions, so the result is proved.

Using a cut-off, we can easily deduce the following. Let $C_{00}(\mathbb{R})$ denote the space of continuous functions on $\mathbb{R}$ with compact support.

Corollary 4.6. *For $1 \leq p < \infty$, the space $C_{00}(\mathbb{R})$ is dense in $L^p(\mathbb{R})$.*

The case $L^2(X, \mu)$ is special. In addition to the L^2-norm, there is an *inner product*, defined by

$$(f, g)_{L^2} = \int_X f(x)\overline{g(x)}\, d\mu(x). \tag{4.17}$$

This makes $L^2(X, \mu)$ into a *Hilbert space*. It is worthwhile to consider the general notion of Hilbert space in some detail. We devote the next few pages to this and then return to the specific consideration of $L^2(X, \mu)$.

Generally, a Hilbert space H is a complete inner product space. That is to say, first the space H is a linear space provided with an inner product, denoted (u, v), for u and v in H, satisfying the following defining conditions:

$$\begin{aligned} (au_1 + u_2, v) &= a(u_1, v) + (u_2, v), \\ (u, v) &= \overline{(v, u)}, \\ (u, u) &> 0 \text{ unless } u = 0. \end{aligned} \tag{4.18}$$

To such an inner product there is assigned a norm, denoted by

$$\|u\| = \sqrt{(u,u)}. \tag{4.19}$$

To establish that the triangle inequality holds for $\|u+v\|$, we can expand $\|u+v\|^2 = (u+v, u+v)$ and deduce that this is $\leq \big[\|u\| + \|v\|\big]^2$, as a consequence of *Cauchy's inequality*:

$$|(u,v)| \leq \|u\| \cdot \|v\|, \tag{4.20}$$

a result that can be proved as follows. The fact that $(u-v, u-v) \geq 0$ implies $2\,\mathrm{Re}\,(u,v) \leq \|u\|^2 + \|v\|^2$; replacing u by $e^{i\theta}u$ with $e^{i\theta}$ chosen so that $e^{i\theta}(u,v)$ is real and positive, we get

$$|(u,v)| \leq \frac{1}{2}\|u\|^2 + \frac{1}{2}\|v\|^2. \tag{4.21}$$

Now in (4.21) we can replace u by tu and v by $t^{-1}v$ to get

$$|(u,v)| \leq \frac{t}{2}\|u\|^2 + \frac{1}{2t}\|v\|^2. \tag{4.22}$$

Minimizing over t gives (4.20). This establishes Cauchy's inequality, so we can deduce the triangle inequality. Thus (4.19) defines a norm on H. Note the parallel between this argument and the proof of (4.7), via (4.10). The completeness hypothesis on H is that, with this norm, H is a Banach space.

The nice properties of Hilbert spaces arise from their similarity with familiar Euclidean space, so a great deal of geometrical intuition is available. For example, we say u and v are orthogonal and write $u \perp v$, provided $(u,v) = 0$. Note that the Pythagorean Theorem holds on a general Hilbert space:

$$u \perp v \Longrightarrow \|u+v\|^2 = \|u\|^2 + \|v\|^2. \tag{4.23}$$

This follows directly from expanding $(u+v, u+v)$.

Another useful identity is the following, called the "parallelogram law," valid for all $u, v \in H$:

$$\|u+v\|^2 + \|u-v\|^2 = 2\|u\|^2 + 2\|v\|^2. \tag{4.24}$$

This also follows directly by expanding $(u+v, u+v) + (u-v, u-v)$, observing some cancellations. One important application of this simple identity is to the following existence result.

Let K be any closed, convex subset of H. Convexity implies $x, y \in K \Rightarrow (x+y)/2 \in K$. Given $x \in H$, we define the distance from x to K to be

$$d(x,K) = \inf\,\{\|x-y\| : y \in K\}. \tag{4.25}$$

Proposition 4.7. *If $K \subset H$ is a nonempty, closed, convex set in a Hilbert space H and if $x \in H$, then there is a unique $z \in K$ such that $d(x, K) = \|x - z\|$.*

Proof. We can pick $y_n \in K$ such that $\|x - y_n\| \to d = d(x, K)$. It will suffice to show that (y_n) must be a Cauchy sequence. Use (4.24) with $u = y_m - x$, $v = x - y_n$, to get

$$\|y_m - y_n\|^2 = 2\|y_n - x\|^2 + 2\|y_m - x\|^2 - 4\Big\|x - \frac{1}{2}(y_n + y_m)\Big\|^2.$$

Since K is convex, $(y_n + y_m)/2 \in K$, so $\|x - (y_n + y_m)/2\| \geq d$. Therefore

$$\limsup_{m,n\to\infty} \|y_n - y_m\|^2 \leq 2d^2 + 2d^2 - 4d^2 \leq 0,$$

which implies convergence.

In particular, this result applies when K is a closed linear subspace of H. In this case, for $x \in H$, denote by $P_K x$ the point in K closest to x. We have

$$x = P_K x + (x - P_K x). \tag{4.26}$$

We claim that $x - P_K x$ belongs to the closed linear space $K^\perp$, called the orthogonal complement of K, defined by

$$K^\perp = \{u \in H : (u, v) = 0 \text{ for all } v \in K\}. \tag{4.27}$$

Indeed, take any $v \in K$. Then

$$\begin{aligned}\Delta(t) &= \|x - P_K x + tv\|^2 \\ &= \|x - P_K x\|^2 + 2t \operatorname{Re}(x - P_K x, v) + t^2\|v\|^2\end{aligned}$$

is minimal at $t = 0$, so $\Delta'(0) = 0$, i.e., $\operatorname{Re}(x - P_K x, v) = 0$, for all $v \in K$. Replacing v by iv shows that $(x - P_K x, v)$ also has vanishing imaginary part for any $v \in K$, so our claim is established. The decomposition (4.26) gives

$$x = x_1 + x_2, \quad x_1 \in K, \ x_2 \in K^\perp, \tag{4.28}$$

with $x_1 = P_K x$, $x_2 = x - P_K x$. Clearly such a decomposition is unique. This implies that H is an orthogonal direct sum of K and $K^\perp$; we write

$$H = K \oplus K^\perp. \tag{4.29}$$

From this it is clear that

$$(K^\perp)^\perp = K, \tag{4.30}$$

that

$$x - P_K x = P_{K^\perp} x, \tag{4.31}$$

and that P_K and $P_{K^\perp}$ are *linear* maps on H. We call P_K the *orthogonal projection* of H on K. Note that $P_K x$ is uniquely characterized by the condition

$$P_K x \in K, \ (P_K x, v) = (x, v) \text{ for all } v \in K. \tag{4.32}$$

We remark that if K is a linear subspace of H that is not closed, then $K^\perp$ coincides with $\overline{K}^\perp$, and (4.30) becomes $(K^\perp)^\perp = \overline{K}$.

Using the orthogonal projection discussed above, we can establish the following result.

Proposition 4.8. *If H is a Hilbert space and $\varphi : H \to \mathbb{C}$ is a continuous linear map, there exists a unique $f \in H$ such that*

$$\varphi(u) = (u, f) \textit{ for all } u \in H. \tag{4.33}$$

Proof. Consider $K = \operatorname{Ker} \varphi = \{u \in H : \varphi(u) = 0\}$, a closed linear subspace of H. If $K = H$, $\varphi = 0$ and we can take $f = 0$. Otherwise, $K^\perp \neq 0$; select a nonzero $x_0 \in K^\perp$, such that $\varphi(x_0) = 1$. We claim $K^\perp$ is one dimensional in this case. Indeed, given any $y \in K^\perp$, $y - \varphi(y)x_0$ is annihilated by φ, so it belongs to K as well as to $K^\perp$, so it is zero. The result is now easily proved by setting $f = a x_0$ with $a \in \mathbb{C}$ chosen so that (4.33) works for $u = x_0$, namely $\overline{a}(x_0, x_0) = 1$, i.e., $a = \|x_0\|^{-2}$.

We note that the correspondence $\varphi \mapsto f$ gives a *conjugate linear* isomorphism

$$H' \to H, \tag{4.34}$$

where H' denotes the space of all continuous linear maps $\varphi : H \to \mathbb{C}$.

Recall that our interest in Hilbert spaces arises from our interest in $L^2(X, \mu)$. Let us record the content of Proposition 4.8 in that case.

Corollary 4.9. *If $\varphi : L^2(X,\mu) \to \mathbb{C}$ is a continuous linear map, there exists a unique $f \in L^2(X,\mu)$ such that*

$$\varphi(u) = \int u(x)f(x)\,d\mu(x), \quad \forall\, u \in L^2(X,\mu). \tag{4.35}$$

We can use orthogonal projection operators to construct an *orthonormal basis* of a Hilbert space H. Let us assume H is *separable*, i.e., H has a countable dense subset $S = \{v_j : j \geq 1\}$. See the exercises for results on separability. In such a case, pick a countable subset $\{w_j : j \geq 1\}$ of S such that each w_k is linearly independent of $\{w_j : j < k\}$ and such that the linear span of $\{w_j\}$ is dense in H. Let $L_k = \operatorname{span}\{w_j : 1 \leq j \leq k\}$. We define an orthonormal set $\{u_j : j \geq 1\}$ inductively, as follows. Set $u_1 = w_1/\|w_1\|$. Suppose you have $\{u_j : 1 \leq j \leq k\}$, an orthonormal basis of L_k. Then set

$$u_{k+1} = \frac{w_{k+1} - P_k w_{k+1}}{\|w_{k+1} - P_k w_{k+1}\|}, \tag{4.36}$$

where P_k is the orthogonal projection onto L_k. One does not need the construction involving Proposition 4.7 to get P_k here. We can simply set

$$P_k f = \sum_{j=1}^{k} (f, u_j) u_j. \tag{4.37}$$

The set $\{u_j : j \geq 1\}$ so constructed is orthonormal, i.e., $(u_k, u_\ell) = \delta_{k\ell}$. Also its linear span is dense in H, since the linear span coincides with that of $\{w_j : j \geq 1\}$.

We claim that, for each $f \in H$, $P_k f \to f$ as $k \to \infty$. To see this, note that

$$\|f\|^2 = \|P_k f\|^2 + \|f - P_k f\|^2 \geq \|P_k f\|^2 = \sum_{j=1}^{k} |(f, u_j)|^2. \tag{4.38}$$

We also see that, if $n \geq k$, then $\|P_n f - P_k f\|^2 = \sum_{k<j\leq n} |(f,u_j)|^2$, and hence that $(P_k f)$ is a Cauchy sequence in H, for each f. We claim that the limit is equal to f, hence that

$$f = \sum_j (f, u_j) u_j. \tag{4.39}$$

Indeed, we now know that the right side of (4.39) defines an element of H; call it g. Then, $f - g$ has inner product 0 with each u_j, hence with all elements of the linear span of $\{u_j\}$, hence with all elements of the closure, i.e., with all elements of H, so $f - g = 0$.

In the case $H = L^2(I, m)$, with $I = [-\pi, \pi]$ and $m = dx/2\pi$, an orthonormal basis is given by

$$e_k(x) = e^{ikx}, \quad k \in \mathbb{Z}. \tag{4.40}$$

See the exercises for a proof of this. In such a case, (4.39) is the expansion of a function in a *Fourier series.*

We use Corollary 4.9 to prove an important result known as the *Radon-Nikodym Theorem.* Let μ and ν be two finite measures on $(X, \mathfrak{F})$. Let

$$\alpha = \mu + 2\nu, \quad \omega = 2\mu + \nu. \tag{4.41}$$

On the Hilbert space $H = L^2(X, \alpha)$, consider the linear functional $\varphi : H \to \mathbb{C}$ given by

$$\varphi(f) = \int_X f(x)\, d\omega(x). \tag{4.42}$$

Note that $|\varphi(f)| \le 2 \int |f|\, d\alpha \le 2\sqrt{\alpha(X)} \|f\|_{L^2(X,\alpha)}$. By Corollary 4.9, there exists $g \in \mathcal{L}^2(X, \alpha)$ such that, for any $f \in \mathcal{L}^2(X, \alpha) = \mathcal{L}^2(X, \mu) \cap \mathcal{L}^2(X, \nu)$,

$$\int_X f(x)\, d\omega(x) = \int_X f(x) g(x)\, d\alpha(x). \tag{4.43}$$

In particular, this holds for any bounded measurable f. Note that this identity is equivalent to

$$\int f(2g - 1)\, d\nu = \int f(2 - g)\, d\mu, \tag{4.44}$$

for all $f \in \mathcal{L}^2(X, \alpha)$. If we let f be the characteristic function of $S_{1\ell} = \{x \in X : g(x) < 1/2 - 1/\ell\}$ or of $S_{2\ell} = \{x \in X : g(x) > 2 + 1/\ell\}$, we see that $\mu(S_{j\ell}) = \nu(S_{j\ell}) = 0$. As a consequence, we can arrange that $1/2 \le g(x) \le 2$, for all $x \in X$. We also see that $Z = \{x \in X : g(x) = 1/2\}$ must have μ-measure zero. (Similarly, $\{x : g(x) = 2\}$ has ν-measure zero.) Also, (4.44) holds for all $f \in \mathcal{M}^+(X)$, by the Monotone Convergence Theorem.

We say that ν is *absolutely continuous* with respect to μ and write $\nu << \mu$, provided

$$\mu(S) = 0 \Longrightarrow \nu(S) = 0. \tag{4.45}$$

In such a case, we see that $Z = \{x \in X : g(x) = 1/2\}$ has ν-measure zero. Given $F \in \mathcal{M}^+(X)$, we can set

$$f(x) = \frac{F(x)}{2g(x) - 1}, \quad h(x) = \frac{2 - g(x)}{2g(x) - 1} \tag{4.46}$$

and apply (4.44) to get

$$\int_X F(x)\, d\nu(x) = \int_X F(x)h(x)\, d\mu(x) \tag{4.47}$$

for all positive measurable F. Note that taking $F = 1$ gives $h \in L^1(X, \mu)$. The result we have just obtained is known as the Radon-Nikodym Theorem. We record a formal statement.

Theorem 4.10. *Let μ and ν be two finite measures on $(X, \mathfrak{F})$. If ν is absolutely continuous with respect to μ, then* (4.47) *holds for some nonnegative $h \in L^1(X, \mu)$ and every positive measurable F.*

We mention that (4.47) also holds for every bounded measurable F.

If we do not assume that $\nu << \mu$, we can still consider

$$h(x) = \begin{cases} \dfrac{2 - g(x)}{2g(x) - 1} & \text{if } g(x) \neq \dfrac{1}{2}, \\ 0 & \text{if } g(x) = \dfrac{1}{2}, \end{cases} \tag{4.48}$$

and we have

$$\int_Y F\, d\nu = \int_Y F h\, d\mu \tag{4.49}$$

for any positive measurable F, where

$$Y = X \setminus Z = \Big\{x \in X : g(x) \neq \frac{1}{2}\Big\}. \tag{4.50}$$

Recall that $\mu(X \setminus Y) = 0$. We can define the measure λ on $(X, \mathfrak{F})$ by

$$\lambda(E) = \nu(Y \cap E). \tag{4.51}$$

Then we have

$$\int_X F\, d\lambda = \int_X F h\, d\mu \tag{4.52}$$

for all positive measurable F. Write

$$\rho(E) = \nu(E \setminus Y) = \nu(E \cap Z), \tag{4.53}$$

so

$$\nu = \lambda + \rho. \tag{4.54}$$

Now the measure λ is *supported* on Y, i.e., $\lambda(X \setminus Y) = 0$. Similarly, ρ is supported on Z. Thus λ and ρ have disjoint supports. Generally, two measures with disjoint supports are said to be *mutually singular*. When two measures λ and ρ are mutually singular, we write $\lambda \perp \rho$.

We have the following result, known as the Lebesgue decomposition of ν with respect to μ.

Theorem 4.11. *If μ and ν are finite measures on $(X, \mathfrak{F})$, then we can write*

$$\nu = \lambda + \rho, \quad \lambda << \mu, \quad \rho \perp \mu. \tag{4.55}$$

This decomposition is unique.

Proof. The measures λ and ρ are given by (4.51) and (4.52). The fact that $\lambda << \mu$ is contained in (4.52). As we have noted, $\mu(X \setminus Y) = 0$, so μ is supported on Y, which is disjoint from Z, on which ρ is supported; hence $\rho \perp \mu$.

If also $\tilde{\lambda}$ and $\tilde{\rho}$ are measures such that $\nu = \tilde{\lambda} + \tilde{\rho}$, $\tilde{\lambda} << \mu$, $\tilde{\rho} \perp \mu$, we have $\widetilde{Z} \in \mathfrak{F}$ such that $\tilde{\rho}$ is supported on $\widetilde{Z}$ and $\mu(\widetilde{Z}) = 0$. Now $\mu(Z \cup \widetilde{Z}) = 0$, so $\lambda(Z \cup \widetilde{Z}) = 0$ and $\tilde{\lambda}(Z \cup \widetilde{Z}) = 0$, and, for $E \in \mathcal{M}$,

$$\begin{aligned} \lambda(E) &= \lambda(E \setminus \widetilde{Z}) = \nu\big(E \setminus (Z \cup \widetilde{Z})\big), \\ \tilde{\lambda}(E) &= \tilde{\lambda}(E \setminus Z) = \nu\big(E \setminus (Z \cup \widetilde{Z})\big). \end{aligned}$$

This gives uniqueness.

We say a measure μ on $(X, \mathfrak{F})$ is σ-finite if we can write X as a countable union $\bigcup_{j \ge 1} X_j$ where $X_j \in \mathfrak{F}$ and $\mu(X_j) < \infty$. A paradigm case is Lebesgue measure on $X = \mathbb{R}$. There are routine extensions of Theorems 4.10–4.11 to the case where μ and ν are σ-finite measures, which we leave to the reader.

Exercises

1. Let V and W be normed linear spaces. Suppose we have linear transformations

$$T_j : V \longrightarrow W, \quad \|T_j v\|_W \le C \|v\|_V, \tag{4.56}$$

with C independent of j. (We say $\{T_j\}$ is uniformly bounded.) Suppose also $T : V \to W$ satisfies this bound. Let L be a dense subspace of V. Then show that

$$T_j v \to Tv, \ \forall\, v \in L \Longrightarrow T_j v \to Tv, \ \forall\, v \in V. \tag{4.57}$$

2. Define

$$\tau_s : L^p(\mathbb{R}) \longrightarrow L^p(\mathbb{R}), \quad \tau_s f(x) = f(x - s). \tag{4.58}$$

Show that, for $p \in [1, \infty)$,

$$f \in L^p(\mathbb{R}) \Longrightarrow \tau_s f \to f \text{ in } L^p\text{-norm, as } s \to 0. \tag{4.59}$$

Hint. Apply Exercise 1, with $V = W = L^p(\mathbb{R})$, $L = C_{00}(\mathbb{R})$, as in Corollary 4.6. Note that $\|\tau_s f\|_{L^p} = \|f\|_{L^p}$.

One says a metric space is *separable* if it has a countable dense subset.

3. If $I = [a, b] \subset \mathbb{R}$ and $a \le \alpha < \beta \le b$, define

$$\varphi_{\alpha\beta}(x) = \operatorname{dist}\big(x, I \setminus [\alpha, \beta]\big).$$

Show that the linear span over $\mathbb{Q}$ of $\{\varphi_{\alpha\beta} : \alpha, \beta \in \mathbb{Q} \cap I\}$ is dense in $C(I)$, and deduce that $C(I)$ is separable. From the denseness and continuity of the inclusion

$$\iota : C(I) \longrightarrow L^p(I),$$

prove that $L^p(I)$ is separable, for $1 \le p < \infty$. Then prove that $L^p(\mathbb{R})$ is separable, for $1 \le p < \infty$.

4. Let X be a compact metric space; X has a countable dense subset $\{z_j : j \ge 1\}$. Given $0 < \rho < (\operatorname{diam} X)/2$, set

$$\psi_{j\rho}(x) = \operatorname{dist}\big(x, X \setminus B_\rho(z_j)\big).$$

Show that the algebra generated by $\{\psi_{j,\rho} : j \in \mathbb{Z}^+, \ \rho \in \mathbb{Q}^+\}$ and 1 is dense in $C(X)$ and deduce that $C(X)$ is separable. Conclude, from Proposition 4.5, that $L^p(X, \mu)$ is separable, for $p \in [1, \infty)$, if μ is a finite measure on the σ-algebra of Borel sets in X.

5. Let $\{u_j : j \ge 1\}$ be a countable orthonormal set in a Hilbert space H. Show that

$$\sum_{j=1}^{\infty} (f, u_j) u_j = P_V f, \tag{4.60}$$

where V is the closure of the linear span of $\{u_j\}$.

The Stone-Weierstrass Theorem states that, if X is a compact Hausdorff space and $\mathcal{A}$ an algebra of functions in $C_{\mathbb{R}}(X)$ (the space of real-valued continuous functions on X), such that $1 \in \mathcal{A}$, and if $\mathcal{A}$ has the property of separating points, i.e., for any two distinct $p, q \in X$, there exists $f \in \mathcal{A}$ such that $f(p) \neq f(q)$, then $\mathcal{A}$ is dense in $C_{\mathbb{R}}(X)$. If $\mathcal{A}$ is an algebra in $C_{\mathbb{C}}(X)$ with these properties, plus the property that $f \in \mathcal{A} \Rightarrow \overline{f} \in \mathcal{A}$, then $\mathcal{A}$ is dense in $C_{\mathbb{C}}(X)$. A proof is given in Appendix A.

6. Use the Stone-Weierstrass Theorem to show that, if $e_k(x) = e^{ikx}$, as in (4.40), then the linear span $\mathcal{E}$ of $\{e_k : k \in \mathbb{Z}\}$ is dense in $C(S^1)$, where $S^1 = \mathbb{R}/2\pi\mathbb{Z}$; hence $\mathcal{E}$ is dense in $L^p(S^1, dx/2\pi)$, for $p \in [1, \infty)$. Hence $\{e_k : k \in \mathbb{Z}\}$ is an orthonormal basis of $L^2(S^1, dx/2\pi)$.

7. For $f, u \in C_0^\infty(\mathbb{R})$, the space of smooth functions with compact support in $\mathbb{R}$, set

$$K_f u(x) = f * u(x) = \int f(y)u(x-y)\, dy. \tag{4.61}$$

Show that, for $1 \le p < \infty$, K_f has a unique bounded extension:

$$K_f : L^p(\mathbb{R}) \longrightarrow L^p(\mathbb{R}), \quad \|K_f u\|_{L^p} \le \|f\|_{L^1} \|u\|_{L^p}. \tag{4.62}$$

The operation in (4.61) is called *convolution*.
Hint. For $f, u \in C_0^\infty(\mathbb{R})$, if f is supported in $[a, b]$, show that

$$f * u(x) = \lim_{n\to\infty} \frac{b-a}{n} \sum_{j=0}^{n} f\Bigl(\frac{j}{n}b + \Bigl(1 - \frac{j}{n}\Bigr)a\Bigr)\, \tau_j u(x),$$

where $\tau_j u(x) = u(x - jb/n - (1 - j/n)a)$. Use the triangle inequality (4.7) to estimate norms.

8. Show that there is a unique extension from $f \in C_0^\infty(\mathbb{R})$ to $f \in L^1(\mathbb{R})$ of $K_f u$, with (4.62) continuing to hold, giving a continuous linear map $K : L^1(\mathbb{R}) \to \mathcal{L}(L^p(\mathbb{R}))$.

9. Let f_j be a sequence of nonnegative functions in $L^1(\mathbb{R})$ such that

$$\int f_j\, dx = 1, \quad \operatorname{supp} f_j \subset \{x \in \mathbb{R} : |x| < 1/j\}. \tag{4.63}$$

Show that, for $1 \le p < \infty$,

$$u \in L^p(\mathbb{R}) \Longrightarrow f_j * u \to u \text{ in } L^p \text{ norm, as } j \to \infty. \tag{4.64}$$

Derive the same conclusion, upon weakening the second hypothesis in (4.63) to

$$\int_{|x|<\varepsilon} f_j\,dx = \beta_j(\varepsilon) \to 1 \text{ as } j \to \infty, \quad \forall\, \varepsilon > 0.$$

Hint. First verify the conclusion for $u \in C_0^\infty(\mathbb{R})$. Then use Exercise 1.

10. Given $f \in L^1(S^1)$, $0 < r < 1$, define

$$P_r f(\theta) = \sum_{n=-\infty}^{\infty} \hat{f}(n) r^{|n|} e^{in\theta}, \quad \hat{f}(n) = \frac{1}{2\pi} \int_0^{2\pi} f(\theta) e^{-in\theta}\,d\theta. \tag{4.65}$$

Show that

$$P_r f(\theta) = p_r * f(\theta) = \frac{1}{2\pi} \int_0^{2\pi} p_r(\theta - \varphi) f(\varphi)\,d\varphi, \tag{4.66}$$

where

$$p_r(\theta) = \sum_{n=-\infty}^{\infty} r^{|n|} e^{in\theta} = \frac{1 - r^2}{1 - 2r\cos\theta + r^2}. \tag{4.67}$$

Show that

$$\frac{1}{2\pi} \int_0^{2\pi} p_r(\theta)\,d\theta = 1. \tag{4.68}$$

11. If $f \in L^p(S^1)$, $1 \le p < \infty$, show that

$$P_r f \to f, \quad \text{as } r \nearrow 1, \tag{4.69}$$

in L^p-norm. If $f \in C(S^1)$, show that you have uniform convergence in (4.69). This is known as Abel convergence of Fourier series.
Hint. Use a variant of the analysis needed for Exercise 9.

12. Show that Exercises 10–11 provide an alternative proof of the conclusion of Exercise 6, that the linear span of $e_k(\theta) = e^{ik\theta}$, $k \in \mathbb{Z}$, is dense in $C(S^1)$ and in $L^p(S^1)$, for $1 \le p < \infty$, and hence that $\{e_k : k \in \mathbb{Z}\}$ is an orthonormal basis of $L^2(S^1, d\theta/2\pi)$.

13. Suppose $(X, \overline{\mathfrak{F}}, \overline{\mu})$ is the completion of $(X, \mathfrak{F}, \mu)$. Show that $L^p(X, \mu)$ and $L^p(X, \overline{\mu})$ are identical.

Hint. Consult Exercise 6 of Chapter 3.

14. Show that if $f \in L^p(X, \mu)$ and $p \in [1, \infty)$, then, for $\lambda \in (0, \infty)$,

$$\mu(\{x \in X : |f(x)| > \lambda\}) \le \lambda^{-p} \|f\|_{L^p}^p. \tag{4.70}$$

Deduce that if $f_k \to 0$ in L^p-norm, for some $p \in [1, \infty)$, then $f_k \to 0$ in measure, as defined in Exercise 10 of Chapter 3.
Hint. Denote the set being measured in (4.70) by E_λ and note that $\int_{E_\lambda} |f|^p \, d\mu \ge \lambda^p \mu(E_\lambda)$.
The inequality (4.70) is called Tchebychev's inequality.

Chapter 5

The Caratheodory Construction of Measures

Recall how our construction of Lebesgue measure in Chapter 2 proceeded from an initial notion of the size of a very restricted class of subsets of $\mathbb{R}$, the intervals. Using this notion, we defined an "outer measure" by the process (2.1) and proceeded from there to obtain a measure.

In general, an outer measure on a set X is a function $\mu^* : \mathcal{P}(X) \to [0, \infty]$ satisfying the following three conditions:

$$\mu^*(\emptyset) = 0, \tag{5.1}$$

$$A \subset B \Longrightarrow \mu^*(A) \le \mu^*(B), \tag{5.2}$$

$$A_j \subset X \text{ countable } \Longrightarrow \mu^*\Bigl(\bigcup_j A_j\Bigr) \le \sum_j \mu^*(A_j). \tag{5.3}$$

Recall that $\mathcal{P}(X)$ is the collection of *all* subsets of X.

Parallel to (2.1), here is a way to construct outer measures. Let $\mathcal{E}$ be some family of subsets of X, such that $\emptyset \in \mathcal{E}$ and $X = \bigcup_{j\ge1} X_j$ for some countable collection $X_j \in \mathcal{E}$. Suppose you have a function

$$\rho : \mathcal{E} \longrightarrow [0, \infty], \quad \rho(\emptyset) = 0. \tag{5.4}$$

For any $S \subset X$, set

$$\mu^*(S) = \inf \Bigl\{ \sum_{j\ge1} \rho(E_j) : E_j \in \mathcal{E},\ S \subset \bigcup_{j\ge1} E_j \Bigr\}. \tag{5.5}$$

Here $\{E_j\}$ is a *countable* cover of S by sets in $\mathcal{E}$.

Proposition 5.1. *Under the hypotheses above, μ^*, defined by (5.5), is an outer measure.*

Proof. Property (5.1) follows from $\rho(\emptyset) = 0$ and (5.2) from the fact that, when $A \subset B$, any countable cover of B by elements of $\mathcal{E}$ is also a cover of A. The proof of (5.3) works the same way as the proof of Proposition 2.1; each A_j has a countable cover $\{E_{jk} : k \geq 1\}$ by elements of $\mathcal{E}$, such that $\mu^*(A_j) \geq \sum_k \rho(E_{jk}) - 2^{-j}\varepsilon$. Then $\{E_{jk} : j, k \geq 1\}$ is a countable cover of $\bigcup_j A_j$ by elements of $\mathcal{E}$, and we obtain (5.3) in the limit as $\varepsilon \to 0$.

If we have an outer measure μ^*, then, as suggested when we stated Proposition 2.12, we say that a set $A \subset X$ is μ^*-measurable if and only if

$$\mu^*(Y) = \mu^*(Y \cap A) + \mu^*(Y \setminus A), \quad \forall\, Y \subset X. \tag{5.6}$$

(By (5.3), $\leq$ always holds, so the condition to check is $\geq$.) Denote by $\mathcal{M}$ the class of μ^*-measurable subsets of X. The following result is known as Caratheodory's Theorem.

Theorem 5.2. *If μ^* is an outer measure on X, then the class $\mathcal{M}$ of μ^*-measurable sets is a σ-algebra, and the restriction of μ^* to $\mathcal{M}$ is a measure.*

Proof. Clearly $\emptyset \in \mathcal{M}$. Also, if $A \in \mathcal{M}$, then, for all $Y \subset X$, $Y \cap A^c = Y \setminus A$ and $Y \setminus A^c = Y \cap A$, so $\mathcal{M}$ is closed under complements.

Next, suppose $A_j \in \mathcal{M}$. We want to show that (5.6) holds with $A = A_1 \cup A_2$. Since μ^* is subadditive, it suffices to establish that

$$\mu^*(Y) \geq \mu^*\big(Y \cap (A_1 \cup A_2)\big) + \mu^*\big(Y \cap (A_1 \cup A_2)^c\big), \tag{5.7}$$

for all $Y \subset X$. Note that $A_1 \cup A_2 = A_1 \cup (A_2 \cap A_1^c)$ is a disjoint union. We see that the right side of (5.7) is

$$\begin{aligned} &\leq \mu^*(Y \cap A_1) + \mu^*(Y \cap A_2 \cap A_1^c) + \mu^*(Y \cap A_1^c \cap A_2^c) \\ &= \mu^*(Y \cap A_1) + \mu^*(Y \cap A_1^c) \\ &= \mu^*(Y), \end{aligned} \tag{5.8}$$

where the last two identities use $A_2 \in \mathcal{M}$ and $A_1 \in \mathcal{M}$, respectively. This yields (5.7) and shows that $\mathcal{M}$ is an algebra, i.e., $\mathcal{M}$ is closed under complements and finite unions.

We next check additivity when $A_1, A_2 \in \mathcal{M}$ are disjoint. Indeed, if we set $Y = A_1 \cup A_2,\ A = A_1$ in (5.6), we get

$$\mu^*(A_1 \cup A_2) = \mu^*(A_1) + \mu^*(A_2), \quad A_j \in \mathcal{M}, \text{ disjoint}. \tag{5.9}$$

Inductively, we have, for $A_j \in \mathcal{M}$, disjoint,

$$\mu^*\Bigl(\bigcup_{j=1}^{N} A_j\Bigr) = \sum_{j=1}^{N} \mu^*(A_j). \tag{5.10}$$

Taking $N \nearrow \infty$ and using monotonicity of μ^* yield

$$\mu^*\Bigl(\bigcup_{j\geq 1} A_j\Bigr) \geq \sum_{j\geq 1} \mu^*(A_j),$$

and since the reverse inequality holds by (5.3), we have for a countable family A_j,

$$\mu^*\Bigl(\bigcup_{j\geq 1} A_j\Bigr) = \sum_{j\geq 1} \mu^*(A_j), \quad A_j \in \mathcal{M}, \text{ disjoint.} \tag{5.11}$$

Compare the proof of Proposition 2.7, leading to Theorem 2.8.

To finish the proof of Theorem 5.2, we need to show that, if $A_j \in \mathcal{M}$ is a countable disjoint family, then $A = \bigcup_{j\geq 1} A_j$ is μ^*-measurable. Let $B_n = \bigcup_{j\leq n} A_j$. The measurability of A_n implies

$$\begin{aligned} \mu^*(Y \cap B_n) &= \mu^*(Y \cap B_n \cap A_n) + \mu^*(Y \cap B_n \cap A_n^c) \\ &= \mu^*(Y \cap A_n) + \mu^*(Y \cap B_{n-1}), \end{aligned}$$

for any $Y \subset X$. Inductively, we obtain

$$\mu^*(Y \cap B_n) = \sum_{j=1}^{n} \mu^*(Y \cap A_j). \tag{5.12}$$

We know $\mathcal{M}$ is an algebra, so $B_n \in \mathcal{M}$, and hence, using (5.12), we have

$$\begin{aligned} \mu^*(Y) &= \mu^*(Y \cap B_n) + \mu^*(Y \cap B_n^c) \\ &\geq \sum_{j=1}^{n} \mu^*(Y \cap A_j) + \mu^*(Y \cap A^c), \end{aligned} \tag{5.13}$$

the last inequality holding because $B_n \subset A$. Taking $n \to \infty$, we have

$$\begin{aligned} \mu^*(Y) &\geq \sum_{j\geq 1} \mu^*(Y \cap A_j) + \mu^*(Y \cap A^c) \\ &\geq \mu^*\Bigl(\bigcup_{j\geq 1} (Y \cap A_j)\Bigr) + \mu^*(Y \cap A^c) \\ &= \mu^*(Y \cap A) + \mu^*(Y \cap A^c) \geq \mu^*(Y). \end{aligned} \tag{5.14}$$

Thus $A \in \mathcal{M}$, and Theorem 5.2 is proved.

While the construction (5.4)–(5.5) is very general, you need extra structure to relate $\mu^*(S)$ to $\rho(S)$ when $S \in \mathcal{E}$. The following is a convenient setting.

Let $\mathcal{A}$ be an algebra of subsets of X, i.e., a nonempty collection of subsets of X, closed under finite unions and under complements, hence under finite intersections. A function $\mu_0 : \mathcal{A} \to [0, \infty]$ is called a *premeasure* if it satisfies the following two conditions:

$$\mu_0(\emptyset) = 0, \tag{5.15}$$

$$S_j \in \mathcal{A} \text{ countable, disjoint, } \bigcup_j S_j = S \in \mathcal{A} \Longrightarrow \mu_0(S) = \sum_j \mu_0(S_j). \tag{5.16}$$

As an example, let $X = I = [a, b]$, and let $\mathcal{A}$ consist of *finite* unions of intervals (open, closed, or half-open) in I. If $S = \bigcup_{k=1}^N J_k$ is a disjoint union of intervals, take $\mu_0(S) = \sum_{k=1}^N \ell(J_k)$, so μ_0 is the restriction of Lebesgue (outer) measure to $\mathcal{A}$. In this case, (5.16) can be demonstrated by an argument similar to that needed for Exercise 1 in Chapter 2.

If μ_0 is a premeasure on $\mathcal{A}$, it induces an outer measure on X via the construction (5.4)–(5.5), i.e.,

$$\mu^*(E) = \inf \Big\{ \sum_{j \ge 0} \mu_0(A_j) : A_j \in \mathcal{A},\ E \subset \bigcup_{j \ge 0} A_j \Big\}. \tag{5.17}$$

Proposition 5.3. *If μ_0 is a premeasure on $\mathcal{A}$ and μ^* is defined by* (5.17), *then*

$$S \in \mathcal{A} \Longrightarrow \mu^*(S) = \mu_0(S), \tag{5.18}$$

and every set in $\mathcal{A}$ is μ^*-measurable.

Proof. To prove (5.18), first note that $\mu^*(S) \le \mu_0(S)$ for $S \in \mathcal{A}$ since S covers itself. Suppose $S \in \mathcal{A}$, and $S \subset \bigcup_{j \ge 1} A_j$, $A_j \in \mathcal{A}$. Then let $B_n = S \cap \big(A_n \setminus \bigcup_{j<n} A_j\big)$, so $\{B_n : n \ge 1\}$ is a disjoint family of members of $\mathcal{A}$, whose union is S. Thus, by (5.16),

$$\mu_0(S) = \sum_{j \ge 1} \mu_0(B_j) \le \sum_{j \ge 1} \mu_0(A_j). \tag{5.19}$$

It follows that $\mu_0(S) \le \mu^*(S)$. This proves (5.18).

To prove that each $A \in \mathcal{A}$ is μ^*-measurable, if $Y \subset X$ and $\varepsilon > 0$, there is a sequence $\{B_j : j \geq 1\} \subset \mathcal{A}$ with $Y \subset \bigcup_{j\geq 1} B_j$ and $\sum \mu_0(B_j) \leq \mu^*(Y) + \varepsilon$. Since μ_0 is additive on $\mathcal{A}$,

$$\begin{aligned} \mu^*(Y) + \varepsilon &\geq \sum_{j\geq 1} \mu_0(B_j \cap A) + \sum_{j\geq 1} \mu_0(B_j \cap A^c) \\ &\geq \mu^*(Y \cap A) + \mu^*(Y \cap A^c), \end{aligned} \tag{5.20}$$

the latter inequality holding, e.g., because $\{B_j \cap A : j \geq 1\}$ is a cover of $Y \cap A$ by elements of $\mathcal{A}$. Taking $\varepsilon \searrow 0$, we obtain (5.6), so any $A \in \mathcal{A}$ is μ^*-measurable.

When we combine the last result with Theorem 5.2, we obtain an extension of the premeasure μ_0 to a measure.

Theorem 5.4. *Let $\mathcal{A} \subset \mathcal{P}(X)$ be an algebra, μ_0 a premeasure on $\mathcal{A}$, and $\mathcal{M} = \sigma(\mathcal{A})$ the σ-algebra generated by $\mathcal{A}$. Then there exists a measure μ on $\mathcal{M}$ whose restriction to $\mathcal{A}$ is μ_0, namely $\mu^*\big|_{\mathcal{M}}$, where μ^* is given by* (5.17).

Proof. In fact, the class of μ^*-measurable sets is a σ-algebra containing $\mathcal{A}$; hence it contains $\sigma(\mathcal{A})$. There is nothing more to prove.

We next examine the extent to which the extension μ of μ_0 to $\sigma(\mathcal{A})$ is unique. Suppose ν is a measure on $\mathcal{M} = \sigma(\mathcal{A})$, which also agrees with μ_0 on $\mathcal{A}$. If $E \in \mathcal{M}$ and $E \subset \bigcup_{j\geq 1} A_j$, $A_j \in \mathcal{A}$, then

$$\nu(E) \leq \sum_{j\geq 1} \nu(A_j) = \sum_{j\geq 1} \mu_0(A_j), \tag{5.21}$$

so, by the construction (5.17) of μ^*,

$$\nu(E) \leq \mu(E), \quad \forall E \in \sigma(\mathcal{A}). \tag{5.22}$$

Also, by the property (3.5),

$$B_j \in \mathcal{A},\ B_j \nearrow B \Longrightarrow \nu(B) = \lim_{n\to\infty} \nu(B_n) = \lim_{n\to\infty} \mu(B_n) = \mu(B). \tag{5.23}$$

Suppose now that $E \in \mathcal{M}$ and $\mu(E) < \infty$. Fix $\varepsilon > 0$. Then we can choose the cover A_j in (5.17) such that $\sum \mu(A_j) < \mu(E) + \varepsilon$. Hence, with

$$B_n = \bigcup_{j\leq n} A_j \nearrow B = \bigcup_{j\geq 1} A_j,$$

we have $\mu(B) < \mu(E) + \varepsilon$, so $\mu(B \setminus E) < \varepsilon$, and hence, by (5.22), $\nu(B \setminus E) < \varepsilon$. Meanwhile, by (5.23), $\nu(B) = \mu(B)$, so

$$\mu(E) \leq \mu(B) = \nu(B) = \nu(E) + \nu(B \setminus E) \leq \nu(E) + \varepsilon. \tag{5.24}$$

Taking $\varepsilon \searrow 0$ gives $\mu(E) \leq \nu(E)$, as long as $\mu(E) < \infty$. In concert with (5.22), this yields the following.

Proposition 5.5. *Let $\mathcal{A}$ be an algebra of subsets of X, generating the σ-algebra $\mathcal{M} = \sigma(\mathcal{A})$. Let μ_0 be a premeasure on $\mathcal{A}$, and let μ be the measure on $\mathcal{M}$ given by Theorem 5.4, extending μ_0. If ν is another measure on $\mathcal{M}$ which agrees with μ_0 on $\mathcal{A}$, then for all $S \in \mathcal{M}$, $\nu(S) \le \mu(S)$, and*

$$\mu(S) < \infty \Longrightarrow \mu(S) = \nu(S). \tag{5.25}$$

Furthermore, if there is a countable family A_j such that

$$X = \bigcup_{j \ge 1} A_j, \quad A_j \in \mathcal{M}, \quad \mu(A_j) < \infty, \tag{5.26}$$

then $\mu(S) = \nu(S)$ for all $S \in \mathcal{M}$.

Proof. The implication (5.25) was established above. If (5.26) holds, we can assume the A_j are disjoint. If $S \in \mathcal{M}$, we have the disjoint union $S = \bigcup S_j$, $S_j = S \cap A_j$. By (5.25), $\mu(S_j) = \nu(S_j)$, and then $\mu(S) = \nu(S)$ follows by countable additivity.

If (5.26) holds, we say $(X, \mathcal{M}, \mu)$ is a σ-finite measure space. See Exercise 15 for another proof of Proposition 5.5.

If Theorem 5.4 is applied to the example mentioned after (5.16), it yields the fact that finite unions of intervals in $I = [a, b]$ are Lebesgue measurable and hence so are sets in the σ-algebra $\sigma(\mathcal{A})$ generated by this algebra of subsets of I. In particular, all open sets in I are measurable, since they are countable unions of open intervals. Hence all closed sets in I are measurable. Of course, this recaptures results obtained in Chapter 2.

Recall that, in our treatment of Lebesgue measure on an interval, we gave a definition of measurability different from (5.6) and then showed in Proposition 2.12 that the definition of measurability given there implied (5.6). We now give a result in counterpoint to Proposition 2.12.

Proposition 5.6. *Suppose we are given an algebra $\mathcal{A}$ of subsets of X and a premeasure μ_0 on $\mathcal{A}$, with associated outer measure μ^*, defined by (5.17). Assume $Z \subset X$ is μ^*-measurable and $\mu^*(Z) < \infty$. Then a set $S \subset Z$ is μ^*-measurable if and only if*

$$\mu^*(S) + \mu^*(Z \setminus S) = \mu^*(Z). \tag{5.27}$$

Proof. We need to show that, if (5.27) holds, then, for any $Y \subset X$,

$$\mu^*(Y) = \mu^*(Y \cap S) + \mu^*(Y \cap S^c). \tag{5.28}$$

Only the "$\geq$" part needs to be established, so we can assume $\mu^*(Y) < \infty$. Write $Y = Y_0 \cup Y_1$, a disjoint union, where $Y_0 = Y \cap Z$, $Y_1 = Y \cap Z^c$. Using the measurability of Z, we have

$$\mu^*(Y) = \mu^*(Y \cap Z) + \mu^*(Y \cap Z^c) = \mu^*(Y_0) + \mu^*(Y_1) \tag{5.29}$$

and

$$\mu^*(Y \cap S^c) = \mu^*(Y \cap S^c \cap Z) + \mu^*(Y \cap S^c \cap Z^c) = \mu^*(Y_0 \cap S^c) + \mu^*(Y_1), \tag{5.30}$$

provided $S \subset Z$, while

$$\mu^*(Y \cap S) = \mu^*(Y_0 \cap S). \tag{5.31}$$

Consequently it suffices to prove (5.28) for $Y = Y_0$, i.e., for $Y \subset Z$. For that, we bring in a lemma

Lemma 5.7. *Let μ^* arise from a premeasure on an algebra $\mathcal{A}$. Let $\mathcal{A}_\sigma$ consist of countable unions of sets in $\mathcal{A}$. Then*

$$S \subset X \Longrightarrow \mu^*(S) = \inf\{\mu^*(E) : E \in \mathcal{A}_\sigma,\ S \subset E\}. \tag{5.32}$$

Proof. We leave this to the reader; see Exercise 3 at the end of this chapter.

We continue the proof of Proposition 5.6. Given $\varepsilon > 0$, there exists $\widetilde{A} \in \mathcal{A}_\sigma$ such that $\widetilde{A} \supset Y$ and $\mu^*(\widetilde{A}) \leq \mu^*(Y) + \varepsilon$. Set $A = \widetilde{A} \cap Z$. Then $A \in \mathcal{M}$ (the σ-algebra of μ^*-measurable sets), $A \supset Y$ (if $Y \subset Z$), and $\mu^*(A) \leq \mu^*(Y) + \varepsilon$. We claim that

$$\mu^*(A) = \mu^*(A \cap S) + \mu^*(A \cap S^c). \tag{5.33}$$

Suppose this is known. Then we have

$$\mu^*(Y) + \varepsilon \geq \mu^*(Y \cap S) + \mu^*(Y \cap S^c), \tag{5.34}$$

for all $\varepsilon > 0$. In the limit $\varepsilon \to 0$, we obtain the $\geq$ part of (5.28), and the problem is solved.

Thus it remains to establish (5.32), given $A \in \mathcal{M},\ A \subset Z,\ \mu^*(Z) < \infty$, and given the hypothesis (5.27). Using the measurability of A, we have

$$\mu^*(Z) = \mu^*(A) + \mu^*(Z \cap A^c). \tag{5.35}$$

Now the left side of (5.35) is equal to $\mu^*(S) + \mu^*(Z \setminus S)$, by hypothesis. By subadditivity of outer measure, the right side of (5.35) is

$$\begin{aligned} &\le \mu^*(A \cap S) + \mu^*(A \cap S^c) + \mu^*\big((Z \setminus A) \cap S\big) + \mu^*\big((Z \setminus A) \cap S^c\big) \\ &= \mu^*(S \cap A) + \mu^*\big((Z \setminus S) \cap A\big) + \mu^*(S \setminus A) + \mu^*\big((Z \setminus S) \setminus A\big) \\ &= \mu^*(S) + \mu^*(Z \setminus S), \end{aligned} \tag{5.36}$$

the last identity following by grouping together the odd terms and the even terms on the second line of (5.36) and using measurability of A. Since the bottom line of (5.36) is equal to the left side of (5.35), the $\le$ in (5.36) must be *equality.* That inequality arose from the sum of two inequalities, and so both of them must be equalities. One of them is the desired result (5.33). This finishes the proof of Proposition 5.6.

We next describe an important class of outer measures on metric spaces, for which all open sets and all closed sets can be shown to be measurable. This result will play an important role in Chapter 12, on Hausdorff measures, and in Chapter 13, on Radon measures.

Let X be a metric space, with distance function $d(x, y)$. An outer measure μ^* on X is called a *metric outer measure* provided

$$\begin{aligned} \rho(S_1, S_2) = \inf\ \{d(x_1, x_2) : x_j \in S_j\} > 0 \\ \Longrightarrow \mu^*(S_1 \cup S_2) = \mu^*(S_1) + \mu^*(S_2). \end{aligned} \tag{5.37}$$

Note that the part (2.15)–(2.16) of Lemma 2.4 is the statement that Lebesgue outer measure is a metric outer measure on $I = [a, b]$. The following is another result of Caratheodory:

Proposition 5.8. *If μ^* is a metric outer measure on a metric space X, then every closed subset of X is μ^*-measurable.*

Proof. We must show that if $F \subset X$ is closed and $Y \subset X$ satisfies $\mu^*(Y) < \infty$, then

$$\mu^*(Y) \ge \mu^*(Y \cap F) + \mu^*(Y \setminus F). \tag{5.38}$$

Consider

$$B_n = \Big\{x \in Y \setminus F : \rho(x, F) \ge \frac{1}{n}\Big\}.$$

Note that $B_n \nearrow Y \setminus F$. Also $\rho(B_n, F) \ge 1/n$, so, by (5.37),

$$\mu^*(Y \cap F) + \mu^*(B_n) = \mu^*\big((Y \cap F) \cup B_n\big) \le \mu^*(Y). \tag{5.39}$$

Thus it will suffice to show that

$$\mu^*(B_n) \longrightarrow \mu^*(Y \setminus F). \tag{5.40}$$

Let $C_n = B_{n+1} \setminus B_n$. Note that

$$|j - k| \geq 2 \Longrightarrow \rho(C_j, C_k) > 0.$$

Hence, for any N, one obtains inductively (using (5.37)) that

$$\begin{aligned} \sum_{j=1}^{N} \mu^*(C_{2j}) &= \mu^*\Big(\bigcup_{j=1}^{N} C_{2j}\Big) \leq \mu^*(Y), \\ \sum_{j=1}^{N} \mu^*(C_{2j+1}) &= \mu^*\Big(\bigcup_{j=1}^{N} C_{2j+1}\Big) \leq \mu^*(Y), \end{aligned} \tag{5.41}$$

and consequently $\sum_{j \geq 1} \mu^*(C_j) < \infty$. Now countable subadditivity implies

$$\mu^*(Y \setminus F) \leq \mu^*(B_n) + \sum_{j \geq n} \mu^*(C_j), \tag{5.42}$$

so as $n \to \infty$, the last sum tends to zero, and we obtain

$$\mu^*(Y \setminus F) \leq \liminf_{n \to \infty} \mu^*(B_n) \leq \limsup_{n \to \infty} \mu^*(B_n) \leq \mu^*(Y \setminus F), \tag{5.43}$$

the last inequality by monotonicity. This implies (5.40) and finishes the proof.

As one may have noticed, at several points in this chapter and previous chapters, we have considered a class of sets with certain desirable properties and have wanted to prove it was a σ-algebra. The following result, called the Monotone Class Lemma, sometimes furnishes a convenient tool for doing this. We define a monotone class on a set X to be a collection $\mathcal{C} \subset \mathcal{P}(X)$ having the properties

$$\begin{aligned} E_j \in \mathcal{C},\ E_j \nearrow E &\Longrightarrow E \in \mathcal{C}, \\ E_j \in \mathcal{C},\ E_j \searrow E &\Longrightarrow E \in \mathcal{C}. \end{aligned} \tag{5.44}$$

The smallest monotone class containing a collection $\mathcal{E} \subset \mathcal{P}(X)$ is called the monotone class generated by $\mathcal{E}$.

Proposition 5.9. *If $\mathcal{A}$ is an algebra of subsets of X and $\mathcal{C}$ is the monotone class generated by $\mathcal{A}$, then $\mathcal{C} = \sigma(\mathcal{A})$.*

Proof. Clearly $\mathcal{C} \subset \sigma(\mathcal{A})$. We will show that $\mathcal{C}$ is a σ-algebra. If $E \in \mathcal{C}$, let

$$\mathcal{C}(E) = \{F \in \mathcal{C} : E \setminus F,\ F \setminus E,\ E \cap F \in \mathcal{C}\}. \tag{5.45}$$

Clearly $\emptyset, E \in \mathcal{C}(E)$, and $\mathcal{C}(E)$ is easily verified to be a monotone class. Also, $E \in \mathcal{C}(F) \Leftrightarrow F \in \mathcal{C}(E)$. Since $\mathcal{A}$ is an algebra, we have

$$E, F \in \mathcal{A} \Longrightarrow F \in \mathcal{C}(E). \tag{5.46}$$

In other words, if $E \in \mathcal{A}$, then $\mathcal{A} \subset \mathcal{C}(E)$; hence $\mathcal{C} \subset \mathcal{C}(E)$. Thus, if $F \in \mathcal{C}$, then $F \in \mathcal{C}(E)$ for all $E \in \mathcal{A}$. This implies

$$E \in \mathcal{C}(F), \quad \forall\, E \in \mathcal{A},\ F \in \mathcal{C}. \tag{5.47}$$

Thus $\mathcal{A} \subset \mathcal{C}(F)$ and hence $\mathcal{C} \subset \mathcal{C}(F)$, for all $F \in \mathcal{C}$. In other words,

$$E, F \in \mathcal{C} \Longrightarrow E \setminus F,\ E \cap F \in \mathcal{C}. \tag{5.48}$$

Since $X \in \mathcal{A} \subset \mathcal{C}$, it follows that $\mathcal{C}$ is an algebra of sets. We finally note that $\mathcal{C}$ must be a σ-algebra. Indeed, if $E_j \in \mathcal{C}$, then $F_j = \bigcup_{n \le j} E_j \in \mathcal{C}$, and (5.44) implies $F_j \nearrow F = \bigcup_{j \ge 1} E_j \in \mathcal{C}$. This finishes the proof.

The proof of Proposition 6.2 in the next chapter will provide a nice application of the Monotone Class Lemma.

Exercises

1. Let $I = [a, b]$ and let $\mathcal{A}$ be the algebra of subsets of I consisting of finite unions of intervals, as in the example after (5.16). Let $\varphi : I \to \mathbb{R}$ be a monotonically increasing function, i.e., assume that $x < y \Rightarrow \varphi(x) \le \varphi(y)$. Set

$$\varphi_-(x) = \lim_{u \nearrow x} \varphi(u), \quad \varphi_+(x) = \lim_{u \searrow x} \varphi(u), \tag{5.49}$$

with the convention that $\varphi_-(0) = \varphi(0)$ and $\varphi_+(1) = \varphi(1)$. Define $\mu_0 : \mathcal{A} \to \mathbb{R}^+$ as follows. If $S = \bigcup_{k=1}^N J_k$ is a union of mutually disjoint intervals, set $\mu_0(S) = \sum_{k=1}^N \mu_0(J_k)$, where μ_0 is defined on a single interval $J \subset I$ as follows:

$$\begin{aligned} \mu_0\big([x, y]\big) &= \varphi_+(y) - \varphi_-(x), \\ \mu_0\big([x, y)\big) &= \varphi_-(y) - \varphi_-(x), \\ \mu_0\big((x, y]\big) &= \varphi_+(y) - \varphi_+(x), \\ \mu_0\big((x, y)\big) &= \varphi_-(y) - \varphi_+(x). \end{aligned} \tag{5.50}$$

Also, set $\mu_0(\emptyset) = 0$. Show that μ_0 is a premeasure. The measure then produced via Theorem 5.4 is called "Lebesgue-Stieltjes measure."
Hint. Say J is a closed interval and $J = \bigcup_{k\ge1} J_k$, with disjoint intervals J_k. First show that $\sum_{k=1}^m \mu_0(J_k) \le \mu_0(J)$ for each m, so $\mu_0(J) \ge \sum_{k\ge1} \mu_0(J_k)$. For the reverse inequality, one can argue as follows. Pick $\varepsilon > 0$. Say the endpoints of J_k are a_k and b_j, so $(a_k, b_k) \subset J_k \subset [a_k, b_k]$. If $a_k \in J_k$, pick $\tilde{a}_k < a_k$ such that $0 \le \varphi_-(a_k) - \varphi_+(\tilde{a}_k) < \varepsilon 2^{-k}$. If $a_k \notin J_k$, let $\tilde{a}_k = a_k$. Make an appropriate analogous choice of $\tilde{b}_k$, and let $\tilde{J}_k = (\tilde{a}_k, \tilde{b}_k)$. Note that $\mu_0(\tilde{J}_k) < \mu_0(J_k) + \varepsilon 2^{-k+1}$. Now $J \subset \bigcup_{k\ge1} \tilde{J}_k$, and this is an open cover. Take a finite subcover.

For Exercises 2–6, suppose we are given an algebra $\mathcal{A}$ of subsets of X and a premeasure μ_0 on $\mathcal{A}$, with associated outer measure μ^*, defined by (5.17). Let $\mathcal{A}_\sigma$ consist of countable unions of sets in $\mathcal{A}$.

2. Show that

$$E \in \mathcal{A}_\sigma \Longrightarrow \mu^*(E) = \sup\ \{\mu_0(A) : A \in \mathcal{A},\ A \subset E\}. \tag{5.51}$$

3. Show that

$$S \subset X \Longrightarrow \mu^*(S) = \inf\ \{\mu^*(E) : E \in \mathcal{A}_\sigma,\ S \subset E\}. \tag{5.52}$$

4. Let $\mathcal{A}_\delta$ consist of countable intersections of sets in $\mathcal{A}$, and set

$$\mu_*(S) = \sup\ \{\mu^*(E) : E \in \mathcal{A}_\delta,\ E \subset S\}, \quad S \subset X. \tag{5.53}$$

Show that $\mu_*(S) \le \mu^*(S)$. If $\mu^*(S) < \infty$, show that S is μ^*-measurable if and only if $\mu_*(S) = \mu^*(S)$.
Hint. Show that, if $S \subset Z \in \mathcal{A}_\sigma$, $\mu^*(Z) < \infty$, then $\mu^*(Z) = \mu_*(S) + \mu^*(Z \setminus S)$. Then use Proposition 5.6.

5. Given $S \subset X$, show that there exist $S_0 \in \mathcal{A}_{\sigma\delta}$ and $S_1 \in \mathcal{A}_{\delta\sigma}$ such that

$$S_0 \subset S \subset S_1, \quad \mu(S_1) = \mu^*(S), \quad \mu(S_0) = \mu_*(S). \tag{5.54}$$

Show that, if $\mu^*(S) < \infty$, then S is μ^*-measurable if and only if $\mu(S_1 \setminus S_0) = 0$. Here, μ is the measure on $\sigma(\mathcal{A})$ given by Theorem 5.4.

6. Let $\mu^\#$ be the outer measure on X obtained via (5.17), with $(\mathcal{A}, \mu_0)$ replaced by $(\sigma(\mathcal{A}), \mu)$. Equivalently,

$$\mu^\#(S) = \inf\ \{\mu(A) : S \subset A \in \sigma(\mathcal{A})\}. \tag{5.55}$$

Clearly $\mu^{\#}(S) \le \mu^*(S)$ for all $S \subset X$. Show that $\mu^{\#} = \mu^*$.
Hint. If $S \subset A \in \sigma(\mathcal{A})$, show that, for any $\varepsilon > 0$, there exists $B \in \mathcal{A}_\sigma$ such that $A \subset B$ and $\mu(B) \le \mu(A) + \varepsilon$.

7. If μ^* is an outer measure on X, then clearly

$$S_1 \subset S_2 \subset S_3 \subset \cdots \nearrow S \Longrightarrow \mu^*(S_j) \nearrow M \le \mu^*(S). \tag{5.56}$$

If μ^* arises from a premeasure, via (5.17), show that $M = \mu^*(S)$.
Hint. First note that $A_n \in \sigma(\mathcal{A})$, $A_n \nearrow A \Rightarrow \mu^*(A_n) \nearrow \mu^*(A)$. Then show that, for any $\varepsilon > 0$, there exist $Z_n \in \sigma(\mathcal{A})$ such that $Z_n \supset S_n$, $Z_n \nearrow$, and $\mu^*(Z_n) = \mu^*(S_n)$. Make use of (5.54).

8. As in Exercise 5 of Chapter 3, a measure μ on $(X, \mathcal{M})$ is said to be complete provided

$$A \in \mathcal{M},\ \mu(A) = 0,\ S \subset A \Longrightarrow S \in \mathcal{M}, \text{ and } \mu(S) = 0. \tag{5.57}$$

Show that the measures provided by Theorem 5.2, from outer measures, are all complete. In particular, Lebesgue measure is complete.

9. If μ is a measure on $(X, \mathfrak{F})$ that is not complete, form the outer measure μ^* by (5.17), with $\mathcal{A}$ replaced by $\mathfrak{F}$. Let $\mathcal{M}$ be the σ-algebra of μ^*-measurable sets. Show that Theorem 5.2 produces a measure $\overline{\mu}$ on $(X, \mathcal{M})$ which is complete and such that $\mathfrak{F} \subset \mathcal{M}$ and $\overline{\mu} = \mu$ on $\mathfrak{F}$. Show that $\mathcal{M} = \overline{\mathfrak{F}}$, given by Exercise 5 of Chapter 3, and that $\overline{\mu}$ coincides with the measure produced there, the "completion" of μ.

In Exercises 10–14, suppose X is a compact metric space, $\mathfrak{B}$ the σ-algebra of Borel subsets of X, and μ a finite measure on $(X, \mathfrak{B})$. Construct the outer measure μ^* via (5.17), with $\mathcal{A}$ replaced by $\mathfrak{B}$. (Note that Proposition 5.3 applies in this case.)

10. Show that μ^* must be a metric outer measure.
Hint. If S_1, S_2 satisfy (5.37), construct compact $K \subset X$ such that $S_1 \subset K \subset X \setminus S_2$. Apply (5.6) with $A = K$, $Y = S_1 \cup S_2$.

11. We know that $\mathfrak{B} = \sigma(\mathcal{A})$ where $\mathcal{A}$ is the algebra generated by the compact sets in X. Show that $\mathcal{A}$ consists of sets of the form

$$\bigcup_{\alpha_\ell=1}^{N_\ell} \bigcap_{\beta_\ell=1}^{M_\ell} \cdots \bigcup_{\alpha_1=1}^{N_1} \bigcap_{\beta_1=1}^{M_1} E_{\alpha\beta}, \tag{5.58}$$

where each $E_{\alpha\beta} \subset X$ is either open or closed.

12. Show that

$$S \subset X \Longrightarrow \mu^*(S) = \inf \{\mu^*(\mathcal{O}) : S \subset \mathcal{O},\ \mathcal{O} \text{ open}\}. \tag{5.59}$$

Hint. By Exercise 6, $\mu^*(S)$ is given by (5.52) or, equivalently, by (5.17) (with some changes in notation). Show that, given $A_j \in \mathcal{A}$, $\varepsilon > 0$, there exist open sets $\mathcal{O}_j$ such that $A_j \subset \mathcal{O}_j$ and $\mu(\mathcal{O}_j) \le \mu(A_j) + 2^{-j}\varepsilon$. It suffices to make such a construction for each $E_{\alpha\beta}$ in (5.58), and we need worry only about the case when $E_{\alpha\beta}$ is compact; i.e., we need to verify (5.59) when S is compact. For this, keep in mind that X is a *metric* space.

13. Show that

$$S \in \mathfrak{B} \Longrightarrow \mu(S) = \sup \{\mu(K) : K \subset S,\ K \text{ compact}\}. \tag{5.60}$$

14. If $f \in \mathcal{L}^1(X, \mu)$ and $\varepsilon > 0$, show that there exists a compact $K \subset X$ such that $\mu(X \setminus K) < \varepsilon$ and $f\big|_K$ is continuous.
Hint. Using Proposition 4.5 (or otherwise), produce $f_\nu \in C(X)$ such that $f_\nu \to f$, μ-a.e. Then use Egoroff's Theorem. Then use (5.60).
This result is Lusin's Theorem. A special case was stated in Exercise 16 of Chapter 3.

15. Use the Monotone Class Lemma to give another proof of Proposition 5.5. Furthermore, establish the following variant of Proposition 5.5.

Proposition 5.5A. *Let $\mathcal{A}$ be an algebra of subsets of X, generating the σ-algebra $\mathcal{M} = \sigma(\mathcal{A})$. Let μ and ν be measures on $\mathcal{M}$. Assume there exist A_j such that*

$$X = \bigcup_{j\ge1} A_j, \quad A_j \in \mathcal{A}, \quad \mu(A_j) < \infty.$$

If $\mu = \nu$ on $\mathcal{A}$, then $\mu = \nu$ on $\mathcal{M}$.

How do these two propositions differ?

16. Suppose $\mathcal{E}$ is a collection of subsets of X having the following properties:
(1) The intersection of any two elements of $\mathcal{E}$ belongs to $\mathcal{E}$.
(2) The collection of finite disjoint unions of elements of $\mathcal{E}$ is an algebra $\mathcal{A}$.

Let $\mu_b : \mathcal{E} \to [0, \infty]$ satisfy
(a) $\mu_b(\emptyset) = 0$,
(b) $E_j \in \mathcal{E}$ countable, disjoint, $\bigcup_j E_j = E \in \mathcal{E} \Rightarrow \mu_b(E) = \sum_j \mu_b(E_j)$.
Show that μ_b extends to a premeasure μ_0 on $\mathcal{A}$, satisfying

$$E_1, \dots, E_K \in \mathcal{E} \text{ disjoint } \Rightarrow \mu_0(E_1 \cup \dots \cup E_K) = \sum_{j=1}^{K} \mu_b(E_j).$$

Show that if μ^* is defined by (5.17), then also, for $E \subset X$,

$$\mu^*(E) = \inf\Big\{ \sum_{j \geq 0} \mu_b(A_j) : A_j \in \mathcal{E},\ E \subset \bigcup_{j \geq 0} A_j \Big\}.$$

Hint. Start with uniqueness; if also $F_1, \dots, F_L \in \mathcal{E}$ are disjoint and $\bigcup_{\ell=1}^{L} F_\ell = \bigcup_{k=1}^{K} E_k$, write this set as $\bigcup_{k,\ell}(E_k \cap F_\ell)$ to show μ_0 is well defined.
Note. An example is $X = [0, 1]$, $\mathcal{E} =$ the collection of intervals in $[0, 1]$, $\mu_b(J) = \ell(J)$, the length. Another family of examples arises at the beginning of Chapter 6.

17. Suppose $\mathcal{E}$ is a collection of subsets of X satisfying property (1) in Exercise 16 and also satisfying
(2′) $E \in \mathcal{E} \Rightarrow X \setminus E$ is a finite disjoint union of elements of $\mathcal{E}$.
Show that $\mathcal{E}$ satisfies property (2) in Exercise 16.

Chapter 6

Product Measures

If $(X, \mathcal{M}, \mu)$ and $(Y, \mathcal{N}, \nu)$ are measure spaces, we construct a σ-algebra $\mathcal{M} \otimes \mathcal{N}$ of subsets of $X \times Y$ and a product measure $\mu \times \nu$ on $\mathcal{M} \otimes \mathcal{N}$, as follows. We say that a *rectangle* is a set of the form $A \times E \subset X \times Y$, with $A \in \mathcal{M}$, $E \in \mathcal{N}$. We define $\mathcal{M} \otimes \mathcal{N}$ to be the σ-algebra generated by the collection $\mathcal{R}$ of rectangles. Note that

$$
\begin{aligned}
(A \times E) \cap (B \times F) &= (A \cap B) \times (E \cap F), \\
(A \times E) \cup (B \times F) &= A \times (E \setminus F) \cup (A \cup B) \times (E \cap F) \\
&\quad \cup B \times (F \setminus E),
\end{aligned} \tag{6.1}
$$

the latter decomposition being a disjoint union of rectangles. Also,

$$
(A \times E)^c = (X \times E^c) \cup (A^c \times E). \tag{6.2}
$$

Hence the collection of finite disjoint unions of rectangles is an algebra, which we denote $\mathcal{M} \boxtimes \mathcal{N}$, and $\mathcal{M} \otimes \mathcal{N}$ is the σ-algebra generated by $\mathcal{M} \boxtimes \mathcal{N}$.

We define the set function $\pi : \mathcal{M} \boxtimes \mathcal{N} \to [0, \infty]$ by

$$
\pi\Big(\bigcup_{j=1}^{N} (A_j \times E_j)\Big) = \sum_{j=1}^{N} \mu(A_j)\nu(E_j), \tag{6.3}
$$

when the rectangles $R_j = A_j \times E_j$ are disjoint. We claim π is a premeasure on $\mathcal{M} \boxtimes \mathcal{N}$. We need to check that

$$
\mu(A)\nu(E) = \sum_{j \ge 1} \mu(A_j)\nu(E_j) \tag{6.4}
$$

when $A \times E$ is a countable disjoint union of rectangles $A_j \times E_j$. Establishing (6.4) will also show that (6.3) is well defined for an element of $\mathcal{M} \boxtimes \mathcal{N}$

which has several different representations as disjoint unions of rectangles. To prove (6.4), write

$$\chi_A(x)\chi_E(y) = \chi_{A\times E}(x,y) = \sum_{j\ge 1} \chi_{A_j\times E_j}(x,y) = \sum_{j\ge 1} \chi_{A_j}(x)\chi_{E_j}(y). \tag{6.5}$$

Integrating with respect to x and applying Proposition 3.5 yield

$$\begin{aligned} \mu(A)\chi_E(y) &= \sum \chi_{E_j}(y) \int \chi_{A_j}(x)\,d\mu(x) \\ &= \sum \mu(A_j)\chi_{E_j}(y), \qquad \forall\, y \in Y. \end{aligned} \tag{6.6}$$

Integrating this with respect to y and using a parallel argument, we get (6.4). We mention that this setup is a special case of that treated in Exercise 16 of Chapter 5. Here $\mathcal{E} = \mathcal{R}$, the collection of rectangles, and $\mathcal{A} = \mathcal{M} \boxtimes \mathcal{N}$.

Using the construction (5.17), we obtain an outer measure π^* on $X \times Y$. Note that, for $S \subset X \times Y$,

$$\pi^*(S) = \inf\Bigl\{\sum_{k\ge 1} \mu(A_k)\nu(E_k) : A_k \in \mathcal{M}, E_k \in \mathcal{N},\ S \subset \bigcup_{k\ge 1}(A_k\times E_k)\Bigr\}. \tag{6.7}$$

By Theorem 5.4, the restriction of π^* to $\mathcal{M}\otimes\mathcal{N}$ is a measure, extending the premeasure on $\mathcal{M}\boxtimes\mathcal{N}$ defined above. We call this measure on $\mathcal{M}\otimes\mathcal{N}$ the product measure $\mu \times \nu$.

We note that, if μ and ν are σ-finite, so is $\mu \times \nu$, and, by Proposition 5.5, in this case $\mu \times \nu$ is the *unique* measure on $\mathcal{M}\otimes\mathcal{N}$ with the property that

$$(\mu \times \nu)(A \times E) = \mu(A)\nu(E), \quad A \in \mathcal{M},\ E \in \mathcal{N}.$$

If f is a measurable function on $(X \times Y, \mathcal{M}\otimes\mathcal{N})$, we want to see when the *iterated integral*

$$\int \Bigl[\int f(x,y)\,d\nu(y)\Bigr]\,d\mu(x) \tag{6.8}$$

is well defined and equal to the "product integral" of f with respect to the product measure $\mu \times \nu$. To begin, we define the x-section f_x and y-section f^y of a function f on $X \times Y$ by

$$f_x(y) = f^y(x) = f(x,y). \tag{6.9}$$

Related concepts are those of the x-section E_x and y-section E^y of a set $E \subset X \times Y$, defined by

$$E_x = \{y \in Y : (x,y) \in E\}, \quad E^y = \{x \in X : (x,y) \in E\}. \tag{6.10}$$

Note that

$$(\chi_E)_x(y) = \chi_{E_x}(y), \quad (\chi_E)^y(x) = \chi_{E^y}(x). \tag{6.11}$$

Lemma 6.1. *If $E \in \mathcal{M}\otimes\mathcal{N}$, then $E_x \in \mathcal{N}$ for all $x \in X$ and $E^y \in \mathcal{M}$ for all $y \in Y$. More generally, if f is $\mathcal{M}\otimes\mathcal{N}$-measurable, then f_x is $\mathcal{N}$-measurable for all $x \in X$ and f^y is $\mathcal{M}$-measurable for all $y \in Y$.*

Proof. Let

$$\mathcal{F} = \{E \subset X \times Y : E_x \in \mathcal{N} \text{ for all } x \text{ and } E^y \in \mathcal{M} \text{ for all } y\}.$$

Then every rectangle is certainly in $\mathcal{F}$. Furthermore,

$$\Big(\bigcup_{j\ge1} E_j\Big)_x = \bigcup_{j\ge1} (E_j)_x, \quad \text{and} \quad (E_x)^c = (E^c)_x.$$

Thus $\mathcal{F}$ is a σ-algebra, so $\mathcal{F} \supset \mathcal{M}\otimes\mathcal{N}$.

Since $(f_x)^{-1}(S) = \big(f^{-1}(S)\big)_x$ and $(f^y)^{-1}(S) = \big(f^{-1}(S)\big)^y$, the rest of the lemma follows.

The next result is a basic special case of the forthcoming Theorem 6.3, when f is a characteristic function.

Proposition 6.2. *Assume μ and ν are σ-finite, and take $E \in \mathcal{M}\otimes\mathcal{N}$. Then,*

$$\varphi(x) = \nu(E_x) \text{ and } \psi(y) = \mu(E^y) \text{ are measurable,} \tag{6.12}$$

and

$$(\mu\times\nu)(E) = \int \nu(E_x)\, d\mu(x) = \int \mu(E^y)\, d\nu(y). \tag{6.13}$$

Proof. First we assume μ and ν are finite. Let $\mathcal{C}$ consist of sets $E \in \mathcal{M}\otimes\mathcal{N}$ such that (6.12) and (6.13) hold; our goal is to prove that $\mathcal{C} = \mathcal{M}\otimes\mathcal{N}$. We will do this by applying the Monotone Class Lemma, Proposition 5.9. Thus, to start, we see that rectangles $A \times B$ clearly belong to $\mathcal{C}$, and so do finite disjoint unions of rectangles, so $\mathcal{M}\boxtimes\mathcal{N} \subset \mathcal{C}$. It remains to show that $\mathcal{C}$ is a monotone class.

Let $E_j \in \mathcal{C}$, $E_j \nearrow E$. Then $E_j^y \nearrow E^y$, so $\psi_j(y) = \mu(E_j^y) \nearrow \psi(y) = \mu(E^y)$, and the Monotone Convergence Theorem implies

$$\int \mu(E^y)\, d\nu(y) = \lim \int \psi_j(y)\, d\nu(y) = \lim\, (\mu\times\nu)(E_j) = (\mu\times\nu)(E). \tag{6.14}$$

Similarly we obain the other identity in (6.13), so $E \in \mathcal{C}$.

On the other hand, if $E_j \in \mathcal{C}$ and $E_j \searrow E$, then $E_j^y \searrow E^y$, and hence $\psi_j(y) \searrow \psi(y)$, provided μ is finite. As long as ν is also finite, we can apply the Dominated Convergence Theorem and again get (6.14). Similarly we get the rest of (6.12)–(6.13) for such E, so $\mathcal{C}$ has been shown to be a monotone class, and the proposition is proved, at least for finite measures μ and ν. The extension to the σ-finite case is routine.

The next two results, known as Tonelli's Theorem and Fubini's Theorem, respectively, are the major results of this chapter.

Theorem 6.3. *Assume $(X, \mathcal{M}, \mu)$ and $(Y, \mathcal{N}, \nu)$ are σ-finite measure spaces. If $f \in \mathcal{M}^+(X \times Y)$, then the functions*

$$g(x) = \int f_x \, d\nu, \quad h(y) = \int f^y \, d\mu \tag{6.15}$$

are measurable, hence elements of $\mathcal{M}^+(X)$ and $\mathcal{M}^+(Y)$, respectively, and

$$\int f(x,y) \, d(\mu \times \nu) = \int g(x) \, d\mu(x) = \int h(y) \, d\nu(y). \tag{6.16}$$

Proof. Take a sequence of simple functions $f_j \in \mathfrak{S}^+(X \times Y)$, $f_j \nearrow f$, such as constructed in (3.20). The results above, for f replaced by f_j, are immediate from Proposition 6.2 and additivity of the integral. Note that, for each $x \in X$, $f_{jx} \nearrow f_x$ and, for each $y \in Y$, $f_j^y \nearrow f^y$. Thus, with obvious notation, $g_j \nearrow g$ and $h_j \nearrow h$, by the Monotone Convergence Theorem. Then (6.16) also follows by the Monotone Convergence Theorem.

Theorem 6.4. *Let X and Y be as in Theorem 6.3. If $f \in \mathcal{L}^1(X \times Y, \mu \times \nu)$, then*

$$f^y \in \mathcal{L}^1(X, \mu) \text{ for } \nu\text{-a.e. } y \in Y, \quad f_x \in \mathcal{L}^1(Y, \nu) \text{ for } \mu\text{-a.e. } x \in X, \tag{6.17}$$

so g and h, given by (6.15), are defined almost everywhere. We have

$$g \in L^1(X, \mu), \quad h \in L^1(Y, \nu), \tag{6.18}$$

and (6.16) holds.

Proof. If we write $f = f^+ - f^-$ with $f^\pm \in \mathcal{M}^+(X \times Y) \cap \mathcal{L}^1(X \times Y, \mu \times \nu)$, then Theorem 6.3 applies to f^+ and to f^-, yielding $g^\pm \in \mathcal{M}^+(X)$ and $h^\pm \in \mathcal{M}^+(Y)$ such that

$$\int f^\pm(x,y) \, d(\mu \times \nu) = \int g^\pm \, d\mu = \int h^\pm \, d\nu. \tag{6.19}$$

The finiteness of the first integral in (6.19) implies that $g^{\pm}$ and $h^{\pm}$ are finite a.e. and belong to $L^1(X,\mu)$ and $L^1(Y,\nu)$, respectively. Thus $g = g^+ - g^-$ and $h = h^+ - h^-$ satisfy (6.18), and (6.16) also follows from (6.19).

Let us write (6.16) in the form

$$\begin{aligned}\int f(x,y)\,d(\mu\times\nu) &= \int\Big[\int f(x,y)\,d\nu(y)\Big]\,d\mu(x)\\ &= \int\Big[\int f(x,y)\,d\mu(x)\Big]\,d\nu(y).\end{aligned}\tag{6.20}$$

The first integral is called a product integral and the other two quantities in the triple identity (6.20) are called iterated integrals.

It is useful to have some sufficient conditions for a function $g : X\times Y \to \mathbb{R}$ to be $\mathcal{M}\otimes\mathcal{N}$-measurable. Clearly if $\pi_X : X\times Y \to X$ and $\pi_Y : X\times Y\to Y$ are the standard projections, then

$$\begin{aligned} f : X\to\mathbb{R}\ \mathcal{M}\text{-measurable} &\Longrightarrow f\circ\pi_X\ \mathcal{M}\otimes\mathcal{N}\text{-measurable},\\ g : Y\to\mathbb{R}\ \mathcal{N}\text{-measurable} &\Longrightarrow g\circ\pi_Y\ \mathcal{M}\otimes\mathcal{N}\text{-measurable}.\end{aligned}\tag{6.21}$$

Also, by Proposition 3.1, the algebra $\mathfrak{A}$ consisting of finite sums of products of functions of the type (6.21) are $\mathcal{M}\otimes\mathcal{N}$-measurable. Also recall that pointwise limits of measurable functions are measurable. Using these facts together with the Stone-Weierstrass Theorem, we can prove the following:

Proposition 6.5. *Let X and Y be compact Hausdorff spaces and $\mathfrak{B}_X$, $\mathfrak{B}_Y$ their σ-algebras of Borel sets. Then any continuous $f : X\times Y\to\mathbb{R}$ is $\mathfrak{B}_X\otimes\mathfrak{B}_Y$-measurable.*

Proof. The observations above imply that f is $\mathfrak{B}_X\otimes\mathfrak{B}_Y$-measurable if f belongs to the algebra $\mathcal{C}$ of finite sums of functions of the form $g(x)h(y)$, with g and h continuous. Also, f is $\mathfrak{B}_X\otimes\mathfrak{B}_Y$-measurable if f belongs to the closure $\overline{\mathcal{C}}$ of $\mathcal{C}$ in $C(X\times Y)$. However, the Stone-Weierstrass Theorem implies $\overline{\mathcal{C}} = C(X\times Y)$, so we have the result.

If X and Y are compact metric spaces, we can say more.

Proposition 6.6. *If X and Y are compact metric spaces, then any compact $K\subset X\times Y$ belongs to $\mathfrak{B}_X\otimes\mathfrak{B}_Y$. Hence*

$$\mathfrak{B}_{X\times Y} = \mathfrak{B}_X\otimes\mathfrak{B}_Y.\tag{6.22}$$

Proof. In this case, $X \times Y$ is a metric space, and, if we denote its distance function by d, then

$$\vartheta_K(z) = d(z, K) \tag{6.23}$$

is a continuous function on $X \times Y$ such that $K = \{z \in X \times Y : \vartheta_K(z) = 0\}$. According to Proposition 6.5, ϑ_K is $\mathfrak{B}_X \otimes \mathfrak{B}_Y$-measurable, so each compact $K \subset X \times Y$ belongs to $\mathfrak{B}_X \otimes \mathfrak{B}_Y$. This implies $\mathfrak{B}_{X \times Y} \subset \mathfrak{B}_X \otimes \mathfrak{B}_Y$.

For the reverse containment, we argue as follows. Let

$$\mathcal{A}_X = \{A \subset X : A \times E \in \mathfrak{B}_{X \times Y},\ \forall \text{compact } E \subset Y\}.$$

Then $\mathcal{A}_X$ is a σ-algebra, containing all compact $A \subset X$, so $\mathcal{A}_X \supset \mathfrak{B}_X$. Hence

$$A \in \mathfrak{B}_X,\ E \subset Y \text{ compact } \Longrightarrow A \times E \in \mathfrak{B}_{X \times Y}.$$

Now, let

$$\mathcal{E}_Y = \{E \subset Y : A \times E \in \mathfrak{B}_{X \times Y},\ \forall A \in \mathfrak{B}_X\}.$$

We see that $\mathcal{E}_Y$ is a σ-algebra, containing all compact $E \subset Y$, by the argument above, so $\mathcal{E}_Y \supset \mathfrak{B}_Y$. Hence

$$A \in \mathfrak{B}_X,\ E \in \mathfrak{B}_Y \Longrightarrow A \times E \in \mathfrak{B}_{X \times Y}.$$

It follows that $\mathfrak{B}_{X \times Y} \supset \mathfrak{B}_X \otimes \mathfrak{B}_Y$. Note that this latter containment holds whenever X and Y are compact Hausdorff spaces; we do not need them to be metric spaces.

The following is an immediate consequence of Proposition 6.6.

Corollary 6.7. *If X and Y are σ-compact metric spaces, then* (6.22) *holds.*

The product measure space $(X \times Y, \mathcal{M} \otimes \mathcal{N}, \mu \times \nu)$ is almost never complete, but one can construct its completion $(X \times Y, \overline{\mathcal{M} \otimes \mathcal{N}}, \overline{\mu \times \nu})$ via the process (3.44). The following is a version of the Fubini-Tonelli Theorem for $\overline{\mathcal{M} \otimes \mathcal{N}}$-measurable functions.

Proposition 6.8. *Assume $(X, \mathcal{M}, \mu)$ and $(Y, \mathcal{N}, \nu)$ are complete, σ-finite measure spaces. Let $f : X \times Y \to \mathbb{R}$ be $\overline{\mathcal{M} \otimes \mathcal{N}}$-measurable and satisfy*

$$(a)\ f \geq 0 \qquad \text{or} \qquad (b)\ f \in \mathcal{L}^1(X \times Y, \overline{\mathcal{M} \otimes \mathcal{N}}).$$

Then

$$\begin{aligned} &f_x \ \text{ is } \mathcal{N}\text{-measurable, for } \mu\text{-a.e. } x, \\ &f^y \ \text{ is } \mathcal{M}\text{-measurable, for } \nu\text{-a.e. } y. \end{aligned} \tag{6.24}$$

In case (b), f_x and f^y are integrable for a.e. x (resp., a.e. y). Also, g and h, defined a.e. by (6.15), *are measurable, and in case (b), integrable, and*

$$\int f \, d(\overline{\mu \times \nu}) = \int g(x)\, d\mu(x) = \int h(y)\, d\nu(y). \tag{6.25}$$

Proof. Using Exercise 6 of Chapter 3, write $f = f_0 + f_1$, where f_0 is $\mathcal{M}\otimes\mathcal{N}$-measurable and satisfies either (a) or (b), while $f_1 = 0$, $\overline{\mu\times\nu}$-a.e. Theorems 6.3–6.4 apply to f_0, so it suffices to show that

$$f_{1x} = 0 \ \nu\text{-a.e., for } \mu\text{-a.e. } x, \quad f_1^y = 0 \ \mu\text{-a.e., for } \nu\text{-a.e. } y. \tag{6.26}$$

As in the proof of Theorems 6.3–6.4, it suffices to show that, for $E \in \overline{\mathcal{M}\otimes\mathcal{N}}$,

$$\begin{aligned}(\overline{\mu\times\nu})(E) = 0 \Longrightarrow \nu(E_x) &= 0 \text{ for } \mu\text{-a.e. } x \\ \text{and } \ \mu(E^y) &= 0 \text{ for } \nu\text{-a.e. } y.\end{aligned} \tag{6.27}$$

Now $E \subset F$ with $(\mu\times\nu)(F) = 0$; hence $E_x \subset F_x$ and $E^y \subset F^y$. Thus (6.27) is a consequence of the statement that, for $F \in \mathcal{M}\otimes\mathcal{N}$,

$$\begin{aligned}(\mu\times\nu)(F) = 0 \Longrightarrow \nu(F_x) &= 0 \text{ for } \mu\text{-a.e. } x \\ \text{and } \ \mu(F^y) &= 0 \text{ for } \nu\text{-a.e. } y.\end{aligned} \tag{6.28}$$

This in turn follows from (6.12)–(6.13), together with (3.59), so the proposition is proved.

One can also consider products of K measure spaces $(X_j, \mathcal{M}_j, \mu_j)$, constructing a σ-algebra $\mathfrak{F} = \mathcal{M}_1 \otimes \cdots \otimes \mathcal{M}_K$ of subsets of $Z = \prod_{j=1}^K X_j$, generated by rectangles of the form $R = A_1 \times \cdots \times A_K$, $A_j \in \mathcal{M}_j$. The product measure $\nu = \mu_1 \times \cdots \times \mu_K$ satisfies

$$\nu(S) = \inf\Big\{\sum_{j\ge1} \mu_1(A_{j1})\cdots\mu_K(A_{jK}) : A_{jk} \in \mathcal{M}_k, S \subset \bigcup_{j\ge1}\Big(\prod_{k=1}^K A_{jk}\Big)\Big\}, \tag{6.29}$$

for $S \in \mathfrak{F}$. Analogues of the results discussed above follow without difficulty; we leave the details to the reader.

There is also a notion of product measure for a countable infinite product of measure spaces, $(X_j, \mathcal{M}_j, \mu_j)$, $j \in \mathbb{Z}^+$, provided $\mu_j(X_j) = 1$ for all j, i.e., each μ_j is a probability measure. Then one takes rectangles of the form

$$R = \prod_{j\ge1} A_j, \quad A_j \in \mathcal{M}_j, \quad A_j = X_j \text{ except for finitely many } j, \tag{6.30}$$

and $\mathfrak{F}$ is the σ-algebra generated by such sets. The product measure ν is a probability measure on $Z = \prod_{j\ge1} X_j$, satisfying

$$\nu(S) = \inf\Big\{\sum_{j\ge1}\prod_{k\ge1} \mu_k(A_{jk}) : \prod_{k\ge1} A_{jk} \in \mathcal{R}, S \subset \bigcup_{j\ge1}\Big(\prod_{k\ge1} A_{jk}\Big)\Big\}, \tag{6.31}$$

for $S \in \mathfrak{F}$, where $\mathcal{R}$ denotes the class of rectangles of the form (6.30). Note that, for each "infinite product" $\prod_{k\geq 1}\mu_k(A_{jk})$ in (6.31), all but finitely many factors are 1.

The proof that setting $\nu(R) = \prod_{j\geq 1}\mu_j(A_j)$, for sets R as in (6.30), yields a premeasure on the algebra of finite disjoint unions of such sets can be given parallel to the argument involving (6.5)–(6.6). Thus, suppose $A_j \in \mathcal{M}_j$, $E_{j\ell} \in \mathcal{M}_j$, and we have a countable disjoint union

$$A_1 \times \cdots \times A_k \times X_{k+1} \times \cdots = \bigcup_{\ell\geq 1} E_{1\ell} \times \cdots \times E_{\kappa(\ell),\ell} \times X_{\kappa(\ell)+1} \times \cdots,$$

so

$$\chi_{A_1}(x_1)\cdots\chi_{A_k}(x_k) = \sum_{\ell\geq 1} \chi_{E_{1\ell}}(x_1)\cdots\chi_{E_{\kappa(\ell),\ell}}(x_{\kappa(\ell)}).$$

One can integrate, first over X_1, then over X_2, etc., obtaining, at the kth step,

$$\mu_1(A_1)\cdots\mu_k(A_k) = \sum_{\ell\geq 1} \mu_1(E_{1\ell})\cdots\mu_k(E_{k\ell})\chi_{E_{k+1,\ell}}(x_{k+1})\cdots\chi_{E_{\kappa(\ell),\ell}}(x_{\kappa(\ell)}).$$

Continue to integrate over X_{k+1}, etc. Any given term in the sum above will stabilize eventually, and we obtain

$$\mu_1(A_1)\cdots\mu_k(A_k) = \sum_{\ell\geq 1} \mu_1(E_{1\ell})\cdots\mu_{\kappa(\ell)}(E_{\kappa(\ell),\ell}),$$

as desired.

With a little more effort, one can define a product measure on an *uncountable* product of probability spaces. For a treatment of this, we refer to [**HS**].

Exercises

Let $\mathcal{A}$ and $\mathcal{B}$ be algebras of subsets of X and Y, respectively, and let μ_0 and ν_0 be premeasures on these algebras, giving rise to measures μ and ν, on $\mathcal{M} = \sigma(\mathcal{A})$ and $\mathcal{N} = \sigma(\mathcal{B})$, respectively. Let us denote by $\mathcal{A} \boxtimes \mathcal{B}$ the algebra of subsets of $X \times Y$ generated by rectangles of the form

$$R = A \times E, \quad A \in \mathcal{A},\ E \in \mathcal{B}. \tag{6.32}$$

1. Show that $\mathcal{A} \boxtimes \mathcal{B}$ consists of finite disjoint unions of rectangles of the form (6.32).

2. Show that $\sigma(\mathcal{A} \boxtimes \mathcal{B}) = \mathcal{M} \otimes \mathcal{N}$.
Hint. Show that $\sigma(\mathcal{A} \boxtimes \mathcal{B}) \supset \mathcal{M} \boxtimes \mathcal{N}$, the latter being the algebra defined after (6.2).

3. Define a set function $\gamma_0 : \mathcal{A} \boxtimes \mathcal{B} \to [0, \infty]$ by

$$\gamma_0\Big(\bigcup_{j=1}^{N}(A_j \times E_j)\Big) = \sum_{j=1}^{N} \mu_0(A_j)\nu_0(E_j),$$

when the rectangles $R_j = A_j \times E_j \in \mathcal{A} \boxtimes \mathcal{B}$ are disjoint. Show that γ_0 is a well-defined premeasure on $\mathcal{A} \boxtimes \mathcal{B}$.

4. Show that the measure γ on $\sigma(\mathcal{A} \boxtimes \mathcal{B})$ obtained from γ_0 via Theorem 5.4 coincides with the product measure $\mu \times \nu$, provided μ and ν are σ-finite. Consequently, for $S \in \mathcal{M} \otimes \mathcal{N}$,

(6.33)
$$(\mu \times \nu)(S) = \inf \Big\{ \sum_{j \geq 1} \mu_0(A_j)\nu_0(E_j) : A_j \in \mathcal{A}, E_j \in \mathcal{B},\ S \subset \bigcup_{j \geq 1} (A_j \times E_j) \Big\}.$$

More generally, for any $S \subset X \times Y$, the outer measure $\pi^*(S)$ is given by the right side of (6.33). (Compare (6.7).)

In Exercises 5–6, assume $(X, \mathcal{M}, \mu)$ and $(Y, \mathcal{N}, \nu)$ are σ-finite measure spaces.

5. Let $\mathcal{S}_{\mathcal{R}}$ consist of simple functions on $(X \times Y, \mathcal{M} \otimes \mathcal{N})$ of the form $f = \sum_{j=1}^{N} a_j \chi_{R_j}$, where each R_j is a rectangle, $R_j = A_j \times E_j$, $A_j \in \mathcal{M}$, $E_j \in \mathcal{N}$, such that $(\mu \times \nu)(R_j) < \infty$. Show that $\mathcal{S}_{\mathcal{R}}$ is dense in $L^p(X \times Y, \mathcal{M} \otimes \mathcal{N})$, for $1 \leq p < \infty$.

6. If $\{u_j : j \geq 1\}$ is an orthonormal basis of $L^2(X, \mu)$ and $\{v_k : k \geq 1\}$ is an orthonormal basis of $L^2(Y, \nu)$, show that $\{u_j(x)v_k(y) : j, k \geq 1\}$ is an orthonormal basis of the Hilbert space $L^2(X \times Y, \mu \times \nu)$.
Hint. Show that the closure of the linear span of this orthonormal set contains $\mathcal{S}_{\mathcal{R}}$.

In Exercises 7–10, let

$$Z = \prod_{k=1}^{\infty} \{0,1\}. \tag{6.34}$$

We put on each factor $\{0,1\}$ the probability measure assigning the measure $\frac{1}{2}$ to $\{0\}$ and to $\{1\}$. On Z we put the product σ-algebra and the product measure, which we denote μ. Given $a = (a_1, a_2, a_3, \dots)$, define

$$F : Z \longrightarrow [0,1], \quad F(a) = \sum_{k=1}^{\infty} a_k 2^{-k}. \tag{6.35}$$

Thus $F(a)$ is a number with dyadic expansion $0.a_1a_2a_3\cdots$.

7. Show that F is measurable.
Hint. Let a dyadic interval be one with endpoints of the form $m2^{-\ell}$ and $(m+1)2^{-\ell}$, with $m,\ \ell \in \mathbb{Z}^+,\ m+1 \le 2^{\ell}$; a dyadic interval may or may not contain its endpoints. Show that the σ-algebra generated by the collection of dyadic intervals in I is equal to the Borel σ-algebra $\mathfrak{B}$.

8. Show that F is onto. Show that $F^{-1}(x)$ consists of one point unless x has a finite dyadic expansion, in which case $F^{-1}(x)$ consists of 2 points.
Hint. $\sum_{k=2}^{\infty} 1 \cdot 2^{-k} = 2^{-1}$.

9. Show that $S \in \mathfrak{B} \Rightarrow \mu\big(F^{-1}(S)\big) = m(S)$, where m is Lebesgue measue on I.
Hint. Start with $S \in \mathcal{A}_0$, the algebra of finite (disjoint) unions of dyadic intervals.

10. Show that you can take a set $Z_0 \subset Z$ of measure zero such that $F : Z \setminus Z_0 \to I$ is one-to-one and onto and measure preserving.

11. For this exercise, make a slight change of notation; set

$$Z = \prod_{k=1}^{\infty} \{0,2\}.$$

We use the same sort of product measure as before. Consider the map $G : Z \to [0,1]$ given by

$$G(a) = \sum_{k=1}^{\infty} a_k 3^{-k}.$$

Show that G is a homeomorphism of Z onto the Cantor middle third set K. Show that the measure μ_K on $[0,1]$, given by $\mu_K(S) = \mu(G^{-1}(S))$, coincides with the Lebesgue-Stieltjes measure on $[0,1]$ given by Exercise 1 of Chapter 5, using the following function φ. On the middle third of $[0,1]$, set $\varphi = 1/2$. On the middle third sets of the two intervals remaining, set $\varphi = 1/4$ and $3/4$, respectively. Continue in the appropriate fashion, to define a continuous monotone function $\varphi : [0,1] \to [0,1]$.

12. Let $(X, \mathfrak{F}, \mu)$ be a measure space and let $f : X \to [0,\infty)$ be measurable. Consider the set

$$U = \{(x,t) \in X \times [0,\infty) : f(x) < t\}.$$

Show that $U \in \mathfrak{F} \otimes \mathfrak{B}$, where $\mathfrak{B}$ is the class of Borel subsets of $\mathbb{R}$, and that

$$(\mu \times m)(U) = \int_X f\,d\mu,$$

where m denotes Lebesgue measure on $\mathbb{R}$. Compare this result with Exercise 18 of Chapter 3.
Hint. First verify this for simple functions, and then consider simple $\varphi_j \nearrow f$, as in (3.26).

As an alternative, what can you deduce from Proposition 6.2?

13. Here is a variant of Exercise 6. For $j \in \mathbb{N}$, let $(X_j, \mathcal{M}_j, \mu_j)$ be measure spaces satisfying $\mu_j(X_j) = 1$. Form the infinite product measure space $(Z, \mathfrak{F}, \nu)$ as in (6.30)–(6.31), with $Z = \prod_{j\ge1} X_j$.

Suppose that for each $j \in \mathbb{N}$, $\{u_{jk} : k \in \mathbb{N}\}$ is an orthonormal basis of $L^2(X_j, \mu_j)$ and $u_{j1} = 1$. Consider the collection of functions on Z,

$$v_\alpha(z) = \prod_{k=1}^{\infty} u_{k\alpha_k}(x_k), \quad z = (x_1, x_2, x_3, \dots),$$

where $\alpha = (\alpha_1, \alpha_2, \alpha_3, \dots)$ runs over the set $\mathcal{A}$ of elements of $\prod_{k=1}^{\infty} \mathbb{N}$ such that $\alpha_k \ne 1$ for only finitely many k. Show that

$$\{v_\alpha : \alpha \in \mathcal{A}\}$$

is an orthonormal basis of $L^2(Z, \nu)$.
Hint. For the denseness of the linear span, consult the hint for Exercise 6.

Chapter 7

Lebesgue Measure on $\mathbb{R}^n$ and on Manifolds

We constructed Lebesgue measure on $\mathbb{R}$ in Chapter 2. The σ-algebra $\mathcal{L}_1$ of Lebesgue measurable subsets of $\mathbb{R}$ contains the σ-algebra $\mathfrak{B}_1$ of Borel subsets of $\mathbb{R}$. The product construction of Chapter 6 applies to $\mathbb{R}^n = \mathbb{R} \times \cdots \times \mathbb{R}$. We have the σ-algebra

$$\mathfrak{B}_n = \mathfrak{B}_1 \otimes \cdots \otimes \mathfrak{B}_1, \tag{7.1}$$

which, by Corollary 6.7, is the σ-algebra of Borel subsets of $\mathbb{R}^n$. We have the larger σ-algebra $\mathcal{L}_1 \otimes \cdots \otimes \mathcal{L}_1$. As indicated in Chapter 6, the product measure $m \times \cdots \times m$ (which we will also denote m) is not complete on this product σ-algebra. Let $(\mathbb{R}^n, \mathcal{L}_n, \overline{m})$ denote the completion. We will also denote the extended measure $\overline{m}$ on $\mathcal{L}_n$ simply by m. It is easy to show that this measure space is also the completion of $(\mathbb{R}^n, \mathfrak{B}_n, m)$.

Frequently, we will omit the subscripts from $\mathfrak{B}_n$ and $\mathcal{L}_n$, denoting them simply $\mathfrak{B}$ and $\mathcal{L}$, respectively.

Recall that $\mathfrak{B}_1$ is the σ-algebra generated by the algebra $\mathcal{A}_1$ of finite (disjoint) unions of intervals. It follows that $\mathfrak{B}_n$ is the σ-algebra generated by the algebra $\mathcal{A}_1 \boxtimes \cdots \boxtimes \mathcal{A}_1$, consisting of finite (disjoint) unions of sets in $\mathbb{R}^n$ which are products of intervals; we call such sets *cells*. Products of intervals of identical length will be called cubes. It is often convenient to use the formula (6.33) specialized to Lebesgue measure:

$$m(S) = \inf \Big\{ \sum_{j\ge1} m(Q_j) : Q_j \text{ cells}, S \subset \bigcup_{j\ge1} Q_j \Big\}, \tag{7.2}$$

for $S \in \mathfrak{B}$, where, for a cell Q,

$$Q = I_1 \times \cdots \times I_n \Longrightarrow m(Q) = \ell(I_1) \cdots \ell(I_n). \tag{7.3}$$

In fact, (7.2) holds for all $S \in \mathcal{L}$, since (via Exercise 9 of Chapter 5) we have

$$m(S) = \inf \{m(E) : E \in \mathfrak{B}, E \supset S\}, \quad \text{for} \quad S \in \mathcal{L}. \tag{7.4}$$

Now $\mathbb{R}^n$ has an important structure in addition to its product structure, namely a linear structure. The next result shows how the measure of a Borel set changes under a linear transformation and, more generally, how the Lebesgue integral on $\mathbb{R}^n$ behaves under such a transformation. Here, $Gl(n, \mathbb{R})$ denotes the group of invertible linear transformations on $\mathbb{R}^n$, and $\det A$ denotes the determinant of A.

Proposition 7.1. *If $A \in Gl(n, \mathbb{R})$, and if $f \in \mathcal{M}^+(\mathbb{R}^n)$ or $f \in \mathcal{L}^1(\mathbb{R}^n, dx)$, then*

$$\int_{\mathbb{R}^n} f(x)\, dx = |\det A| \int_{\mathbb{R}^n} f(Ax)\, dx. \tag{7.5}$$

Proof. Let $\mathcal{G}$ be the set of elements $A \in Gl(n, \mathbb{R})$ for which (7.5) is true. Clearly $I \in \mathcal{G}$. Using $\det A^{-1} = (\det A)^{-1}$ and $\det AB = (\det A)(\det B)$, it is easy to show that $\mathcal{G}$ is a subgroup of $Gl(n, \mathbb{R})$. Thus, to prove the proposition, it suffices to show that $\mathcal{G}$ contains all elements of the following three forms, since it is an exercise in linear algebra to show that these elements generate $Gl(n, \mathbb{R})$. Here, $\{e_j : 1 \le j \le n\}$ denotes the standard basis of $\mathbb{R}^n$ and σ a permutation of $\{1, \ldots, n\}$.

$$\begin{aligned} &A_1 e_j = e_{\sigma(j)}, \\ &A_2 e_j = c_j e_j, \quad c_j \neq 0, \\ &A_3 e_1 = e_1 + c e_2, \quad A_3 e_j = e_j \text{ for } j \ge 2. \end{aligned} \tag{7.6}$$

For the case $A = A_1$ we can work directly. If $f = \chi_R$ is the characteristic function of a rectangle $E_1 \times \cdots \times E_n$, the result is obvious. By the construction of product measure the identity (7.5) then holds for χ_S, the characteristic function of a Borel set. Then (7.5) follows for $f \in \mathfrak{S}^+(\mathbb{R}^n)$, by additivity, and for $f \in \mathcal{M}^+(\mathbb{R}^n)$, by (3.25).

A similar argument shows that (7.5) holds for transformations of the form A_2. In such a case, a cell of size $\ell(I_1), \ldots, \ell(I_n)$ is mapped by A_2 to a cell of size $|c_1|\ell(I_1), \ldots, |c_n|\ell(I_n)$, so its measure is multiplied by $|c_1 \cdots c_n| = |\det A_2|$. Thus, (7.5) holds for $f = \chi_S$, $S \in \mathfrak{B}$, by (7.2)–(7.3), and hence, as before, for more general f.

We will show that (7.5) holds for each of the transformations of the type A_3, using the Fubini-Tonelli Theorem. We note that the identity

$$\int_{\mathbb{R}} f(x)\,dx = \int_{\mathbb{R}} f(x+c)\,dx, \quad f \in \mathcal{M}^+(\mathbb{R}), \tag{7.7}$$

is elementary. To keep the formulas short, take $n = 2$. For the case $A = A_3$, we have

$$\begin{aligned}\int f(x_1 + cx_2, x_2)\,dx &= \int \Big[\int f(x_1 + cx_2, x_2)\,dx_1\Big]\,dx_2 \\ &= \int \Big[\int f(x_1, x_2)\,dx_1\Big]\,dx_2,\end{aligned} \tag{7.8}$$

the second identity by (7.7). Again we get (7.5). It is clear how to extend the analysis of A_3 to $n \geq 3$, so the proposition is proved.

REMARK 1. One consequence of Proposition 7.1 is the invariance of Lebesgue measure under rotations.

REMARK 2. At the end of this chapter there are exercise sets on determinants and on row reduction, deriving the results on matrices used in the proof of Proposition 7.1.

Proposition 7.1 generalizes to a change of variables formula, which we establish next, an extension to Lebesgue measure of one of the fundamental results of multi-variable calculus. Let $F : \mathcal{O} \to \Omega$ be a C^1 diffeomorphism, where $\mathcal{O}$ and Ω are open in $\mathbb{R}^n$. Let $G = F^{-1}$, and form the absolute values of the Jacobian determinants:

$$J(x) = |\det DF(x)|, \quad K(y) = |\det DG(y)|. \tag{7.9}$$

Theorem 7.2. *Let $F : \mathcal{O} \to \Omega$ be a C^1 diffeomorphism. If either $u \in \mathcal{M}^+(\Omega)$ or $u \in \mathcal{L}^1(\Omega, dx)$, then*

$$\int_{\Omega} u(x)\,dx = \int_{\mathcal{O}} u\big(F(x)\big)J(x)\,dx. \tag{7.10}$$

Proof. Take a Borel set $S \subset\subset \mathcal{O}$. Fix $\varepsilon > 0$, and cover S with a countable family of cells $\{Q_j\}$ such that

$$m(S) \geq \sum m(Q_j) - \varepsilon. \tag{7.11}$$

Subdividing each of the cells if necessary, we can assume that, if x_j is the center of Q_j and $y_j = F(x_j)$,

$$y \in F(Q_j) \Longrightarrow K(y_j) \geq (1-\varepsilon)K(y) \tag{7.12}$$

and

$$F(Q_j) \subset y_j + (1+\varepsilon)DF(x_j)(Q_j^0), \tag{7.13}$$

where we set $Q_j^0 = Q_j - x_j$, centered at 0. From Proposition 7.1, we deduce that

$$m\big(F(Q_j)\big) \leq (1+\varepsilon)^n J(x_j) m(Q_j). \tag{7.14}$$

It follows that

$$\begin{aligned} m(S) + \varepsilon \geq \sum m(Q_j) &\geq \frac{1}{(1+\varepsilon)^n} \sum K(y_j) m\big(F(Q_j)\big) \\ &\geq \frac{1-\varepsilon}{(1+\varepsilon)^n} \int\limits_{\Omega} K(y)\chi_{F(S)}(y)\,dy. \end{aligned} \tag{7.15}$$

Taking $\varepsilon \to 0$, we have $m(S) \geq \int K(y)\chi_{F(S)}(y)\,dy$ or, equivalently,

$$\int\limits_{\mathcal{O}} \varphi(x)\,dx \geq \int\limits_{\Omega} \varphi\big(G(y)\big)K(y)\,dy, \tag{7.16}$$

for $\varphi = \chi_S$. By additivity, we have (7.16) for any $\varphi \in \mathfrak{S}^+(K)$, any compact $K \subset \mathcal{O}$. The Monotone Convergence Theorem then gives (7.16) for any $\varphi \in \mathcal{M}^+(\mathcal{O})$. Setting $\varphi(x) = \psi(x)J(x)$, we have $\int \psi(x)J(x)\,dx \geq \int \psi\big(G(y)\big)\,dy$ or, equivalently, setting $v(y) = \psi(G(y))$,

$$\int\limits_{\Omega} v(y)\,dy \leq \int\limits_{\mathcal{O}} v\big(F(x)\big)J(x)\,dx, \tag{7.17}$$

for all $v \in \mathcal{M}^+(\Omega)$. Reversing the roles of $\mathcal{O}$ and Ω, we hence have (7.10), for $u \in \mathcal{M}^+(\Omega)$, from which the identity for $u \in \mathcal{L}^1(\Omega, dx)$ follows.

There are various more refined change of variables formulas, some of which we will discuss below. First, we apply the results obtained so far to show how to integrate functions on a Riemannian manifold, i.e., a smooth manifold equipped with a metric tensor. Background material on manifolds and metric tensors is given in Appendix B.

Let M be a C^1 manifold of dimension n. A continuous metric tensor on M gives a continuous inner product on tangent vectors to M. In a local coordinate system $(x_1, \dots, x_n)$, identifying an open subset of M with an open set $\mathcal{O} \subset \mathbb{R}^n$, the metric tensor is given by a positive definite $n \times n$ matrix $G(x) = (g_{jk}(x))$, and the inner product of vectors U and V is given by

$$\langle U, V \rangle = U \cdot G(x)V = \sum_{j,k} g_{jk}(x) u_j(x) v_k(x), \tag{7.18}$$

where $U = \sum_j u_j(x) e_j$, $V = \sum_k v_k(x) e_k$, and $\{e_1, \dots, e_n\}$ is the standard basis of $\mathbb{R}^n$. If we change coordinates by a C^1 diffeomorphism $F : \mathcal{O} \to \Omega$, the metric tensor $H(y) = (h_{jk}(y))$ in the coordinate system $y = F(x)$ is related to $G(x)$ by

$$DF(x)U \cdot H(y)DF(x)V = \langle U, V \rangle = U \cdot G(x)V, \tag{7.19}$$

at $y = F(x)$, i.e.,

$$G(x) = DF(x)^t H(y) DF(x), \tag{7.20}$$

or

$$g_{jk}(x) = \sum_{i,\ell} \frac{\partial F_i}{\partial x_j} \frac{\partial F_\ell}{\partial x_k} h_{i\ell}(y).$$

Now, for a Borel measurable function u supported on a coordinate patch, the integral is given by

$$\int u \, dV = \int u(x) \sqrt{g} \, dx, \quad g(x) = \det G(x). \tag{7.21}$$

(One often uses dS instead of dV when $\dim M = 2$ or when one is integrating over the boundary of a manifold of dimension $(n+1)$.) To see that (7.21) is well defined, note that under the change of coordinates $y = F(x)$ we have by (7.20) that $\det G(x) = (\det DF(x))^2 \det H(y)$. Hence

$$\sqrt{h} = |\det DF|^{-1} \sqrt{g}, \quad h = \det H, \tag{7.22}$$

so, by (7.10),

$$\begin{aligned} \int u(y) \sqrt{h} \, dy &= \int u(F(x)) \, |\det DF|^{-1} \sqrt{g} \, |\det DF| \, dx \\ &= \int u(F(x)) \sqrt{g} \, dx. \end{aligned} \tag{7.23}$$

More generally $\int_M u\,dV$ is defined by writing u as a sum of terms supported on coordinate charts. We see that we have a well-defined measure on the σ-algebra of Borel subsets of a C^1 manifold with a C^0 metric tensor.

In case M is an n-dimensional submanifold of $\mathbb{R}^m$ and a local coordinate chart arises via a C^1 map $\varphi : \mathcal{O} \to \mathbb{R}^m$, the metric tensor $G(x)$ is given by $G(x) = D\varphi(x)^t D\varphi(x)$, i.e.,

$$g_{jk}(x) = \sum_{\ell=1}^{m} \frac{\partial \varphi_\ell}{\partial x_j}\frac{\partial \varphi_\ell}{\partial x_k} = \frac{\partial \varphi}{\partial x_j} \cdot \frac{\partial \varphi}{\partial x_k}, \tag{7.24}$$

using the dot product on $\mathbb{R}^m$.

An example is the graph of a real-valued C^1 function u on $\Omega \subset \mathbb{R}^n$. If M is the graph of $z = u(x)$, then M is an n-dimensional C^1 manifold in $\mathbb{R}^{n+1}$. The map $\varphi(x) = \big(u(x), x\big)$ provides a natural coordinate system, in which the metric tensor is given by

$$g_{jk}(x) = \delta_{jk} + \frac{\partial u}{\partial x_j}\frac{\partial u}{\partial x_k}. \tag{7.25}$$

If u is C^1, we see that the metric tensor is C^0. To calculate $g = |\det(g_{jk})|$ at a given point $x_0 \in \Omega$, if $\nabla u(x_0) \neq 0$, rotate coordinates so $\nabla u(x_0)$ is parallel to the x_1-axis. We get

$$\sqrt{g} = \big(1 + |\nabla u|^2\big)^{1/2}. \tag{7.26}$$

In particular, the n-dimensional measure of the surface M is given by

$$V(M) = \int_M dV = \int_\Omega \sqrt{g}\,dx = \int_\Omega \big(1 + |\nabla u(x)|^2\big)^{1/2}\,dx.$$

Particularly important examples of Riemannian manifolds are the unit spheres S^{n-1} in $\mathbb{R}^n$,

$$S^{n-1} = \{x \in \mathbb{R}^n : |x| = 1\}.$$

In case $n = 2$, we have S^1, the unit circle. Since the time of the ancient Greek geometers, the length of the unit circle has been used to define π; $\pi =$ half the length of S^1.

Spherical polar coordinates on $\mathbb{R}^n$ are defined in terms of a smooth diffeomorphism

$$R : (0, \infty) \times S^{n-1} \longrightarrow \mathbb{R}^n \setminus 0, \quad R(r, \omega) = r\omega. \tag{7.27}$$

Let $(h_{\ell m})$ denote the metric tensor on S^{n-1} (induced from its inclusion in $\mathbb{R}^n$) with respect to some coordinate chart $\varphi : \mathcal{O} \subset U \subset S^{n-1}$. Then, with respect to the coordinate chart $\Phi : (0,\infty) \times \mathcal{O} \to \mathcal{U} \subset \mathbb{R}^n$ given by $\Phi(r,y) = r\varphi(y)$, the Euclidean metric tensor can be written

$$(e_{jk}) = \begin{pmatrix} 1 & \\ & r^2 h_{\ell m} \end{pmatrix}. \tag{7.28}$$

(Compare (7.24).) To see that the blank terms vanish, i.e., $\partial_r \Phi \cdot \partial_{x_j} \Phi = 0$, note that $\varphi(x) \cdot \varphi(x) \equiv 1 \Rightarrow \partial_{x_j}\varphi(x) \cdot \varphi(x) = 0$. Now (7.28) yields

$$\sqrt{e} = r^{n-1}\sqrt{h}. \tag{7.29}$$

We therefore have the following result for integrating a function in spherical polar coordinates. Let us denote by dS the measure on S^{n-1} produced via its Riemannian metric.

Proposition 7.3. *If $f \in \mathcal{M}^+(\mathbb{R}^n)$ or $f \in \mathcal{L}^1(\mathbb{R}^n, dx)$, then*

$$\int_{\mathbb{R}^n} f(x)\,dx = \int_{S^{n-1}} \left[\int_0^\infty f(r\omega) r^{n-1}\,dr\right] dS(\omega). \tag{7.30}$$

See the exercises for applications of Proposition 7.3.

We return to the change of variable formula (7.10), approaching it from another point of view, which yields more general results. In fact, we start with a very general setting. Let $(\mathcal{O}, \mathfrak{F}, \mu)$ and $(\Omega, \mathfrak{G}, \nu)$ be measure spaces and $F : \mathcal{O} \to \Omega$ a measurable map, i.e., $S \in \mathfrak{G} \Rightarrow F^{-1}(S) \in \mathfrak{F}$. If $u : \Omega \to \mathbb{R}$ is measurable and ≥ 0, we have

$$\int_{\mathcal{O}} u(F(x))\,d\mu(x) = \int_{\Omega} u(x)\,d\mu_{F^{-1}}(x), \tag{7.31}$$

where, for $S \in \mathfrak{G}$,

$$\mu_{F^{-1}}(S) = \mu\big(F^{-1}(S)\big). \tag{7.32}$$

Now, if we have the property

$$\nu(S) = 0 \Longrightarrow \mu\big(F^{-1}(S)\big) = 0, \tag{7.33}$$

we can apply the Radon-Nikodym Theorem, Theorem 4.10, to conclude that there is a $\mathfrak{G}$-measurable function $J_{F^{-1}} \geq 0$ such that

$$\int_{\mathcal{O}} u\big(F(x)\big)\,d\mu(x) = \int_{\Omega} u(x) J_{F^{-1}}(x)\,d\nu(x). \tag{7.34}$$

Now we specialize to $\mathcal{O}, \Omega$ open in $\mathbb{R}^n$, $\mathfrak{F}, \mathfrak{G}$ the Borel sets, $\mu = \nu =$ Lebesgue measure. Assume $F : \mathcal{O} \to \Omega$ is continuous, hence measurable. Suppose F is one-to-one and onto. The property (7.33) certainly holds if F^{-1} is Lipschitz. If we interchange the roles of $\mathcal{O}$ and Ω and of F and F^{-1}, we have the following.

Proposition 7.4. *Let $F : \mathcal{O} \to \Omega$ be both a Lipschitz map and a homeomorphism. There exists a Borel function $J_F \ge 0$ such that, for any $u \in \mathcal{M}^+(\Omega)$,*

$$\int_{\Omega} u(x)\,dx = \int_{\mathcal{O}} u\big(F(x)\big) J_F(x)\,dx. \tag{7.35}$$

In Theorem 7.2, this result was established for a C^1 diffeomorphism F, and the factor $J_F(x)$ was identified with the absolute value of the Jacobian determinant:

$$J_F(x) = |\det DF(x)|. \tag{7.36}$$

We will now establish this identity for a larger class of maps F.

Note that, if Q is a cell in $\mathcal{O}$, then

$$m\big(F(Q)\big) = \int_{\Omega} \chi_{F(Q)}(x)\,dx = \int_{Q} J_F(x)\,dx. \tag{7.37}$$

Let us assume that Ω is bounded, so $m(\Omega) < \infty$, and hence $J_F \in L^1(\mathcal{O}, dx)$. Extend J_F to be 0 on $\mathbb{R}^n \setminus \mathcal{O}$. For $\varepsilon > 0$, set

$$J_\varepsilon(x) = \frac{1}{m\big(Q_\varepsilon(x)\big)} \int_{Q_\varepsilon(x)} J_F(y)\,dy = \int \varphi_\varepsilon(x-y) J_F(y)\,dy, \tag{7.38}$$

where Q_ε is the cube of edge ε, centered at x, so $\varphi_\varepsilon(x) = \varepsilon^{-n}\chi_{Q_\varepsilon(0)}(x)$. We have

$$J_\varepsilon \longrightarrow J_F \ \text{ in } \ L^1(\mathbb{R}^n), \ \text{ as } \ \varepsilon \to 0. \tag{7.39}$$

See Exercise 10 in the first exercise set below. Consequently, by Exercise 11 of Chapter 3, there is a sequence $\varepsilon_\nu \to 0$ such that

$$J_{\varepsilon_\nu}(x) \longrightarrow J_F(x) \ \text{ for a.e. } \ x \in \mathbb{R}^n. \tag{7.40}$$

Consequently, by (7.37), we have

$$\frac{m\big(F(Q_\varepsilon(x))\big)}{m\big(Q_\varepsilon(x)\big)} \longrightarrow J_F(x) \ \text{ for a.e. } \ x \in \mathcal{O}, \quad \varepsilon = \varepsilon_\nu \to 0. \tag{7.41}$$

We next establish a lemma which will take the place of (7.14) in our extension of the range of validity of (7.10).

Lemma 7.5. *If $F : \mathcal{O} \to \Omega$ is a homeomorphism that is differentiable at $x_0 \in \mathcal{O}$, then*

$$\lim_{\varepsilon \to 0} \frac{m\big(F(Q_\varepsilon(x_0))\big)}{m\big(Q_\varepsilon(x_0)\big)} = |\det DF(x_0)|. \tag{7.42}$$

Proof. First assume $DF(x_0)$ is invertible. Composing F with a linear transformation (and recalling Proposition 7.1), we can assume $DF(x_0) = I$. Thus

$$F(x_0 + y) = F(x_0) + y + R(y), \quad |R(y)| = o(|y|). \tag{7.43}$$

Fix $\delta \in (0, 1/2)$, and pick ε_0 so small that $|y| \le \varepsilon_0 \Rightarrow |R(y)| \le \delta|y|$. Then, as in (7.13), we have, for $\varepsilon \le \varepsilon_0$,

$$F\big(Q_\varepsilon(x_0)\big) \subset Q_{(1+\delta)\varepsilon}\big(F(x_0)\big). \tag{7.44}$$

Hence

$$m\big(F\big(Q_\varepsilon(x_0)\big)\big) \le (1+\delta)^n \, m\big(Q_\varepsilon(x_0)\big). \tag{7.45}$$

This implies that the lim sup of the left side of (7.42) is $\le |\det DF(x_0)|$ in this case.

We now need to establish the reverse inequality for the lim inf. This is not hard, if one uses some fundamental results from topology. The same reasoning that gives (7.44) shows that the *boundary* $\partial Q_\varepsilon(x_0)$ of $Q_\varepsilon(x_0)$ is mapped by F onto a set Σ, contained in

$$\mathcal{C} = \{x \in \mathcal{O} : \operatorname{dist}(x, \partial Q_\varepsilon(x_0)) \le \delta\varepsilon\} = Q_{(1+\delta)\varepsilon}(x_0) \setminus \overset{\circ}{Q}_{(1-\delta)\varepsilon}(x_0). \tag{7.46}$$

Furthermore, for each $\tau \in [0,1]$, we can define $F_\tau(x_0+y) = f(x_0)+y+\tau R(y)$, and each map F_τ maps $\partial Q_\varepsilon(x_0)$ into $\mathcal{C}$. Recall that F is assumed to be a homeomorphism. Thus, by degree theory, $F\big(\overset{\circ}{Q}_\varepsilon(x_0)\big)$ contains a point x_1 if and only if the map $F\big|_{\partial Q_\varepsilon(x_0)}$ has degree 1 about x_1. However, the maps $F_\tau\big|_{\partial Q_\varepsilon(x_0)}$ all have the same degree about any point $x_1 \in \overset{\circ}{Q}_{(1-\delta)\varepsilon}(x_0)$, for $\tau \in [0,1]$, so we see that $F\big|_{\partial Q_\varepsilon(x_0)}$ has degree 1 about each such point, since F_0 obviously has this property. Thus we conclude that

$$F\big(Q_\varepsilon(x_0)\big) \supset Q_{(1-\delta)\varepsilon}(x_0); \tag{7.47}$$

hence

$$m\big(F\big(Q_\varepsilon(x_0)\big)\big) \ge (1-\delta)^n m\big(Q_\varepsilon(x_0)\big). \tag{7.48}$$

This establishes the lemma when $DF(x_0)$ is invertible.

It remains to treat the case when $\det DF(x_0) = 0$, i.e., $A = DF(x_0)$ is not invertible. In this case, we have

$$F(x_0 + y) = F(x_0) + Ay + R(y), \quad |R(y)| = o(|y|). \tag{7.49}$$

Consequently, if we fix $\delta > 0$ and pick ε_0 so small that $|y| \le \varepsilon_0 \Rightarrow |R(y)| \le \delta|y|$, we have

$$F\big(Q_\varepsilon(x_0)\big) \subset \{x \in \mathbb{R}^n : \operatorname{dist}\big(x - F(x_0), AQ_\varepsilon(0)\big) \le \varepsilon\delta\}, \tag{7.50}$$

for $\varepsilon \le \varepsilon_0$. Now $AQ_\varepsilon(0)$ is contained in a set $\{x \in V : |x| \le K\varepsilon\}$, where $V = \mathcal{R}(A)$ is a linear subspace of $\mathbb{R}^n$ of dimension $\le n - 1$. Hence, in this case,

$$m\Big(F\big(Q_\varepsilon(x_0)\big)\Big) \le K\varepsilon^n\delta, \tag{7.51}$$

so the lim sup of the left side of (7.42) is $\le K\delta$ in this case. Letting $\delta \to 0$, we finish the proof.

Using (7.41), we see that, if $F : \mathcal{O} \to \Omega$ is as in Proposition 7.4 and if also F is differentiable for almost all $x \in \mathcal{O}$, then the identity (7.36) holds for almost all x. We have the following extension of Theorem 7.2.

Proposition 7.6. *If $F : \mathcal{O} \to \Omega$ is both a Lipschitz map and a homeomorphism and if F is differentiable for almost all $x \in \mathcal{O}$, then the conclusion of Theorem 7.2 holds.*

It is a result of Rademacher that every Lipschitz function on $\mathcal{O}$ is differentiable almost everywhere. We will prove this in Chapter 11.

To end this chapter, we show that Lebesgue measure on $\mathbb{R}^n$ is uniquely characterized by *translation invariance*, together with a simple finiteness condition.

Proposition 7.7. *Let λ be a measure on $\mathfrak{B}_n$. Assume that λ is translation invariant:*

$$S \in \mathfrak{B}_n \Longrightarrow \lambda(S) = \lambda(S + y), \quad \forall\, y \in \mathbb{R}^n. \tag{7.52}$$

Also assume there exists a bounded open $U \subset \mathbb{R}^n$ such that

$$0 < \lambda(U) < \infty. \tag{7.53}$$

Then $\lambda = cm$ for some constant $c > 0$, where m is Lebesgue measure on $\mathbb{R}^n$.

Proof. Let S be any bounded Borel set in $\mathbb{R}^n$. By (7.52), for each $y \in \mathbb{R}^n$,

$$\int \chi_S(x)\, d\lambda(x) = \int \chi_S(x-y)\, d\lambda(x). \tag{7.54}$$

Integrate this identity over a ball B with respect to $dm(y)$. Since λ is σ-finite, one can apply Fubini's Theorem to write

$$m(B)\lambda(S) = \int \left[\int_B \chi_S(x-y)\, dm(y) \right] d\lambda(x). \tag{7.55}$$

Now $m(S) = 0 \Rightarrow$ the right side of (7.55) is zero, so $\lambda << m$. Hence by the Radon-Nikodym Theorem we have $f \in \mathcal{M}^+(\mathbb{R}^n)$ such that

$$\lambda(S) = \int_S f(x)\, dm(x), \quad \forall\, S \in \mathfrak{B}_n, \tag{7.56}$$

and translation invariance (7.52) implies

$$\int_S f(x+y)\, dm(x) = \int_S f(x)\, dm(x), \quad \forall\, y \in \mathbb{R}^n,\ S \in \mathfrak{B}_n. \tag{7.57}$$

Note that (7.53) implies $\int_U f\, dm < \infty$, so f is integrable on bounded subsets of $\mathbb{R}^n$.

Consider the positive number

$$c = \frac{\lambda(U)}{m(U)} = \frac{1}{m(U)} \int_U f(x)\, dm(x) = \frac{1}{m(U)} \int_U f(x+y)\, dm(x). \tag{7.58}$$

If we write (7.56)–(7.57) as $\lambda(S) = \int_S f(x+y)\, dm(x)$ and integrate over U with respect to $dm(y)$, we have by Fubini's Theorem that, for each $S \in \mathfrak{B}_n$,

$$m(U)\lambda(S) = \int_S \int_U f(x+y)\, dm(y)\, dm(x) = c m(U) m(S), \tag{7.59}$$

the last identity by (7.58). This proves that $\lambda = cm$.

Exercises

1. Let $a = \int_{-\infty}^{\infty} e^{-x^2}\, dx$. This is called a Gaussian integral. Show that

$$a^n = \int_{\mathbb{R}^n} e^{-|x|^2}\, dx.$$

2. Show that
$$a^2 = 2\pi \int_0^\infty e^{-r^2} r\, dr = \pi.$$
Hint. Apply Proposition 7.3 and use the fact that 2π is defined to be the length of S^1.

3. Let A_n denote the n-dimensional measure of the unit sphere S^n. Show that
$$a^n = A_{n-1} \int_0^\infty e^{-r^2} r^{n-1}\, dr = \frac{1}{2} A_{n-1} \int_0^\infty e^{-s} s^{n/2-1}\, ds,$$
and show that
$$A_{n-1} = \frac{2\pi^{n/2}}{\Gamma(\frac{n}{2})}, \tag{7.60}$$
where $\Gamma(z) = \int_0^\infty e^{-s} s^{z-1}\, ds$, for $z > 0$. The function $\Gamma(z)$ is called Euler's gamma function.
Hint. Apply Proposition 7.3.

4. Show that $z\Gamma(z) = \Gamma(z+1)$.
Hint. Integrate by parts.

5. Show that $\Gamma(1) = 1$ and $\Gamma(1/2) = \pi^{1/2}$. Deduce that, for $k \in \mathbb{Z}^+$,
$$\Gamma(k) = (k-1)!, \quad \Gamma\Big(k + \frac{1}{2}\Big) = \Big(k - \frac{1}{2}\Big) \cdots \frac{1}{2}\,\pi^{1/2}.$$

6. Let B^n be the unit ball in $\mathbb{R}^n$, $B^n = \{x \in \mathbb{R}^n : |x| < 1\}$. Show that
$$m(B^n) = \frac{\pi^{n/2}}{\Gamma(\frac{n}{2}+1)}.$$
Hint. Apply (7.30) to $f = \chi_{B^n}$.

7. Show that the arc of the circle lying above $0 \le x \le 1/2$ has $1/12$ the length of the entire circle S^1.
Hint. Find the equilateral triangles in Figure 7.1.

8. Deduce from (7.26) that
$$\frac{\pi}{6} = \int_0^{1/2} \frac{dx}{\sqrt{1-x^2}} = \sum_{n=0}^\infty \frac{a_n}{2n+1}\Big(\frac{1}{2}\Big)^{2n+1}, \tag{7.61}$$

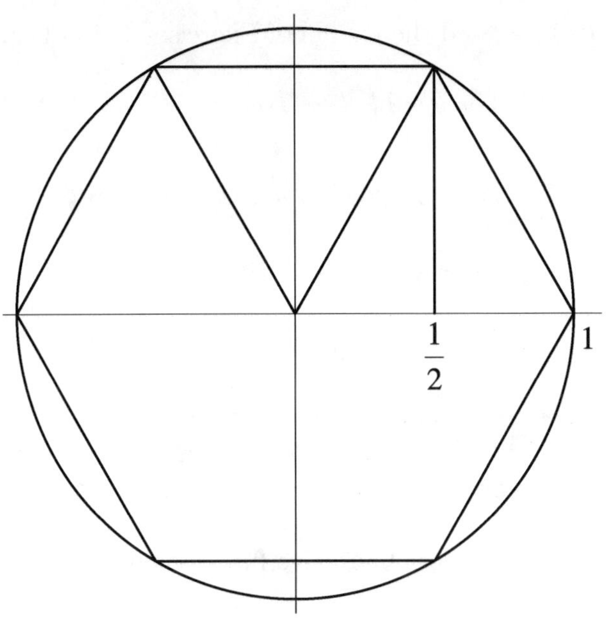

Figure 7.1

where $a_0 = 1,\ a_{n+1} = a_n(2n+1)/(2n+2)$. Show that

$$\frac{\pi}{6} - \sum_{n=0}^{k} \frac{a_n}{2n+1}\left(\frac{1}{2}\right)^{2n+1} < \frac{4^{-k}}{3(2k+3)}.$$

Using a calculator, sum the series over $0 \le n \le 20$ to approximate π to 13 digits after the decimal point.

9. Consider a 2D surface $M \subset \mathbb{R}^3$, with a coordinate chart $\varphi : \mathcal{O} \to U \subset M$. For integrable $f : M \to \mathbb{R}$ supported on U, show that

$$\int_M f\, dS = \int_{\mathcal{O}} f \circ \varphi(x)\, |\partial_1\varphi \times \partial_2\varphi|\, dx_1\, dx_2.$$

Hint. Apply (7.21). Note that

$$g = \det \begin{pmatrix} \partial_1\varphi \cdot \partial_1\varphi & \partial_1\varphi \cdot \partial_2\varphi \\ \partial_2\varphi \cdot \partial_1\varphi & \partial_2\varphi \cdot \partial_2\varphi \end{pmatrix} = |\partial_1\varphi|^2|\partial_2\varphi|^2 - (\partial_1\varphi \cdot \partial_2\varphi)^2.$$

Use $|u \times v| = |u| \cdot |v| \cdot |\sin\theta|$ and $u \cdot v = |u| \cdot |v| \cos\theta$ to show that $|u \times v|^2 = |u|^2|v|^2 - (u \cdot v)^2$, and hence $g = |\partial_1\varphi \times \partial_2\varphi|^2$.

Exercises 10-11 extend the results of Exercises 7–8 of Chapter 4 to $\mathbb{R}^n$.

10. Let $f \in C_0^\infty(\mathbb{R}^n)$. For $u \in L^1_{\text{loc}}(\mathbb{R}^n)$, set

$$K_f u(x) = f * u(x) = \int f(y)u(x-y)\,dy. \tag{7.62}$$

Show that, for $1 \le p < \infty$,

$$K_f : L^p(\mathbb{R}^n) \longrightarrow L^p(\mathbb{R}^n), \quad \|K_f u\|_{L^p} \le \|f\|_{L^1}\|u\|_{L^p}. \tag{7.63}$$

Show that there is a unique extension from $f \in C_0^\infty(\mathbb{R}^n)$ to $f \in L^1(\mathbb{R}^n)$ of $K_f u$, with (7.63) continuing to hold, giving a continuous linear map $K : L^1(\mathbb{R}^n) \to \mathcal{L}(L^p(\mathbb{R}^n))$. The operator (7.62) is called convolution.

11. Let f_j be a sequence of nonnegative functions in $L^1(\mathbb{R}^n)$ such that

$$\int f_j\,dx = 1, \quad \operatorname{supp} f_j \subset \{x \in \mathbb{R}^n : |x| < 1/j\}. \tag{7.64}$$

Show that, for $1 \le p < \infty$,

$$u \in L^p(\mathbb{R}^n) \Longrightarrow f_j * u \to u \text{ in } L^p \text{ norm, as } j \to \infty. \tag{7.65}$$

Hint. See the hint for Exercise 9 of Chapter 4.
The sequence (f_j) is called an *approximate identity.*

12. Given $f \in \mathcal{L}^1(\mathbb{R}^n, dx)$ and $\delta > 0$, show that there is a Lebesgue measurable $S \subset \mathbb{R}^n$ such that $m(S) < \delta$ and such that $f\big|_{\mathbb{R}^n \setminus S}$ is continuous. This is Lusin's Theorem for $\mathbb{R}^n$; compare Exercise 16 of Chapter 3 and Exercise 14 of Chapter 5.

13. Let u be a C^1 function on $\mathbb{R}^n$ with compact support; we write $u \in C_0^1(\mathbb{R}^n)$. Show that, for $1 \le j \le n$,

$$\int_{\mathbb{R}^n} \frac{\partial u}{\partial x_j}(x)\,dx = 0. \tag{7.66}$$

Deduce that, if $v \in C^1(\mathbb{R}^n)$,

$$\int_{\mathbb{R}^n} u(x)\frac{\partial v}{\partial x_j}\,dx = -\int_{\mathbb{R}^n} \frac{\partial u}{\partial x_j} v(x)\,dx. \tag{7.67}$$

Hint. If, say, $j = 1$, write the left side of (7.66) as the iterated integral

$$\int_{\mathbb{R}^{n-1}} \Big(\int_{\mathbb{R}} \frac{\partial u}{\partial x_1}(x_1, \dots, x_n)\, dx_1 \Big)\, dx_2 \cdots dx_n.$$

Use the fact that the inner Lebesgue integral coincides with the Riemann integral, in combination with the Fundamental Theorem of Calculus, as established in Chapter 1.

14. Consult a topology text for material about degree theory, used to establish (7.47). Sufficient material can be found in [**Dug**] or [**Spa**].

Exercises relating the Riemann and Lebesgue integrals on $\mathbb{R}^n$

Let $R = I_1 \times \cdots \times I_n$ be a cell in $\mathbb{R}^n$, with $I_\nu = [a_\nu, b_\nu]$ intervals in $\mathbb{R}$. The notion of the Riemann integral of a bounded function $f : R \to \mathbb{R}$ is defined similarly to the one-dimensional case treated in Chapter 1. If one has a partition of each I_ν into $J_{\nu 1} \cup \cdots \cup J_{\nu, N(\nu)}$, then a partition $\mathcal{P}$ of R consists of the cells

$$R_\alpha = J_{1\alpha_1} \times J_{2\alpha_2} \times \cdots \times J_{n\alpha_n}, \tag{7.68}$$

where $0 \le \alpha_\nu \le N(\nu)$. Each cell has n-dimensional volume

$$V(R_\alpha) = \ell(J_{1\alpha_1}) \cdots \ell(J_{n\alpha_n}). \tag{7.69}$$

Then we set

$$\overline{I}_{\mathcal{P}}(f) = \sum_\alpha \sup_{R_\alpha} f(x)\, V(R_\alpha), \quad \underline{I}_{\mathcal{P}}(f) = \sum_\alpha \inf_{R_\alpha} f(x)\, V(R_\alpha). \tag{7.70}$$

Note that $\underline{I}_{\mathcal{P}}(f) \le \overline{I}_{\mathcal{P}}(f)$. As in Chapter 1, we pass to limits via finer partitions.

To be precise, if $\mathcal{P}$ and $\mathcal{Q}$ are two partitions of R, we say $\mathcal{P}$ refines $\mathcal{Q}$ and write $\mathcal{P} \succ \mathcal{Q}$, if each partition of each interval factor I_ν of R involved in the definition of $\mathcal{Q}$ is further refined in order to produce the partitions of the factors I_ν, used to define $\mathcal{P}$, via (7.68). It is an exercise to show that any two partitions of R have a common refinement. Note also that, as in (1.4),

$$\mathcal{P} \succ \mathcal{Q} \Longrightarrow \overline{I}_{\mathcal{P}}(f) \le \overline{I}_{\mathcal{Q}}(f) \ \text{ and } \ \underline{I}_{\mathcal{P}}(f) \ge \underline{I}_{\mathcal{Q}}(f). \tag{7.71}$$

Consequently, if $\mathcal{P}_j$ are any two partitions of R and $\mathcal{Q}$ is a common refinement, we have

$$\underline{I}_{\mathcal{P}_1}(f) \le \underline{I}_{\mathcal{Q}}(f) \le \overline{I}_{\mathcal{Q}}(f) \le \overline{I}_{\mathcal{P}_2}(f). \tag{7.72}$$

Now, whenever $f : R \to \mathbb{R}$ is bounded, the following quantities are well defined:

$$\overline{I}(f) = \inf_{\mathcal{P}\in\Pi(R)} \overline{I}_{\mathcal{P}}(f), \quad \underline{I}(f) = \sup_{\mathcal{P}\in\Pi(R)} \underline{I}_{\mathcal{P}}(f), \tag{7.73}$$

where $\Pi(R)$ is the set of all partitions of R, as defined above. Clearly, by (7.72), $\underline{I}(f) \le \overline{I}(f)$. We then say that f is Riemann integrable (on R) provided $\overline{I}(f) = \underline{I}(f)$, and in such a case, we write $f \in \mathcal{R}(R)$ and set

$$I(f) = \overline{I}(f) = \underline{I}(f). \tag{7.74}$$

The following exercises deal with the Riemann integral and its relation to the Lebesgue integral.

1. Show that $f \in C(R) \Rightarrow f \in \mathcal{R}(R)$.

2. Show that $f, g \in \mathcal{R}(R) \Rightarrow f + g \in \mathcal{R}(R)$ and $I(f+g) = I(f) + I(g)$. *Hint.* Consult the proof of Proposition 1.1.

3. Show that $f, g \in \mathcal{R}(R) \Rightarrow fg \in \mathcal{R}(R)$.

4. Given a partition $\mathcal{P}$ of R, define maxsize $(\mathcal{P})$. Extend Corollary 1.4 and (1.42) to the multi-dimensional case.

5. Extend the proof of Proposition 3.10 to the following result.

   Proposition 7.8. Let R be a cell in $\mathbb{R}^n$, and let $\varphi : R \to \mathbb{R}$ be bounded. Then φ is Riemann integrable if and only if the set of points $x \in R$ at which φ is discontinuous has Lebesgue measure zero. If this condition holds, then the Riemann integral and the Lebesgue integral of φ coincide.

6. For a bounded set $S \subset \mathbb{R}^n$, with characteristic function χ_S, set

   $$\text{cont}^+(S) = \overline{I}(\chi_S), \quad \text{cont}^-(S) = \underline{I}(\chi_S),$$

   as in (1.23), and define S to "have content" if and only if $\text{cont}^+(S) = \text{cont}^-(S)$. Show that S has content if and only if the set of boundary points of S has Lebesgue measure zero.

When doing Exercises 7–8, look back at Exercises 10–14 of Chapter 5.

7. If $\mathcal{O} \subset \mathbb{R}^n$ is open and bounded, show that its Lebesgue measure is given by
$$m(\mathcal{O}) = \text{cont}^-(\mathcal{O}).$$
If $K \subset \mathbb{R}^n$ is compact, show that its Lebesgue measure is given by
$$m(K) = \text{cont}^+(K).$$

8. If $S \subset \mathbb{R}^n$ is a bounded set, show that its Lebesgue outer measure is given by
$$m^*(S) = \inf \{m(\mathcal{O}) : S \subset \mathcal{O}, \text{open}\}.$$
Remove the hypothesis that S be bounded.

Exercises on determinants

Let $M(n, \mathbb{R})$ denote the space of $n \times n$ real matrices. These exercises investigate the existence, uniqueness, and basic properties of the determinant,

$$\det : M(n, \mathbb{R}) \to \mathbb{R}, \tag{7.75}$$

which will be uniquely specified as a function $\vartheta : M(n, \mathbb{R}) \to \mathbb{R}$ satisfying

(a) ϑ is *linear* in each column a_j of A,
(b) $\vartheta(\widetilde{A}) = -\vartheta(A)$ if $\widetilde{A}$ is obtained from A by interchanging two columns,
(c) $\vartheta(I) = 1$.

1. Let $A = (a_1, \dots, a_n)$, where a_j are column vectors; $a_j = (a_{1j}, \dots, a_{nj})^t$. Show that, if (a) holds, we have the expansion
$$\begin{aligned} \det A &= \sum_j a_{j1} \det (e_j, a_2, \dots, a_n) \\ &= \cdots = \sum_{j_1, \cdots, j_n} a_{j_1 1} \cdots a_{j_n n} \det (e_{j_1}, e_{j_2}, \dots, e_{j_n}), \end{aligned} \tag{7.76}$$
where $\{e_1, \dots, e_n\}$ is the standard basis of $\mathbb{R}^n$.

2. Show that, if (b) and (c) also hold, then
$$\det A = \sum_{\sigma \in S_n} (\text{sgn } \sigma)\, a_{\sigma(1)1} a_{\sigma(2)2} \cdots a_{\sigma(n)n}, \tag{7.77}$$

where S_n is the set of *permutations* of $\{1, \dots, n\}$ and

$$\text{sgn}\ \sigma = \det\left(e_{\sigma(1)}, \dots, e_{\sigma(n)}\right) = \pm 1. \tag{7.78}$$

To define sgn σ, the "sign" of a permutation σ, we note that every permutation σ can be written as a product of transpositions: $\sigma = \tau_1 \cdots \tau_\nu$, where a transposition of $\{1, \dots, n\}$ interchanges two elements and leaves the rest fixed. We say sgn $\sigma = 1$ if ν is even and sgn $\sigma = -1$ if ν is odd. It is necessary to show that sgn σ is independent of the choice of such a product representation. (Referring to (7.78) begs the question until we know that det is well defined.)

3. Let $\sigma \in S_n$ act on a function of n variables by

$$(\sigma f)(x_1, \dots, x_n) = f(x_{\sigma(1)}, \dots, x_{\sigma(n)}). \tag{7.79}$$

Let P be the polynomial

$$P(x_1, \dots, x_n) = \prod_{1 \le j < k \le n} (x_j - x_k). \tag{7.80}$$

Show that

$$(\sigma P)(x) = (\text{sgn}\ \sigma)\ P(x) \tag{7.81}$$

and that this implies that sgn σ is well defined.

4. Deduce that there is a unique determinant satisfying (a)–(c) and that it is given by (7.77). If (c) is replaced by $\vartheta(I) = r$, show that $\vartheta(A) = r \det A$.

5. Show that, if (a)–(c) hold, it follows that

(d) $\vartheta(\widetilde{A}) = \vartheta(A)$ if $\widetilde{A}$ is obtained from A by adding ca_ℓ to a_k, for some $c \in \mathbb{R}$, where $a_1, \dots, a_n$ are the columns of A.

6. Show that

$$\det(AB) = (\det A)(\det B). \tag{7.82}$$

Hint. For fixed $A \in M(n, \mathbb{R})$, apply Exercise 4 to $\vartheta_1(B) = \det(AB)$. Note that if we represent B by its columns, as $B = (b_1, \dots, b_n)$, then $AB = (Ab_1, \dots, Ab_n)$. To verify rule (b) for $\vartheta_1(B)$, note that if $\widetilde{B} =$

$(b_{\sigma(1)}, \dots, b_{\sigma(n)})$ is obtained from B by permuting its columns, then $A\widetilde{B}$ has columns $(Ab_{\sigma(1)}, \dots, Ab_{\sigma(n)})$, obtained by permuting the columns of AB.

7. Show that (7.77) implies

$$\det A = \det A^t, \tag{7.83}$$

where A^t denotes the *transpose* of A, i.e., $A = (a_{jk}) \Rightarrow A^t = (a_{kj})$. Conclude that one can replace columns by rows in the characterization (a)–(d) of determinants.
Hint. $a_{\sigma(j)j} = a_{\ell\tau(\ell)}$ with $\ell = \sigma(j)$, $\tau = \sigma^{-1}$. Also, sgn $\sigma =$ sgn τ.

8. Show that

$$\det \begin{pmatrix} 1 & a_{12} & \cdots & a_{1n} \\ 0 & a_{22} & \cdots & a_{2n} \\ \vdots & \vdots & & \vdots \\ 0 & a_{n2} & \cdots & a_{nn} \end{pmatrix} = \det \begin{pmatrix} 1 & 0 & \cdots & 0 \\ 0 & a_{22} & \cdots & a_{2n} \\ \vdots & \vdots & & \vdots \\ 0 & a_{n2} & \cdots & a_{nn} \end{pmatrix} = \det A_{11} \tag{7.84}$$

where $A_{11} = (a_{jk})_{2 \le j,k \le n}$.

Hint. Do the first identity using (d), from Exercise 5. Then exploit uniqueness for det on $M(n-1, \mathbb{R})$.

9. Deduce that $\det(e_j, a_2, \dots, a_n) = (-1)^{j-1} \det A_{1j}$ where A_{kj} is formed by deleting the kth column and the jth row from A.

10. Deduce from the first sum in (7.76) that

$$\det A = \sum_{j=1}^n (-1)^{j-1} a_{j1} \det A_{1j}. \tag{7.85}$$

More generally, for any $k \in \{1, \dots, n\}$,

$$\det A = \sum_{j=1}^n (-1)^{j-k} a_{jk} \det A_{kj}. \tag{7.86}$$

This is called an expansion of det A by minors, down the kth column.

11. Let $c_{kj} = (-1)^{j-k} \det A_{kj}$. Show that

$$\sum_{j=1}^n a_{j\ell} c_{kj} = 0, \quad \text{if } \ell \ne k. \tag{7.87}$$

Deduce that $C = (c_{jk})$ satisfies

$$CA = (\det A)I. \tag{7.88}$$

Hint. Reason as in Exercises 8–10 that the left side of (7.87) is equal to

$$\det (a_1, \dots, a_\ell, \dots, a_\ell, \dots, a_n),$$

with a_ℓ in the kth column as well as in the ℓth column. The identity (7.88) is known as Cramer's formula.

12. Show that

$$\det \begin{pmatrix} a_{11} & a_{12} & \cdots & a_{1n} \\ & a_{22} & \cdots & a_{2n} \\ & & \ddots & \vdots \\ & & & a_{nn} \end{pmatrix} = a_{11}a_{22}\cdots a_{nn}. \tag{7.89}$$

Hint. Use (7.84) and induction.

Exercises on row reduction and matrix products

We consider the following three types of row operations on an $n \times n$ matrix $A = (a_{jk})$. If σ is a permutation of $\{1, \dots, n\}$, let

$$\rho_\sigma(A) = (a_{\sigma(j)k}).$$

If $c = (c_1, \dots, c_j)$ and all c_j are nonzero, set

$$\mu_c(A) = (c_j^{-1} a_{jk}).$$

Finally, if $c \in \mathbb{R}$ and $\mu \neq \nu$, define

$$\varepsilon_{\mu\nu c}(A) = (b_{jk}), \quad b_{\nu k} = a_{\nu k} - c a_{\mu k},\ b_{jk} = a_{jk} \ \text{ for } \ j \neq \nu.$$

We relate these operations to left multiplication by matrices P_σ, M_c, and $E_{\mu\nu c}$, defined by the following actions on the standard basis $\{e_1, \dots, e_n\}$ of $\mathbb{R}^n$:

$$P_\sigma e_j = e_{\sigma(j)}, \quad M_c e_j = c_j e_j, \tag{7.90}$$

and

$$E_{\mu\nu c} e_\mu = e_\mu + c e_\nu, \quad E_{\mu\nu c} e_j = e_j \ \text{ for } \ j \neq \mu. \tag{7.91}$$

1. Show that

$$A = P_\sigma \rho_\sigma(A), \quad A = M_c \mu_c(A), \quad A = E_{\mu\nu c} \varepsilon_{\mu\nu c}(A)$$

and that

$$\det P_\sigma = \operatorname{sgn} \sigma, \quad \det M_c = c_1 \cdots c_n, \quad \det E_{\mu\nu c} = 1.$$

2. Show that $P_{\sigma\tau} = P_\sigma P_\tau$ and $P_\sigma^{-1} = P_{\sigma^{-1}}$.

3. Show that, if $\mu \neq \nu$, then $E_{\mu\nu c} = P_\sigma^{-1} E_{21c} P_\sigma$, for some permutation σ.

4. If $B = \rho_\sigma(A)$ and $C = \mu_c(B)$, show that $A = P_\sigma M_c C$. Generalize this to other cases where a matrix C is obtained from a matrix A via a sequence of row operations.

5. If A is an invertible, real $n \times n$ matrix (i.e., $A \in Gl(n, \mathbb{R})$), then the rows of A form a basis of $\mathbb{R}^n$. Use this to show that A can be transformed to the identity matrix via a sequence of row operations. Deduce that any $A \in Gl(n, \mathbb{R})$ can be written as a finite product of matrices of the form P_σ, M_c and $E_{\mu\nu c}$, hence as a finite product of matrices of the form listed in (7.6).

Further exercises on $Gl(n, \mathbb{R})$

The set $Gl(n, \mathbb{R})$ is the subset of the space $M(n, \mathbb{R})$ of $n \times n$ real matrices consisting of invertible matrices, where we say $A \in M(n, \mathbb{R})$ is invertible if and only if there exists $B \in M(n, \mathbb{R})$ such that $AB = BA = I$. (Then we write $A^{-1} = B$.)

1. Show that

$$Gl(n, \mathbb{R}) = \{A \in M(n, \mathbb{R}) : \det A \neq 0\}. \tag{7.92}$$

Hint. Use (7.82) and (7.88).

2. Show that

$$Gl(n, \mathbb{R}) \text{ is open in } M(n, \mathbb{R}). \tag{7.93}$$

Hint. Use Exercise 1.

Note that $Gl(n,\mathbb{R}) = Gl_+(n,\mathbb{R}) \cup Gl_-(n,\mathbb{R})$, where

$$Gl_\pm(n,\mathbb{R}) = \{A \in M(n,\mathbb{R}) : \pm \det A > 0\}. \tag{7.94}$$

The goal of the next six exercises is to show that

$$Gl_+(n,\mathbb{R}) \ \text{ is connected.} \tag{7.95}$$

3. Show that (7.95) is a consequence of the following three assertions, regarding the matrices $E_{\mu\nu c}$, M_c, and P_σ, defined by (7.90)–(7.91):

   (a) There is a path from I to $E_{\mu\nu c}$ in $Gl_+(n,\mathbb{R})$.
   (b) If $c_1 \cdots c_n > 0$, there is a path from I to M_c in $Gl_+(n,\mathbb{R})$.
   (c) If $\operatorname{sgn} \sigma > 0$, there is a path from I to P_σ in $Gl_+(n,\mathbb{R})$.
   Hint. Use Exercise 5 from the previous exercise set.

4. Show that assertion (a) of Exercise 3 holds.
   Hint. Evaluate $\det E_{\mu\nu(tc)}$.

5. Show that assertion (b) of Exercise 3 holds.
   Hint. Reduce to the case where each $c_j = \pm 1$. The number of -1's is even. There is a path from I to $-I$ in $Gl_+(2,\mathbb{R})$ given by

$$\begin{pmatrix} \cos t & -\sin t \\ \sin t & \cos t \end{pmatrix}, \quad 0 \le t \le \pi.$$

   That is, $-I \in Gl_+(2,\mathbb{R})$ is a rotation by π.

6. Reduce the proof of assertion (c) in Exercise 3 to the following assertion:

   There is a path from I to $P_{\sigma\tau}$ in $Gl_+(n,\mathbb{R})$ whenever σ and τ are transpositions in S_n.

   In turn, reduce the proof of this assertion to the following two assertions:
   (a) There is a path from

$$I = \begin{pmatrix} 1 & & \\ & 1 & \\ & & 1 \end{pmatrix} \quad \text{to } A = \begin{pmatrix} 0 & 0 & 1 \\ 1 & 0 & 0 \\ 0 & 1 & 0 \end{pmatrix} \quad \text{in } Gl_+(3,\mathbb{R}).$$

   (b) There is a path from

$$I = \begin{pmatrix} 1 & & & \\ & 1 & & \\ & & 1 & \\ & & & 1 \end{pmatrix} \quad \text{to } B = \begin{pmatrix} 0 & 1 & & \\ 1 & 0 & & \\ & & 0 & 1 \\ & & 1 & 0 \end{pmatrix} \quad \text{in } Gl_+(4,\mathbb{R}).$$

7. Prove assertion (a) of Exercise 6.
   Hint. Show that A is a rotation by $2\pi/3$ about the axis through $(1,1,1)^t$.

8. Prove assertion (b) of Exercise 6.
   Hint. Show that B is the identity on the linear span of $(1,1,0,0)^t$ and $(0,0,1,1)^t$ and that it is rotation by π on the 2D orthogonal complement of this space.

Exercises on matrix integrals

Let $M(n,\mathbb{R})$ denote the space of $n \times n$ real matrices, $M(n,\mathbb{C})$ the space of $n \times n$ complex matrices. Set

$$O(n) = \{A \in M(n,\mathbb{R}) : A^*A = I\}, \tag{7.96}$$

$$U(n) = \{A \in M(n,\mathbb{C}) : A^*A = I\}, \tag{7.97}$$

where

$$A = (a_{jk}) \Longrightarrow A^* = (\overline{a}_{kj}).$$

Also set

$$SO(n) = \{A \in O(n) : \det A = 1\}, \tag{7.98}$$

$$SU(n) = \{A \in U(n) : \det A = 1\}. \tag{7.99}$$

1. Use Proposition B.7 from Appendix B to show that each of the four matrix groups (7.96)–(7.99) is a smooth, compact submanifold of $M(n,\mathbb{R})$ or $M(n,\mathbb{C})$.

2. For each group G of the form (7.96)–(7.99), define, for $g \in G$,

$$R_g,\ L_g : G \to G, \quad R_g(x) = xg, \quad L_g(x) = gx.$$

   Show that, for each $g \in G$, R_g and L_g preserve the metric tensor on G induced from $M(n,\mathbb{R})$ or $M(n,\mathbb{C})$.
   Hint. Show that $R_g, L_g : M(n,\mathbb{F}) \to M(n,\mathbb{F})$ are isometries, with $\mathbb{F} = \mathbb{R}$ or $\mathbb{C}$, as appropriate.

3. Let dV denote the volume element on G associated to the metric tensor discussed above, via (7.21). Show that for each $f \in L^1(G, dV)$,

$$\int_G f(x)\,dV(x) = \int_G f(gx)\,dV(x) = \int_G f(xg)\,dV(x), \quad \forall\, g \in G. \tag{7.100}$$

We set $dg = V(G)^{-1}\,dV$ and call this *Haar measure* on G.

4. Show that

$$\int_G f(g^{-1})\,dg = \int_G f(g)\,dg.$$

5. Let V be a finite-dimensional Hilbert space and assume $\pi : G \to Gl(V)$ is a continuous homomorphism such that $\pi(g) : V \to V$ is unitary for each $g \in G$. (One says π is a unitary representation of G on V.) For $v \in V$, set

$$Pv = \int_G \pi(g)v\,dg.$$

Show that P is the orthogonal projection of V onto

$$V_0 = \{v \in V : \pi(g)v = v, \ \forall\, g \in G\}.$$

Hint. Show that

$$\pi(g)v = v\ \forall\, g \Longrightarrow Pv = v, \quad \pi(h)Pv = Pv, \quad \forall\, h \in G,$$

$$P^2 v = \int \pi(h)Pv\,dh = Pv, \quad P^* v = \int \pi(g^{-1})v\,dg = Pv.$$

6. Also let λ be a unitary representation of G on a finite-dimensional Hilbert space W. Give $\mathcal{L}(W,V)$ the Hilbert-Schmidt inner product, $(A,B) = \operatorname{Tr} B^*A$, and define a unitary representation ν of G on $\mathcal{L}(W,V)$ by

$$\nu(g)A = \pi(g)A\lambda(g^{-1}).$$

Define $Q : \mathcal{L}(W,V) \to \mathcal{L}(W,V)$ by

$$QA = \int_G \nu(g)A\,dg = \int_G \pi(g)A\lambda(g^{-1})\,dg.$$

Show that Q is the orthogonal projection of $\mathcal{L}(W,V)$ onto

$$\mathcal{I}(\pi,\lambda) = \{A \in \mathcal{L}(W,V) : \pi(g)A = A\lambda(g), \ \forall\, g \in G\}.$$

7. Assume the representations π and λ are *irreducible*, i.e., V and W have no nontrivial G-invariant linear subspaces. Schur's Lemma (cf. [**T1**], Appendix B, §§7–8) implies that $\mathcal{I}(\pi,\lambda)$ is either zero- or one-dimensional. Using Exercise 6, what can you conclude about

$$\int_G \pi_{ij}(g)\overline{\lambda_{k\ell}(g)}\,dg?$$

Here $\pi_{ij}(g)$ are the matrix entries of $\pi(g)$ in an orthonormal basis of V, and $\lambda_{k\ell}(g)$ are similarly defined. The resulting formulas are called the Weyl orthogonality relations.

Chapter 8

Signed Measures and Complex Measures

As opposed to the measures we have considered so far, a signed measure is allowed to take on both positive and negative values. To be precise, if $\mathcal{M}$ is a σ-algebra of subsets of X, a signed measure ν on $\mathcal{M}$ is a function

$$\nu : \mathcal{M} \longrightarrow [-\infty, \infty], \tag{8.1}$$

allowed to assume at most *one* of the values $\pm\infty$ and satisfying

$$\nu(\emptyset) = 0 \tag{8.2}$$

and

$$E_j \in \mathcal{M} \text{ disjoint sequence } \Longrightarrow \nu\Big(\bigcup_j E_j\Big) = \sum_j \nu(E_j), \tag{8.3}$$

the series $\sum_j \nu(E_j)$ being absolutely convergent. A signed measure taking values in $[0, \infty]$ is what we have dealt with in Chapters 2–7; sometimes we call this a *positive* measure. If μ_1 and μ_2 are positive measures and one of them is finite, then $\mu_1 - \mu_2$ is a signed measure.

The following result is easy to prove but useful.

Proposition 8.1. *If ν is a signed measure on $(X, \mathcal{M})$, then for a sequence $\{E_j\} \subset \mathcal{M}$,*

$$E_j \nearrow E \Longrightarrow \nu(E) = \lim_{j\to\infty} \nu(E_j), \tag{8.4}$$

and

$$E_j \searrow E,\ \nu(E_1) \text{ finite } \Longrightarrow \nu(E) = \lim_{j\to\infty} \nu(E_j). \tag{8.5}$$

Proof. Exercise.

If ν is a signed measure on $(X, \mathcal{M})$ and $E \in \mathcal{M}$, we say ν is E-positive if $\nu(F) \geq 0$ for every $F \in \mathcal{M}$, $F \subset E$. Similarly we say ν is E-negative if $\nu(F) \leq 0$ for every such F. We say ν is null on E if $\nu(F) = 0$ for all such F. Note that, if P_j is a sequence of sets in $\mathcal{M}$, then

$$\nu \ P_j\text{-positive} \Longrightarrow \nu \ P\text{-positive, for } P = \bigcup_j P_j. \tag{8.6}$$

To see this, let $Q_n = P_n \setminus \bigcup_{j \leq n-1} P_j$. Then $Q_n \subset P_n$, so ν is Q_n-positive. Now, if $F \subset \bigcup P_j$, then $\nu(F) = \sum \nu(F \cap Q_n) \geq 0$, as desired.

The following key result is known as the Hahn Decomposition Theorem.

Theorem 8.2. *If ν is a signed measure on $(X, \mathcal{M})$, then there exist $P, N \in \mathcal{M}$ such that ν is P-positive and N-negative and $P \cup N = X$, $P \cap N = \emptyset$.*

Let us assume that ν does not take on the value $+\infty$. The theorem can be proved via the following:

Lemma 8.3. *If $\nu(A) > -\infty$, then there exists a measurable $P \subset A$ such that ν is P-positive and $\nu(P) \geq \nu(A)$.*

Given Lemma 8.3, we can prove Theorem 8.2 as follows. Let

$$s = \sup \{\nu(A) : A \in \mathcal{M}\}.$$

By (8.2), $s \geq 0$. Take $A_j \in \mathcal{M}$ such that $\nu(A_j) \nearrow s$, $\nu(A_j) > -\infty$. By the lemma, ν is P_j-positive on a sequence of sets $P_j \subset A_j$, such that $\nu(P_j) \to s$. Set $P = \bigcup P_j$. By (8.6), ν is P-positive. Then we deduce that $\nu(P) = s$. In particular, under our hypothesis, $s < \infty$.

Set $N = X \setminus P$. We claim that ν is N-negative. If not, there is a measurable $S \subset N$ with $\nu(S) > 0$, and then $\nu(P \cup S) = s + \nu(S) > s$, which is not possible. Thus we have the desired partition of X into P and N.

To prove Lemma 8.3, it is convenient to start with a weaker result:

Lemma 8.4. *If $\nu(A) > -\infty$, then for all $\varepsilon > 0$, there exist measurable $B \subset A$ such that $\nu(B) \geq \nu(A)$ and $\nu(E) \geq -\varepsilon$, for all measurable $E \subset B$.*

We show how Lemma 8.4 yields Lemma 8.3. We define a sequence of sets $A_j \in \mathcal{M}$ inductively. Let $A_1 = A$. Given A_j, $j \leq n-1$, take $A_n \subset A_{n-1}$ such that $\nu(A_n) \geq \nu(A_{n-1})$ and $\nu(E) \geq -1/n$, for all measurable $E \subset A_n$. Lemma 8.4 says you can do this. Then let $P = \bigcap A_j$. Clearly ν is P-positive, and (8.5) implies that $\nu(P) = \lim \nu(A_j) \geq \nu(A)$.

It remains to prove Lemma 8.4. We use a proof by contradiction. If Lemma 8.4 is false, then (considering $B = A$), we see that, for some $\varepsilon > 0$, there is a measurable $E_1 \subset A$ such that $\nu(E_1) \leq -\varepsilon$. Thus $\nu(A \setminus E_1) \geq \nu(A) + \varepsilon$. Next, considering $B = A \setminus E_1$, we have a measurable $E_2 \subset A \setminus E_1$ such that $\nu(E_2) \leq -\varepsilon$, so $\nu\big(A \setminus (E_1 \cap E_2)\big) \geq \nu(A) + 2\varepsilon$. Continue, producing a disjoint sequence of measurable $E_j \subset A$ with $\nu(E_j) \leq -\varepsilon$. Then $A_j = A \setminus (E_1 \cup \cdots \cup E_j)$ has $\nu(A_j) \geq \nu(A) + j\varepsilon$, and hence, by (8.5), we must have

$$A_j \searrow F, \quad \nu(F) = +\infty.$$

But we are working under the hypothesis that ν does not take on the value $+\infty$, so this gives a contradiction, and the proof is complete.

We call the pair P, N produced by Theorem 8.2 a "Hahn partition" for ν. It is essentially unique, as the following result shows.

Proposition 8.5. *If ν, P and N are as in Theorem 8.2 and if also $\widetilde{P}, \widetilde{N}$ is a Hahn partition for ν, then ν is null on*

$$P \vartriangle \widetilde{P} = N \vartriangle \widetilde{N} = (P \cap \widetilde{N}) \cup (N \cap \widetilde{P}).$$

Proof. Let $E \in \mathcal{M}$, $E \subset P \vartriangle \widetilde{P}$. Write $E = E_0 \cup E_1$, a disjoint union, where we take $E_0 = E \cap (P \cap \widetilde{N})$, $E_1 = E \cap (N \cap \widetilde{P})$. Then $\nu(E) = \nu(E_0) + \nu(E_1)$. But each $\nu(E_j)$ is simultaneously ≥ 0 and ≤ 0, so we have $\nu(E_j) = 0$.

If ν is a signed measure on $(X, \mathcal{M})$ and we have a Hahn partition of X into $P \cup N$, we define two positive measures ν_+ and ν_- by

$$\nu_+(E) = \nu(P \cap E), \quad \nu_-(E) = -\nu(N \cap E), \tag{8.7}$$

for $E \in \mathcal{M}$. We have

$$\nu = \nu_+ - \nu_-. \tag{8.8}$$

Now ν_+ is *supported* on P, i.e., ν_+ is null on $X \setminus P$. Similarly ν_- is supported on N, so

(8.9) $\qquad \nu_+$ and ν_- have disjoint supports.

This is called the Hahn decomposition of the signed measure ν. By Proposition 8.5, it is unique. We can also form the positive measure

$$|\nu| = \nu_+ + \nu_-, \tag{8.10}$$

called the total variation measure of ν.

Given a signed measure ν, we can define $\int f\,d\nu$ for a suitable class of measurable functions f. We can use the Hahn decomposition (8.8) to do this, setting

$$\int f\,d\nu = \int f\,d\nu_+ - \int f\,d\nu_-, \tag{8.11}$$

when ν satisfies (8.8). In particular this is well defined for $f \in L^1(X, |\nu|)$.

Let μ be a positive measure on $(X, \mathcal{M})$. We say that a signed measure ν on $(X, \mathcal{M})$ is absolutely continuous with respect to μ (and write $\nu << \mu$) provided that, for $E \in \mathcal{M}$,

$$\mu(E) = 0 \Longrightarrow \nu(E) = 0. \tag{8.12}$$

This definition was made in (4.45) in the case when ν is also a positive measure. It is useful to have the following.

Proposition 8.6. *If μ is a positive measure and ν a signed measure on $(X, \mathcal{M})$, with Hahn decomposition* (8.8), *then*

$$\nu << \mu \Longrightarrow \nu_+ << \mu \text{ and } \nu_- << \mu. \tag{8.13}$$

Proof. If $E \in \mathcal{M}$ and $\mu(E) = 0$, then $\mu(E \cap P) = \mu(E \cap N) = 0$, so $\nu_+(E) = \nu(E \cap P) = 0$ and $\nu_-(E) = -\nu(E \cap N) = 0$.

We can therefore apply Theorem 4.10 to ν_+ and ν_- to obtain the following extension of the Radon-Nikodym Theorem.

Theorem 8.7. *Let μ be a positive finite measure and ν a finite signed measure on $(X, \mathcal{M})$, such that $\nu << \mu$. Then there exists $h \in \mathcal{L}^1(X, \mu)$ such that*

$$\int_X F\,d\nu = \int_X Fh\,d\mu, \tag{8.14}$$

for all $F \in \mathcal{L}^1(X, |\nu|)$.

Proof. Note that $\nu_+(X) = \nu(P) < \infty$ and $\nu_-(X) = -\nu(N) < \infty$. Apply Theorem 4.10 to $\nu_\pm$, obtaining $h_\pm \in \mathcal{L}^1(X, \mu)$, and let $h = h_+ - h_-$.

The only difference between (4.47) and (8.14) is that this time h need not be ≥ 0.

It is also useful to consider the concept of a complex measure on $(X, \mathcal{M})$, which is defined to be a function

$$\rho : \mathcal{M} \longrightarrow \mathbb{C}, \tag{8.15}$$

satisfying

$$\rho(\emptyset) = 0 \tag{8.16}$$

and

$$E_j \in \mathcal{M} \text{ disjoint sequence } \Longrightarrow \rho\Bigl(\bigcup_j E_j\Bigr) = \sum_j \rho(E_j). \tag{8.17}$$

In this case, we do not allow ρ to assume any "infinite" values. It is clear that we can set

$$\nu_0(S) = \text{ Re } \rho(S), \quad \nu_1(S) = \text{ Im } \rho(S), \tag{8.18}$$

and ν_j are signed measures, which do not take on either $+\infty$ or $-\infty$ as a value. We have

$$\rho(S) = \nu_0(S) + i\nu_1(S), \tag{8.19}$$

and we can define

$$\int F\, d\rho = \int F\, d\nu_0 + i \int F\, d\nu_1, \tag{8.20}$$

where in turn $\int F\, d\nu_j$ are defined as in (8.11). There is an obvious extension of the Radon-Nikodym Theorem: if μ is a finite positive measure and ν a complex measure on $(X, \mathcal{M})$, such that $\nu << \mu$, then (8.14) continues to hold for some complex-valued h.

Exercises

1. Write down a proof of Proposition 8.1.

In Exercises 2–3, let μ and ν be finite positive measures on $(X, \mathfrak{F})$. For $\tau \in [0, \infty)$, set $\varphi_\tau = \nu - \tau\mu$. For each such τ, φ_τ is a signed measure.

2. Show that there exists a family P_τ and a family N_τ of elements of $\mathfrak{F}$ such that $P_0 = X$, for each $\tau \in [0, \infty)$, P_τ, N_τ is a partition of X,
$$\tau_1 < \tau_2 \Longrightarrow P_{\tau_1} \supset P_{\tau_2},$$
and
$$\varphi_\tau \text{ is } P_\tau\text{-positive and } N_\tau\text{-negative}, \ \forall\, \tau \in [0, \infty).$$
Hint. To get the nesting property, make a preliminary Hahn decomposition $\widetilde{P}_\alpha, \widetilde{N}_\alpha$ for $\alpha \in \mathbb{Q}^+ = \mathbb{Q} \cap [0, \infty)$, and set
$$P_\tau = \bigcap \{\widetilde{P}_\alpha : \alpha \in \mathbb{Q}^+, \alpha \le \tau\}.$$

3. For each $k \in \mathbb{Z}^+ = \{0, 1, 2, \dots\}$, set
$$h_k(x) = \sup\ \{\tau \in 2^{-k} \cdot \mathbb{Z}^+ : x \in P_\tau\}.$$
Show that $h_k \nearrow h \in \mathcal{M}^+(X)$ and
$$\nu = h\mu + \rho, \quad \rho \perp \mu,$$
obtaining another proof of the Lebesgue-Radon-Nikodym Theorems 4.10–4.11.

4. Suppose μ_1 and μ_2 are finite positive measures on $(X, \mathcal{M})$, and form the signed measure $\nu = \mu_1 - \mu_2$. Show that
$$|\nu|(E) \le \mu_1(E) + \mu_2(E), \quad \forall\, E \in \mathcal{M}.$$
Hint. First look at $E \subset P$ and $E \subset N$, with P and N as in Theorem 8.2.

5. Take ν as in Exercise 4. With $\int f\, d\nu$ defined by (8.11), show that for $f \in L^1(X, \mu_1 + \mu_2)$,
$$\int f\, d\nu = \int f\, d\mu_1 - \int f\, d\mu_2.$$
Hint. $\nu_+ + \mu_2 = \nu_- + \mu_1$.

6. Let $\mathfrak{M}(X, \mathfrak{F})$ denote the set of finite signed measures on $(X, \mathfrak{F})$. Show that this is a linear space and, in fact, a Banach space, with norm
$$\|\nu\| = |\nu|(X). \tag{8.21}$$

7. Suppose ν_j are finite signed measures on $(X_j, \mathfrak{F}_j)$. Show that $\nu_1 \times \nu_2$ is a well-defined, finite signed measure on $(X_1 \times X_2, \mathfrak{F}_1 \otimes \mathfrak{F}_2)$, and
$$\|\nu_1 \times \nu_2\| = \|\nu_1\| \cdot \|\nu_2\|.$$

8. Let E be a Banach space. Define the notion of an E-valued measure on $(X, \mathcal{M})$, and develop some basic properties.

Chapter 9

L^p Spaces, II

In Chapter 4 we defined L^p spaces and studied some of their basic properties as Banach spaces. We also studied special properties of L^2 spaces, as Hilbert spaces. In particular, we characterized continuous linear functionals $\varphi : L^2(X, \mu) \to \mathbb{C}$ as always having the form $\varphi(u) = (u, f)_{L^2}$, for some $f \in L^2(X, \mu)$. We illustrated the usefulness of this result by applying it to establish the Radon-Nikodym Theorem. Exercises at the end of Chapter 4 took a brief look at some bounded linear transformations on L^p spaces, such as translations and convolutions.

We will develop these topics further here. We begin with a more general look at linear functionals.

Associated to any Banach space V is an important space, the *dual* of V, denoted V', which consists of continuous linear maps

$$\omega : V \longrightarrow \mathbb{C} \tag{9.1}$$

($\omega : V \to \mathbb{R}$ if V is a real vector space). Elements $\omega \in V'$ are called linear functionals on V. Sometimes one finds the following notation for the action of $\omega \in V'$ on $v \in V$:

$$\langle v, \omega \rangle = \omega(v). \tag{9.2}$$

If V has norm $\| \quad \|$, the condition for the map (9.1) to be continuous is the following: the set of $v \in V$ such that $|\omega(v)| \le 1$ must be a neighborhood of $0 \in V$. Thus this set must contain a ball $B_R = \{v \in V : \|v\| \le R\}$, for some $R > 0$. With $C = 1/R$, it follows that ω must satisfy

$$|\omega(v)| \le C\|v\| \tag{9.3}$$

for some $C < \infty$. The infimum of the C's for which this holds is defined to be $\|\omega\|$; equivalently,

$$\|\omega\| = \sup \{|\omega(v)| : \|v\| \leq 1\}. \tag{9.4}$$

It is easy to verify that V', with this norm, is also a Banach space.

As shown in Chapter 4, if H is a Hilbert space, the inner product produces a conjugate linear isomorphism of H' with H; see (4.34). In particular, as noted in Corollary 4.9, $L^2(X, \mu)$ is its own dual. We next identify the dual of $L^p(X, \mu)$, for general $p \in [1, \infty)$, using the Radon-Nikodym Theorem.

Proposition 9.1. *Let (X, μ) be a σ-finite measure space. Let $1 \leq p < \infty$. Then the dual space $L^p(X, \mu)'$, with norm given by (9.4), is naturally isomorphic to $L^q(X, \mu)$, with $1/p + 1/q = 1$.*

Note that Hölder's inequality and its refinement (4.9) show that there is a natural inclusion $\iota : L^q(X, \mu) \to L^p(X, \mu)'$, which is an isometry. It remains to show that ι is surjective. We give a proof in the case when $\mu(X)$ is finite, from which the general case is easily deduced. If $\omega \in L^p(X, \mu)'$, define a set function ν on measurable sets $E \subset X$ by $\nu(E) = \langle \chi_E, \omega \rangle$, where χ_E is the characteristic function of E; ν is readily verified to be countably additive, as long as $p < \infty$. Thus ν is a complex measure. We have

$$\int f \, d\nu = \langle f, \omega \rangle,$$

first for $f = \chi_E$, then for simple f. Furthermore, ν annihilates sets of μ-measure zero, so the Radon-Nikodym Theorem, in the form of Theorem 8.7, implies

$$\int f \, d\nu = \int f w \, d\mu$$

for some measurable function w. We hence have

$$\langle f, \omega \rangle = \int f w \, d\mu, \tag{9.5}$$

for simple f. A variant of the proof of (4.9) gives $w \in L^q(X, \mu)$, $\|w\|_{L^q} = \|\omega\|$, and (9.5) holds for all $f \in L^p(X, \mu)$.

We mention that countable additivity of ν fails for $p = \infty$; in fact $L^\infty(X, \mu)'$ is identified with the space of *finitely additive* bounded set functions on the σ-algebra of μ-measurable sets that annihilate sets of μ-measure zero; see [**Yo**] for a proof.

Given a Banach space V, since V' is a Banach space, one can construct its dual, V''. Note that the action (9.2) produces a natural linear map

$$\kappa : V \longrightarrow V'', \tag{9.6}$$

and it is obvious that $\|\kappa(v)\| \le \|v\|$. In fact, $\|\kappa(v)\| = \|v\|$, i.e., κ is an isometry. More precisely, for any $v \in V$, there exists $\omega \in V'$, $\|\omega\| = 1$, such that $\omega(v) = \|v\|$. One can check directly that this property holds when $V = L^p(X,\mu)$, for a σ-finite measure space, with $1 \le p < \infty$. More generally, this property is a consequence of the Hahn-Banach Theorem; see, e.g., Chapter 3 of [**RS**] or §4 in Appendix A of [**T1**].

Sometimes κ in (9.6) is surjective, so it gives an isometric isomorphism of V with V''. In this case we say V is *reflexive*. Clearly any Hilbert space is reflexive. Also, in view of Proposition 9.1, we see that $L^p(X,\mu)$ is reflexive provided $1 < p < \infty$. On the other hand, $L^1(X,\mu)$ is not reflexive; $L^\infty(X,\mu)'$ is strictly larger than $L^1(X,\mu)$, except for the trivial cases where $L^1(X,\mu)$ is finite dimensional.

One of the values of identifying a certain Banach space Y as the dual V' of another Banach space V is the use of the "weak* topology" on V', in which a neighborhood basis of an element $\omega \in V'$ consists of sets of the form

$$U_\omega(v_1,\dots,v_K,\varepsilon) = \{w \in V' : |\langle v_j, w - \omega\rangle| < \varepsilon\}. \tag{9.7}$$

In particular, for a sequence $w_\nu \in V'$, we say $w_\nu \to \omega$ in the weak* topology provided $\langle v, w_\nu\rangle \to \langle v, \omega\rangle$ for each $v \in V$. We have the following important result called Alaoglu's Theorem:

Proposition 9.2. *If V is a Banach space, then the closed unit ball $B \subset V'$ is compact in the weak* topology.*

This result is readily deduced from the following fundamental result in topology:

Tychonov's Theorem. *If $\{X_\alpha : \alpha \in A\}$ is any family of compact Hausdorff spaces, then the Cartesian product $\prod_{\alpha\in A} X_\alpha$, with the product topology, is a compact Hausdorff space.*

For a proof of Tychonov's Theorem, see [**Dug**] or [**RS**]. To see how this yields Proposition 9.2, note that $B \in V'$ above, with the weak* topology, is homeomorphic to a closed subset of the Cartesian product $\prod\{X_v : v \in V\}$, where $X_v = \{z \in \mathbb{C} : |z| \le \|v\|\}$ and $\kappa : B \to \prod X_v$ is given by $\kappa(\omega) = \{\omega(v) : v \in V\}$.

If V is *separable*, we can say a little more. Let $\{v_j : j \in \mathbb{Z}^+\}$ be a dense subset of B_1, and, for $\omega, \sigma \in B \subset V'$, set

$$d(\omega,\sigma) = \sum_{j\ge 0} 2^{-j} |\langle v_j, \omega - \sigma\rangle|. \tag{9.8}$$

This defines a *metric* on B, and it is clear that the identity map on B gives a continuous map $\mathrm{id} : (B, w^*) \to (B, d)$, from B with the weak* topology to B

with the metric topology. But a bijective continuous map from a compact Hausdorff space onto a Hausdorff space must be a *homeomorphism*, as is shown in Appendix A, so the two topologies must coincide. Thus we have

Corollary 9.3. *If V is a separable Banach space, then the unit ball B in V' is a compact metric space, with the weak* topology.*

REMARK. In case V is separable, one has $\kappa : B \to \prod X_{v_j}$, mapping (B, w^*) homeomorphically onto a closed subset of $\prod X_{v_j}$, for a countable dense subset $\{v_j\}$ of V. The compactness of $\prod X_{v_j}$ follows from the special case of Tychonov's Theorem proven in Appendix A, Proposition A.17.

Applying Proposition 9.2 and Corollary 9.3 to Proposition 9.1, we have

Proposition 9.4. *If (X, μ) is a σ-finite measure space, then, for $1 < p \leq \infty$, the closed unit ball B in $L^p(X, \mu)$ is compact in the weak* topology, relative to $L^q(X, \mu)$, where $p^{-1} + q^{-1} = 1$. If $L^q(X, \mu)$ is separable, then B is a compact metric space with this topology.*

We turn now to more general mappings. If V and W are two Banach spaces, we denote by $\mathcal{L}(V, W)$ the space of continuous linear transformations from V to W. We set $\mathcal{L}(V) = \mathcal{L}(V, V)$. As in the derivation of (9.3), it is easy to see that, when V and W are Banach spaces, a linear map $T : V \to W$ is continuous if and only if there exists a constant $C < \infty$ such that

$$\|Tv\| \leq C\|v\| \tag{9.9}$$

for all $v \in V$. Thus we call T a *bounded* operator. The infimum of all the C's for which this holds is defined to be $\|T\|$; equivalently,

$$\|T\| = \sup \{\|Tv\| : \|v\| \leq 1\}. \tag{9.10}$$

It is clear that $\mathcal{L}(V, W)$ is a linear space. If V and W are Banach spaces and $T_j \in \mathcal{L}(V, W)$, then $\|T_1 + T_2\| \leq \|T_1\| + \|T_2\|$; completeness is also easy to establish in this case, so $\mathcal{L}(V, W)$ is also a Banach space. If X is a third Banach space and $S \in \mathcal{L}(W, X)$, it is clear that $ST \in \mathcal{L}(V, X)$ and

$$\|ST\| \leq \|S\| \cdot \|T\|. \tag{9.11}$$

If V and W are Banach spaces and $T \in \mathcal{L}(V, W)$, then the *adjoint* $T' \in \mathcal{L}(W', V')$ is uniquely defined to satisfy

$$\langle Tv, w \rangle = \langle v, T'w \rangle, \quad v \in V, \ w \in W'. \tag{9.12}$$

Granted that κ is an isometry in (9.6), it is easy to see that

$$\|T\| = \|T'\| = \sup \{|\langle Tv, w\rangle| : \|v\| = 1, \|w\| = 1\}. \tag{9.13}$$

When V and W are reflexive, it is clear that $T'' = T$.

In case V and W are Hilbert spaces, $T \in \mathcal{L}(V, W)$, then we also have an adjoint $T^* \in \mathcal{L}(W, V)$, given by

$$(Tv, w) = (v, T^*w), \quad v \in V, \ w \in W, \tag{9.14}$$

using the inner products on W and V, respectively. As in (9.13) we have $\|T\| = \|T^*\|$. Also it is clear that $T^{**} = T$.

We consider some simple examples of bounded linear operators. If (X, μ) is a measure space, $f \in L^\infty(X, \mu)$, then the multiplication operator M_f, defined by $M_f u = fu$, is bounded on $L^p(X, \mu)$ for each $p \in [1, \infty]$, with $\|M_f\| = \|f\|_{L^\infty}$. If X is a compact Hausdorff space and $f \in C(X)$, then $M_f \in \mathcal{L}(C(X))$, with $\|M_f\| = \|f\|_{\sup}$.

Another class of examples, a little more elaborate than above, is given by *integral operators* of the form

$$Ku(x) = \int_Y k(x, y)\, u(y)\, d\nu(y). \tag{9.15}$$

We have the following result:

Proposition 9.5. *Let $(X, \mathcal{M}, \mu)$ and $(Y, \mathcal{N}, \nu)$ be σ-finite measure spaces. Suppose k is measurable on $X \times Y$ and*

$$\int_X |k(x, y)|\, d\mu(x) \le C_1, \qquad \int_Y |k(x, y)|\, d\nu(y) \le C_2, \tag{9.16}$$

for all y and for all x, respectively. Then (9.15) defines K as a bounded operator from $L^p(Y, \nu)$ to $L^p(X, \mu)$, for each $p \in [1, \infty]$, with

$$\|K\| \le C_1^{1/p}\, C_2^{1/q}, \quad \frac{1}{p} + \frac{1}{q} = 1. \tag{9.17}$$

Proof. For $p \in (1, \infty)$, we estimate

$$\Bigl| \int_X \int_Y k(x, y) f(y) g(x)\, d\mu(x)\, d\nu(y) \Bigr| \tag{9.18}$$

via the estimate $ab \le a^p/p + b^q/q$ of (4.11), used to prove Hölder's inequality. Apply this to $|f(y)g(x)|$. Then (9.18) is dominated by

$$(C_1/p)\|f\|_{L^p}^p + (C_2/q)\|g\|_{L^q}^q \tag{9.19}$$

provided (9.16) holds. Replacing f, g by $tf, t^{-1}g$, we see that (9.18) is dominated by $(C_1 t^p/p)\|f\|_{L^p}^p + (C_2/qt^q)\|g\|_{L^q}^q$, and minimizing over $t \in (0,\infty)$, via elementary calculus, we see that (9.18) is dominated by

$$C_1^{1/p} C_2^{1/q} \|f\|_{L^p} \|g\|_{L^q}, \tag{9.20}$$

proving the result. The exceptional cases $p = 1$ and $p = \infty$ are easily handled.

We call $k(x,y)$ the integral kernel of K. Note that K' is an integral operator, with kernel $k'(x,y) = k(y,x)$. In the case of the Hilbert space $L^2(X,\mu)$, K^* is an integral operator, with kernel $k^*(x,y) = \overline{k(y,x)}$.

We next take a look at an important class of bounded operators on Hilbert spaces, the *Hilbert-Schmidt operators*, defined as follows. Let H_1 and H_2 be separable Hilbert spaces and $A \in \mathcal{L}(H_1, H_2)$. Let $\{u_j\}$ be an orthonormal basis of H_1. We say A is a Hilbert-Schmidt operator, or an HS operator for short, provided

$$\sum_j \|Au_j\|^2 < \infty, \tag{9.21}$$

equivalently, if

$$\sum_{j,k} |a_{jk}|^2 < \infty, \quad a_{jk} = (Au_j, v_k), \tag{9.22}$$

where $\{v_k\}$ is an orthonormal basis of H_2. The class of HS operators in $\mathcal{L}(H_1, H_2)$ will be denoted $\mathrm{HS}(H_1, H_2)$. If $H_1 = H_2 = H$, we denote the class $\mathrm{HS}(H)$. We define the Hilbert-Schmidt norm of an HS operator:

$$\|A\|_{\mathrm{HS}}^2 = \sum_j \|Au_j\|^2 = \sum_{j,k} |a_{jk}|^2. \tag{9.23}$$

By the second identity in (9.23), we see that $A^* \in \mathcal{L}(H_2, H_1)$ is HS if A is and that

$$\|A^*\|_{\mathrm{HS}} = \|A\|_{\mathrm{HS}}. \tag{9.24}$$

Since $\|A\|_{\mathrm{HS}}$ is clearly independent of the choice of orthonormal basis of H_2, $\|A^*\|_{\mathrm{HS}}$ is independent of the choice of orthonormal basis of H_1 and hence so is $\|A\|_{\mathrm{HS}}$.

The first identity in (9.23) makes it clear that $\|BA\|_{\mathrm{HS}} \le \|B\| \cdot \|A\|_{\mathrm{HS}}$ if A is HS and B is bounded on H_2. Since $AB = (B^*A^*)^*$, we deduce via (9.24) that an analogous bound holds on $\|AB\|_{\mathrm{HS}}$ when B is bounded on H_1. In other words,

$$\|B_2 A B_1\|_{\mathrm{HS}} \le \|B_2\| \cdot \|A\|_{\mathrm{HS}} \cdot \|B_1\|, \tag{9.25}$$

if $B_j \in \mathcal{L}(H_j)$. In particular, recapitulating the observation made following (9.24), we have

$$\|U_2 A U_1\|_{\mathrm{HS}} = \|A\|_{\mathrm{HS}}, \tag{9.26}$$

if $U_j \in \mathcal{L}(H_j)$ are unitary transformations, i.e., invertible isometries of H_j onto H_j.

We note that, if A is a Hilbert-Schmidt operator, then

$$\|A\| \le \|A\|_{\mathrm{HS}}. \tag{9.27}$$

To see this, pick unit u_1 such that $\|Au_1\| \ge \|A\| - \varepsilon$ and make that part of an orthonormal basis.

The following classical result might be called the Hilbert-Schmidt Kernel Theorem.

Proposition 9.6. *If $T : L^2(X_1, \mu_1) \to L^2(X_2, \mu_2)$ is Hilbert-Schmidt, then there exists $K \in L^2(X_1 \times X_2, \mu_1 \times \mu_2)$ such that*

$$(Tu, v)_{L^2} = \iint K(x_1, x_2) u(x_1) \overline{v(x_2)} \, d\mu_1(x_1) \, d\mu_2(x_2). \tag{9.28}$$

Proof. Pick orthonormal bases $\{f_j\}$ for $L^2(X_1)$ and $\{g_k\}$ for $L^2(X_2)$ and set

$$K(x_1, x_2) = \sum_{j,k} a_{jk} \overline{f_j(x_1)} g_k(x_2) \tag{9.29}$$

where $a_{jk} = (Tf_j, g_k)$. The hypothesis that T is HS is precisely what is necessary to guarantee that $K \in L^2(X_1 \times X_2)$, and then (9.28) is obvious. It is also clear that

$$\|T\|^2_{\mathrm{HS}} = \|K\|^2_{L^2}. \tag{9.30}$$

Also of interest is the converse, proved simply by reversing the argument and noting that $\{f_j g_k\}$ is an orthonormal *basis* of $L^2(X_1 \times X_2, \mu_1 \times \mu_2)$ (see Exercise 6 of Chapter 6).

Proposition 9.7. *If $K \in L^2(X_1 \times X_2, \mu_1 \times \mu_2)$, then* (9.28) *defines an HS operator T, satisfying* (9.30).

We now look at a particular linear operator of singular importance, the Fourier transform, defined by

$$\mathcal{F}f(\xi) = \hat{f}(\xi) = (2\pi)^{-n/2} \int f(x) e^{-ix\cdot\xi} \, dx \tag{9.31}$$

when $f \in L^1(\mathbb{R}^n)$. It is clear that

$$\mathcal{F} : L^1(\mathbb{R}^n) \longrightarrow L^\infty(\mathbb{R}^n). \tag{9.32}$$

It is convenient to bring in the Schwartz space of rapidly decreasing functions:

$$\mathcal{S}(\mathbb{R}^n) = \big\{u \in C^\infty(\mathbb{R}^n) : x^\beta D^\alpha u \in L^\infty(\mathbb{R}^n) \text{ for all } \alpha, \beta \geq 0\big\}, \tag{9.33}$$

where $x^\beta = x_1^{\beta_1} \cdots x_n^{\beta_n}$, $D^\alpha = D_1^{\alpha_1} \cdots D_n^{\alpha_n}$, with $D_j = -i\partial/\partial x_j$. Using an argument involving integration by parts (cf. (7.67)), it is easy to verify that

$$\mathcal{F} : \mathcal{S}(\mathbb{R}^n) \longrightarrow \mathcal{S}(\mathbb{R}^n) \tag{9.34}$$

and

$$\xi^\alpha D_\xi^\beta \mathcal{F}f(\xi) = (-1)^{|\beta|} \mathcal{F}(D^\alpha x^\beta f)(\xi). \tag{9.35}$$

We define $\mathcal{F}^*$ by

$$\mathcal{F}^* f(\xi) = \check{f}(\xi) = (2\pi)^{-n/2} \int f(x) e^{ix\cdot\xi} \, dx, \tag{9.36}$$

which differs from (9.31) only in the sign of the exponent. It is clear that $\mathcal{F}^*$ satisfies the mapping properties (9.32), (9.34), and

$$(\mathcal{F}u, v) = (u, \mathcal{F}^* v) \tag{9.37}$$

for $u, v \in \mathcal{S}(\mathbb{R}^n)$, where (u, v) denotes the usual L^2-inner product, $(u, v) = \int_{\mathbb{R}^n} u(x) \, \overline{v(x)} \, dx$.

The first major result is the Fourier inversion formula. The following is our first version.

Proposition 9.8. *We have the inversion formula*

$$\mathcal{F}^* \mathcal{F} = \mathcal{F}\mathcal{F}^* = I \quad \textit{on} \quad \mathcal{S}(\mathbb{R}^n). \tag{9.38}$$

We will sneak up on the inversion formula by throwing in a convergence factor which will allow interchange of orders of integration. So, let us write, for $f \in \mathcal{S}(\mathbb{R}^n)$,

$$
\begin{aligned}
\mathcal{F}^*\mathcal{F}f(x) &= (2\pi)^{-n} \int \Big[\int f(y) e^{-iy\cdot\xi}\, dy \Big] e^{ix\cdot\xi}\, d\xi \\
&= (2\pi)^{-n} \lim_{\varepsilon \searrow 0} \iint f(y)\, e^{-\varepsilon|\xi|^2}\, e^{i(x-y)\cdot\xi}\, dy\, d\xi.
\end{aligned} \tag{9.39}
$$

We can interchange the order of integration on the right for any $\varepsilon > 0$, to obtain

$$
\mathcal{F}^*\mathcal{F}f(x) = \lim_{\varepsilon \searrow 0} \int f(y) p(\varepsilon, x-y)\, dy, \tag{9.40}
$$

where

$$
p(\varepsilon, x) = (2\pi)^{-n} \int e^{-\varepsilon|\xi|^2 + ix\cdot\xi}\, d\xi. \tag{9.41}
$$

Note that

$$
p(\varepsilon, x) = \varepsilon^{-n/2}\, q(x/\sqrt{\varepsilon}), \tag{9.42}
$$

where $q(x) = p(1, x)$. In a moment we will show that

$$
p(\varepsilon, x) = (4\pi\varepsilon)^{-n/2}\, e^{-|x|^2/4\varepsilon}. \tag{9.43}
$$

The derivation of this identity will also show that

$$
\int_{\mathbb{R}^n} q(x)\, dx = 1. \tag{9.44}
$$

From this, it follows, as in the proof of (4.64) and (4.69), that

$$
\lim_{\varepsilon \searrow 0} \int f(y) p(\varepsilon, x-y)\, dy = f(x) \tag{9.45}
$$

for any $f \in \mathcal{S}(\mathbb{R}^n)$, even for f bounded and continuous, so we have proved $\mathcal{F}^*\mathcal{F} = I$ on $\mathcal{S}(\mathbb{R}^n)$; the proof that $\mathcal{F}\mathcal{F}^* = I$ on $\mathcal{S}(\mathbb{R}^n)$ is identical.

It remains to verify (9.43). We observe that $p(\varepsilon, x)$, defined by (9.41), is an entire holomorphic function of $x \in \mathbb{C}^n$, for any $\varepsilon > 0$. It is convenient to verify that

$$
p(\varepsilon, ix) = (4\pi\varepsilon)^{-n/2}\, e^{|x|^2/4\varepsilon}, \quad x \in \mathbb{R}^n, \tag{9.46}
$$

from which (9.43) follows by analytic continuation. Now

$$
\begin{aligned}
p(\varepsilon, ix) &= (2\pi)^{-n} \int e^{-x\cdot\xi-\varepsilon|\xi|^2}\, d\xi \\
&= (2\pi)^{-n} e^{|x|^2/4\varepsilon} \int e^{-|x/2\sqrt{\varepsilon}+\sqrt{\varepsilon}\xi|^2}\, d\xi \\
&= (2\pi)^{-n} \varepsilon^{-n/2} e^{|x|^2/4\varepsilon} \int_{\mathbb{R}^n} e^{-|\xi|^2}\, d\xi.
\end{aligned} \tag{9.47}
$$

To prove (9.46), it remains to show that

$$
\int_{\mathbb{R}^n} e^{-|\xi|^2}\, d\xi = \pi^{n/2}. \tag{9.48}
$$

Indeed, for this see Exercises 1–2 in the first exercise set of Chapter 7. This completes the proof of the identity (9.46) and hence of (9.43).

REMARK. See Exercises 7–8 for another demonstration of (9.43), not using analytic continuation.

In light of (9.37) and the Fourier inversion formula (9.38), we see that, for $u, v \in \mathcal{S}(\mathbb{R}^n)$,

$$
(\mathcal{F}u, \mathcal{F}v) = (u, v) = (\mathcal{F}^*u, \mathcal{F}^*v). \tag{9.49}
$$

Thus $\mathcal{F}$ and $\mathcal{F}^*$ extend uniquely from $\mathcal{S}(\mathbb{R}^n)$ to isometries on $L^2(\mathbb{R}^n)$, and are inverses to each other, i.e., they are unitary. Thus we have the Plancherel Theorem:

Proposition 9.9. *The Fourier transform*

$$
\mathcal{F} : L^2(\mathbb{R}^n) \longrightarrow L^2(\mathbb{R}^n) \tag{9.50}
$$

is unitary, with inverse $\mathcal{F}^*$.

We now make some comments on the relation between the Fourier transform and convolutions. The convolution $u * v$ of two functions on $\mathbb{R}^n$ is defined by

$$
\begin{aligned}
u * v(x) &= \int u(y)v(x-y)\, dy \\
&= \int u(x-y)v(y)\, dy.
\end{aligned} \tag{9.51}
$$

Note that $u * v = v * u$. If $u, v \in \mathcal{S}(\mathbb{R}^n)$, so is $u * v$. As in (4.62) and (7.63) and also as a special case of Proposition 9.5,

$$\|u * v\|_{L^p(\mathbb{R}^n)} \le \|u\|_{L^1} \|v\|_{L^p}, \tag{9.52}$$

so the convolution has a unique continuous extension to a bilinear map

$$L^1(\mathbb{R}^n) \times L^p(\mathbb{R}^n) \longrightarrow L^p(\mathbb{R}^n), \tag{9.53}$$

for $1 \le p < \infty$; one can directly perceive this also works for $p = \infty$. Note that the right side of (9.40), for any $\varepsilon > 0$, is an example of a convolution. Computing the Fourier transform of (9.51) leads immediately to the formula

$$\mathcal{F}(u * v)(\xi) = (2\pi)^{n/2}\, \hat{u}(\xi)\hat{v}(\xi). \tag{9.54}$$

Via Proposition 9.9, we have the estimate

$$\|u * v\|_{L^2(\mathbb{R}^n)} \le (2\pi)^{n/2}\, \|\hat{u}\|_{L^\infty} \|v\|_{L^2}, \tag{9.55}$$

which is stronger than the $p = 2$ case of (9.52). The estimate (9.55) does not hold if L^2 is replaced by L^p for $p \neq 2$. Useful strengthenings of (9.52) for other values of $p \in (1, \infty)$ lead one to the important topic of singular integral operators. See [**St**] for a treatment of this and also Chapter 13 of [**T1**].

It is useful to extend the Fourier transform from $L^1(\mathbb{R}^n) \to L^\infty(\mathbb{R}^n)$ to $\mathfrak{M}(\mathbb{R}^n) \to L^\infty(\mathbb{R}^n)$, where $\mathfrak{M}(\mathbb{R}^n)$ denotes the space of finite (signed or, more generally, complex) Borel measures on $\mathbb{R}^n$, and to extend the convolution product from $L^1(\mathbb{R}^n) \times L^1(\mathbb{R}^n) \to L^1(\mathbb{R}^n)$ to $\mathfrak{M}(\mathbb{R}^n) \times \mathfrak{M}(\mathbb{R}^n) \to \mathfrak{M}(\mathbb{R}^n)$. The formula (9.31) for $\hat{f}(\xi)$ makes it natural to set

$$\hat{\mu}(\xi) = (2\pi)^{-n/2} \int e^{-ix\cdot\xi}\, d\mu(x), \quad \mu \in \mathfrak{M}(\mathbb{R}^n). \tag{9.56}$$

It readily follows that $|\hat{\mu}(\xi)| \le (2\pi)^{-n/2}\|\mu\|$, the norm on $\mathfrak{M}(\mathbb{R}^n)$ defined by (8.21). Also the Dominated Convergence Theorem implies $\hat{\mu}$ is continuous. Note that if $u \in \mathcal{S}(\mathbb{R}^n)$, then, by Fubini's Theorem,

$$\begin{aligned} \int u(\xi)\hat{\mu}(\xi)\, d\xi &= (2\pi)^{-n/2} \iint u(\xi) e^{-ix\cdot\xi}\, d\mu(x)\, d\xi \\ &= \int \hat{u}(x)\, d\mu(x). \end{aligned} \tag{9.57}$$

Applying Proposition 9.8, we deduce that, given $\mu \in \mathfrak{M}(\mathbb{R}^n)$,

$$\int v(x)\, d\mu(x) = \int \tilde{v}(\xi)\hat{\mu}(\xi)\, d\xi, \quad \forall\, v \in \mathcal{S}(\mathbb{R}^n). \tag{9.58}$$

A natural way to define the convolution $\mu * \nu$ of two finite measures is suggested by the following computation. If $f, g \in L^1(\mathbb{R}^n)$ and $u \in L^\infty(\mathbb{R}^n)$,

$$\begin{aligned} \int u(x)\,(f*g)(x)\,dx &= \iint u(x) f(x-y) g(y)\,dy\,dx \\ &= \iint u(x+y) f(x) g(y)\,dx\,dy. \end{aligned} \tag{9.59}$$

Thus it is natural to define $\mu * \nu$ by

$$\mu * \nu(S) = \iint \chi_S(x+y)\,d\mu(x)\,d\nu(y), \tag{9.60}$$

when $S \subset \mathbb{R}^n$ is a Borel set and $\mu, \nu \in \mathfrak{M}(\mathbb{R}^n)$. An equivalent formula is

$$\mu * \nu(S) = (\mu \times \nu)(\varphi^{-1}(S)), \quad \varphi : \mathbb{R}^n \times \mathbb{R}^n \to \mathbb{R}^n, \quad \varphi(x,y) = x+y. \tag{9.61}$$

It readily follows that $\mu * \nu \in \mathfrak{M}(\mathbb{R}^n)$ and

$$\|\mu * \nu\| \le \|\mu\| \cdot \|\nu\|. \tag{9.62}$$

Note also that if u is a bounded Borel function on $\mathbb{R}^n$,

$$\int u\,d(\mu * \nu) = \iint u(x+y)\,d\mu(x)\,d\nu(y). \tag{9.63}$$

In fact, (9.63) follows by linearity from (9.60) if u is a finite linear combination of χ_{S_j}, $S_j \in \mathfrak{B}(\mathbb{R}^n)$, and then by the Dominated Convergence Theorem for a general bounded Borel function u. A particular consequence of (9.63) is the following analogue of (9.54):

$$\begin{aligned} (\mu * \nu)\hat{}(\xi) &= (2\pi)^{-n/2} \int e^{-i(x+y)\cdot\xi}\,d\mu(x)\,d\nu(y) \\ &= (2\pi)^{n/2} \hat{\mu}(\xi) \hat{\nu}(\xi). \end{aligned} \tag{9.64}$$

Exercises

1. Extend the $p = 2$ case of Proposition 9.5 to the following result of Schur. Let (X, μ) and (Y, ν) be measure spaces, and let $k(x,y)$ be measurable on $(X \times Y, \mu \times \nu)$. Assume that there are measurable functions $p(x)$, $q(y)$, *positive* a.e. on X and Y, respectively, such that

$$\int_X |k(x,y)| p(x)\,d\mu(x) \le C_1 q(y), \qquad \int_Y |k(x,y)| q(y)\,d\nu(y) \le C_2 p(x). \tag{9.65}$$

Show that $Ku(x) = \int_Y k(x,y)u(y)\,d\nu(y)$ defines a bounded operator

$$K : L^2(Y,\nu) \longrightarrow L^2(X,\mu), \quad \|K\|^2 \le C_1C_2.$$

Give an appropriate modification of the hypothesis (9.65) in order to obtain an operator bound $K : L^p(Y,\nu) \to L^p(X,\mu)$.
Hint. To estimate (9.18), set

$$|f(y)g(x)| = \Bigl|\frac{B(x)}{A(y)}f(y)\Bigr| \cdot \Bigl|\frac{A(y)}{B(x)}g(x)\Bigr|,$$

and apply the estimate $ab \le (a^p/p) + (b^q/q)$ to this.

2. Show that $k(x,y)$ is the integral kernel of a bounded map $K : L^2(\mathbb{R}^n) \to L^2(\mathbb{R}^n_+)$ provided it satisfies the estimate

$$\text{(9.66)} \quad |k(x,y)| \le C\Bigl(|x'-y'|^2 + x_1^2 + y_1^2\Bigr)^{-n/2}, \quad x = (x_1,x'),\ y = (y_1,y').$$

Hint. Construct $p(x)$ and $q(y)$ so that (9.65) holds. Look for $p(x)$ as a function of x_1 alone.

3. Given $u \in L^p(\mathbb{R}^n)$, $1 \le p < \infty$, and $\varphi \in C_0^\infty(\mathbb{R}^n)$, set $v = \varphi * u$.
(a) Show that v is continuous on $\mathbb{R}^n$.
Hint. Apply dominated convergence.
(b) Show that $h^{-1}\bigl[v(x+he_j) - v(x)\bigr] \to (\partial\varphi/\partial x_j) * u$ as $h \to 0$, and conclude that $v \in C^1(\mathbb{R}^n)$.
(c) Show that $v \in C^\infty(\mathbb{R}^n)$ and $D^\alpha v = (D^\alpha\varphi) * u$.

4. If $u \in L^p(\mathbb{R}^n)$ and $w \in L^{p'}(\mathbb{R}^n)$, $1/p + 1/p' = 1$, show that $u * w$ is continuous on $\mathbb{R}^n$.

In Exercises 5–6, let $(X,\mathfrak{F},\mu)$ be a measure space and $\mathcal{A}$ an algebra of subsets of X such that $\sigma(\mathcal{A}) = \mathfrak{F}$. Let

$$\Lambda = \text{ span } \{\chi_A : A \in \mathcal{A}\}.$$

5. If $f \in L^1(X,\mu)$ and $\int_A f\,d\mu = 0$ for all $A \in \mathcal{A}$, show that $f = 0$.
Hint. Show that $\{B \in \mathfrak{F} : \int_B f\,d\mu = 0\}$ is a monotone class. Apply Proposition 5.9.

6. If $\mu(X) < \infty$, show that Λ is dense in $L^2(X,\mu)$.
Hint. Show that the orthogonal complement $\Lambda^\perp$ of Λ in $L^2(X,\mu)$ is 0.

Deduce that Λ is dense in $L^p(X,\mu)$, for $1 \le p \le 2$.
Note. There is a general result known as the Hahn-Banach Theorem, which implies that if V is a Banach space and L a linear subspace which is not dense in V, then there exists a nonzero $\omega \in V'$ such that $\omega(v) = 0$ for all $v \in L$. (See [**RS**], [**T1**], [**Yo**].) Using this, show that, when $\mu(X) < \infty$, Λ is dense in $L^p(X,\mu)$ for all $p \in [1,\infty)$.

Exercises 7–8 give another approach to the identity (9.43), i.e.,

$$(2\pi)^{-n}\int e^{-\varepsilon|\xi|^2+ix\cdot\xi}\,d\xi = (4\pi\varepsilon)^{-n/2}e^{-|x|^2/4\varepsilon}, \tag{9.67}$$

one that does not use analytic continuation.

7. Show that to demonstrate (9.67) it suffices to treat the case $n = 1$. Furthermore, it suffices to show that

$$g(x) = e^{-x^2/2} \Longrightarrow \hat{g}(\xi) = e^{-\xi^2/2}. \tag{9.68}$$

8. Note that $g \in \mathcal{S}(\mathbb{R})$ and g satisfies the differential equation

$$\Big(\frac{d}{dx} + x\Big)g(x) = 0.$$

Use (9.34)–(9.35) to deduce that $\hat{g} \in \mathcal{S}(\mathbb{R})$ and

$$\Big(\frac{d}{d\xi} + \xi\Big)\hat{g}(\xi) = 0.$$

Then establish (9.68). First show $\hat{g}(\xi) = Ce^{-\xi^2/2}$; then evaluate C.

9. Let $p(t,x)$ be given by (9.43) with $t = \varepsilon$, i.e.,

$$p(t,x) = (4\pi t)^{-n/2}e^{-|x|^2/4t}. \tag{9.69}$$

Show that

$$\mathcal{F}^{-1}e^{-t|\xi|^2}\mathcal{F}f(x) = \int f(y)p(t,x-y)\,dy. \tag{9.70}$$

Show that, if $f \in L^p(\mathbb{R}^n)$, $1 \le p < \infty$, then $u(t,x) = \int f(y)p(t,x-y)\,dy$ satisfies the "diffusion equation" (also called the "heat equation"):

$$\frac{\partial u}{\partial t} = \Delta u, \quad \Delta = \frac{\partial^2}{\partial x_1^2} + \cdots + \frac{\partial^2}{\partial x_n^2}, \tag{9.71}$$

for $t > 0$, and that

$$u(t,\cdot) \to f \text{ in } L^p(\mathbb{R}^n), \text{ as } t \searrow 0.$$

Hint. Use (9.35).

10. Verify that the map κ in (9.6) is an isometry when $V = L^p(X,\mu)$, $1 \le p < \infty$, by showing directly that, given $v \in L^p(X,\mu)$, there exists $w \in L^{p'}(X,\mu)$ such that $\|w\|_{L^{p'}} = 1$ and $\int vw\,d\mu = \|v\|_{L^p}$.

11. Assume $f \in C^2([0,1])$, $f(0) = f(1) = 0$. Show that

$$f(x) = \int_0^1 G(x,y) f''(y)\,dy,$$

where

$$\begin{aligned} G(x,y) &= -x(1-y) \quad \text{for } 0 \le x \le y \le 1, \\ &\quad -(1-x)y \quad \text{for } 0 \le y \le x \le 1. \end{aligned}$$

12. If f is as in Exercise 11, use Proposition 9.5 to show that

$$\int_0^1 |f(x)|\,dx \le \frac{1}{8} \int_0^1 |f''(x)|\,dx. \tag{9.72}$$

Exercises 13–14 deal with the *trapezoidal rule*, defined as follows. Let an interval $I = [a,b]$ be divided into subintervals, all of length h. Given $f \in C(I)$, $\int_I f\,dx$ is approximated by

$$\mathcal{T}_I^h f = \sum_{j=1}^{N} \frac{f(a_j) + f(b_j)}{2}\, h, \tag{9.73}$$

where $[a_j, b_j]$ are the intervals into which I is subdivided, with $b_j - a_j = h = (b-a)/N$.

13. Show that if $I = [0,1]$, then

$$\Bigl| \int_0^1 f\,dx - \mathcal{T}_I^1 f \Bigr| \le \frac{1}{8} \int_0^1 |f''(x)|\,dx. \tag{9.74}$$

Hint. Subtract from f the linear function agreeing with $f(x)$ at $x = 0$ and 1.

14. Show that, if $I = [a, b]$ and $f \in C^2(I)$, then

$$\Big| \int_I f\, dx - \mathcal{T}_I^h f \Big| \le \frac{h^2}{8} \int_I |f''(x)|\, dx. \tag{9.75}$$

Hint. Derive (9.75) from (9.74) by scaling when $I = [0, h]$. Then sum this result over the intervals of length h into which $I = [a, b]$ is subdivided.

15. If $f \in L^p(\mathbb{R}^n)$ and $\mu \in \mathfrak{M}(\mathbb{R}^n)$, define

$$f * \mu(x) = \int f(x - y)\, d\mu(y).$$

Show that $f * \mu \in L^p(\mathbb{R}^n)$ and

$$\|f * \mu\|_{L^p} \le \|f\|_{L^p} \|\mu\|.$$

Show that if also $\nu \in \mathfrak{M}(\mathbb{R}^n)$, then

$$(f * \mu) * \nu = f * (\mu * \nu).$$

16. Suppose P is a bounded linear operator on a Hilbert space H satisfying

$$P^2 = P, \quad P^* = P.$$

Let $K = \operatorname{Range} P$. Show that $K = \operatorname{Ker}(I - P)$ and $K^\perp = \operatorname{Ker} P = \operatorname{Range}(I - P)$. Deduce that K is closed and that P is the orthogonal projection of H onto K.

17. Let H be a Hilbert space and let $T \in \mathcal{H}$. Show that T is an isometry if and only if $T^*T = I$. Show that if T is an isometry, then TT^* is the orthogonal projection of H onto the range of T.

Chapter 10

Sobolev Spaces

We now define spaces $H^{1,p}(\mathbb{R}^n)$, known as Sobolev spaces. For u to belong to $H^{1,p}(\mathbb{R}^n)$, we require that $u \in L^p(\mathbb{R}^n)$ and that u have *weak derivatives* of first order in $L^p(\mathbb{R}^n)$:

$$\partial_j u = f_j \in L^p(\mathbb{R}^n), \tag{10.1}$$

where (10.1) means

$$-\int \frac{\partial \varphi}{\partial x_j}\, u\, dx = \int \varphi f_j\, dx, \quad \forall\, \varphi \in C_0^\infty(\mathbb{R}^n). \tag{10.2}$$

If $u \in C_0^\infty(\mathbb{R}^n)$, we see that $\partial_j u = \partial u / \partial x_j$, by integrating by parts, using (7.67). We define a norm on $H^{1,p}(\mathbb{R}^n)$ by

$$\|u\|_{H^{1,p}} = \|u\|_{L^p} + \sum_j \|\partial_j u\|_{L^p}. \tag{10.3}$$

We claim that $H^{1,p}(\mathbb{R}^n)$ is complete, hence a Banach space. Indeed, let (u_ν) be a Cauchy sequence in $H^{1,p}(\mathbb{R}^n)$. Then (u_ν) is Cauchy in $L^p(\mathbb{R}^n)$; hence it has an L^p-norm limit $u \in L^p(\mathbb{R}^n)$. Also, for each j, $\partial_j u_\nu = f_{j\nu}$ is Cauchy in $L^p(\mathbb{R}^n)$, so there is a limit $f_j \in L^p(\mathbb{R}^n)$, and it is easily verified from (10.2) that $\partial_j u = f_j$.

We can consider convolutions and products of elements of $H^{1,p}(\mathbb{R}^n)$ with elements of $C_0^\infty(\mathbb{R}^n)$ and readily obtain identities

$$\begin{aligned} \partial_j(\varphi * u) &= (\partial_j \varphi) * u = \varphi * (\partial_j u), \\ \partial_j(\psi u) &= \frac{\partial \psi}{\partial x_j} u + \psi(\partial_j u), \end{aligned} \tag{10.4}$$

and estimates

$$\tag{10.5} \begin{aligned} \|\varphi * u\|_{H^{1,p}} &\le \|\varphi\|_{L^1} \|u\|_{H^{1,p}}, \\ \|\psi u\|_{H^{1,p}} &\le \|\psi\|_{L^\infty} \|u\|_{H^{1,p}} + \sum \|\partial_j \psi\|_{L^\infty} \|u\|_{L^p}, \end{aligned}$$

for $\varphi,\ \psi \in C_0^\infty(\mathbb{R}^n),\ u \in H^{1,p}(\mathbb{R}^n)$. For example, the first identity in (10.4) is equivalent to

$$\begin{aligned} &-\iint \frac{\partial \psi}{\partial x_j}(x)\, \varphi(x-y) u(y)\, dy\, dx \\ &= \iint \psi(x)\, \frac{\partial \varphi}{\partial x_j}(x-y)\, u(y)\, dy\, dx, \quad \forall\, \psi \in C_0^\infty(\mathbb{R}^n), \end{aligned}$$

an identity that can be established by using Fubini's Theorem (to first do the x-integral) and integration by parts, via (7.67).

If $p < \infty,\ u \in H^{1,p}(\mathbb{R}^n)$, and (φ_j) is an approximate identity of the form (7.64), with $\varphi_j \in C_0^\infty(\mathbb{R}^n)$, then we can show that

$$\tag{10.6} \varphi_j * u \longrightarrow u \ \text{ in } \ H^{1,p}\text{-norm},$$

using (10.4) and (7.65). Given $\varepsilon > 0$, we can take j such that

$$\|\varphi_j * u - u\|_{H^{1,p}} < \varepsilon.$$

Then we can pick $\psi \in C_0^\infty(\mathbb{R}^n)$ such that $\|\psi(\varphi_j * u) - \varphi_j * u\|_{H^{1,p}} < \varepsilon$. Of course, $\varphi_j * u$ is smooth, so $\psi(\varphi_j * u) \in C_0^\infty(\mathbb{R}^n)$. We have established

Proposition 10.1. *For $p \in [1,\infty)$, the space $C_0^\infty(\mathbb{R}^n)$ is dense in $H^{1,p}(\mathbb{R}^n)$.*

Sobolev spaces are very useful in analysis, particularly in the study of partial differential equations. We will establish just a few results here, some of which will be useful in Chapter 11. More material can be found in [**EG**], [**Fol**], [**T1**], and [**Yo**].

The following result is known as a Sobolev Imbedding Theorem.

Proposition 10.2. *If $p > n$ or if $p = n = 1$, then*

$$\tag{10.7} H^{1,p}(\mathbb{R}^n) \subset C(\mathbb{R}^n) \cap L^\infty(\mathbb{R}^n).$$

For now we concentrate on the case $p \in (n,\infty)$. Since $C_0^\infty(\mathbb{R}^n)$ is then dense in $H^{1,p}(\mathbb{R}^n)$, it suffices to establish the estimate

$$\tag{10.8} \|u\|_{L^\infty} \le C\|u\|_{H^{1,p}}, \ \text{for } u \in C_0^\infty(\mathbb{R}^n).$$

In turn, it suffices to establish

$$|u(0)| \le C\|u\|_{H^{1,p}}, \text{ for } u \in C_0^\infty(\mathbb{R}^n). \tag{10.9}$$

To get this, it suffices to show that, for a given $\varphi \in C_0^\infty(\overset{\circ}{B}_1)$ with $\varphi(0) = 1$,

$$|u(0)| \le C\|\nabla(\varphi u)\|_{L^p}, \tag{10.10}$$

where $\nabla v = (\partial_1 v, \dots, \partial_n v)$ or, equivalently, that

$$|u(0)| \le C\|\nabla u\|_{L^p}, \quad u \in C_0^\infty(\overset{\circ}{B}_1). \tag{10.11}$$

In turn, this will follow from an estimate of the form

$$|u(0) - u(\omega)| \le C\|\nabla u\|_{L^p(B_1)}, \quad u \in C^\infty(\mathbb{R}^n), \tag{10.12}$$

given $\omega \in \mathbb{R}^n$, $|\omega| = 1$. Thus we turn to a proof of (10.12).

Without loss of generality, we can take $\omega = e_n = (0, \dots, 0, 1)$. We will work with the set $\Sigma = \{z \in \mathbb{R}^{n-1} : |z| \le \sqrt{3}/2\}$. For $z \in \Sigma$, let γ_z be the path from 0 to e_n consisting of a line segment from 0 to $(z, 1/2)$, followed by a line segment from $(z, 1/2)$ to e_n, as illustrated in Figure 10.1. Then (with $A = \text{Area } \Sigma$)

$$u(e_n) - u(0) = \int_\Sigma \Big(\int_{\gamma_z} du \Big) \frac{dz}{A} = \int_{B_1} \nabla u(x) \cdot \psi(x)\, dx, \tag{10.13}$$

where the last identity applies the change of variable formula. The behavior of the Jacobian determinant of the map $(t, z) \mapsto \gamma_z(t)$ yields

$$|\psi(x)| \le C|x|^{-(n-1)} + C|x - e_n|^{-(n-1)}. \tag{10.14}$$

Thus

$$\int_{B_{1/2}} |\psi(x)|^q\, dx \le C \int_0^{1/2} r^{-nq+q} r^{n-1}\, dr. \tag{10.15}$$

It follows that

$$\psi \in L^q(B_1), \quad \forall\, q < \frac{n}{n-1}. \tag{10.16}$$

Thus

$$|u(e_n) - u(0)| \le \|\nabla u\|_{L^p(B_1)} \|\psi\|_{L^{p'}(B_1)} \le C\|\nabla u\|_{L^p(B_1)}, \tag{10.17}$$

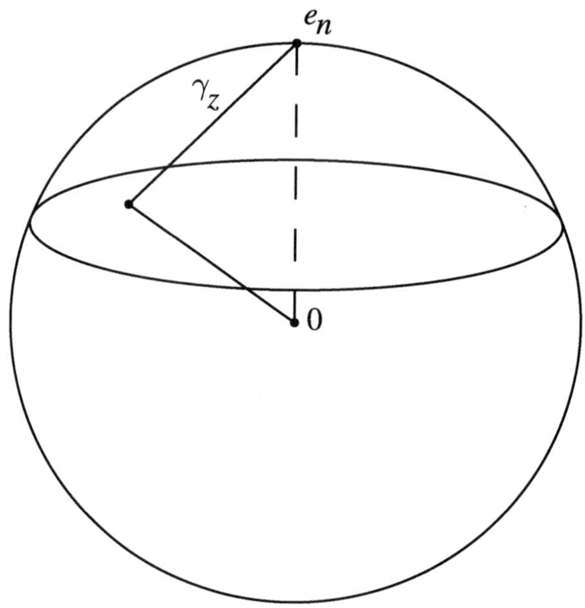

Figure 10.1

as long as $p' < n/(n-1)$, which is the same as $p > n$.

This proves (10.12) and hence Proposition 10.2, for $p \in (n, \infty)$. For $n = 1$, (10.13) simplifies to $u(1) - u(0) = \int_0^1 u'(x)\,dx$, which immediately gives the estimate (10.12) for $p = n = 1$.

We can refine Proposition 10.2 to the following.

Proposition 10.3. *If $p \in (n, \infty)$, then every $u \in H^{1,p}(\mathbb{R}^n)$ satisfies a Hölder condition:*

$$H^{1,p}(\mathbb{R}^n) \subset C^s(\mathbb{R}^n), \quad s = 1 - \frac{n}{p}. \tag{10.18}$$

Proof. Applying (10.12) to $v(x) = u(rx)$, we have, for $|\omega| = 1$,

$$|u(r\omega) - u(0)|^p \le Cr^p \int_{B_1} |\nabla u(rx)|^p\,dx = Cr^{p-n} \int_{B_r} |\nabla u(x)|^p\,dx. \tag{10.19}$$

This implies

$$|u(x) - u(y)| \le C'|x-y|^{1-n/p} \Big(\int_{B_r(x)} |\nabla u(z)|^p\,dz \Big)^{1/p}, \quad r = |x-y|, \tag{10.20}$$

which gives (10.18).

If $u \in H^{1,\infty}(\mathbb{R}^n)$ and if $\varphi \in C_0^\infty(\mathbb{R}^n)$, then $\varphi u \in H^{1,p}(\mathbb{R}^n)$ for all $p \in [1,\infty)$, so Proposition 10.3 applies. We next show that in fact $H^{1,\infty}(\mathbb{R}^n)$ coincides with the space of Lipschitz functions:

$$\text{Lip}(\mathbb{R}^n) = \{u \in L^\infty(\mathbb{R}^n) : |u(x) - u(y)| \le K|x-y|\}. \tag{10.21}$$

Proposition 10.4. *We have the identity*

$$H^{1,\infty}(\mathbb{R}^n) = \text{Lip}(\mathbb{R}^n). \tag{10.22}$$

Proof. First, suppose $u \in \text{Lip}(\mathbb{R}^n)$. Thus

$$h^{-1}\big[u(x+he_j) - u(x)\big] \text{ is bounded in } L^\infty(\mathbb{R}^n). \tag{10.23}$$

Hence, by Proposition 9.4, there is a sequence $h_\nu \to 0$ and $f_j \in L^\infty(\mathbb{R}^n)$ such that

$$h_\nu^{-1}\big[u(x+h_\nu e_j) - u(x)\big] \to f_j \text{ weak}^* \text{ in } L^\infty(\mathbb{R}^n). \tag{10.24}$$

In particular, for all $\varphi \in C_0^\infty(\mathbb{R}^n)$,

$$h_\nu^{-1}\int \varphi(x)\big[u(x+h_\nu e_j) - u(x)\big]\, dx \longrightarrow \int \varphi(x) f_j(x)\, dx. \tag{10.25}$$

But the left side of (10.25) is equal to

$$h_\nu^{-1}\int \big[\varphi(x-h_\nu e_j) - \varphi(x)\big]u(x)\, dx \longrightarrow -\int \frac{\partial\varphi}{\partial x_j}u(x)\, dx. \tag{10.26}$$

This shows that $\partial_j u = f_j$. Hence $\text{Lip}(\mathbb{R}^n) \subset H^{1,\infty}(\mathbb{R}^n)$.

Next, suppose $u \in H^{1,\infty}(\mathbb{R}^n)$. Let $\varphi_j(x) = j^n\varphi(jx)$ be an approximate identity as in (10.6), with $\varphi \in C_0^\infty(\mathbb{R}^n)$. We do not get $\varphi_j * u \to u$ in $H^{1,\infty}$-norm, but we do have $u_j = \varphi_j * u$ bounded in $H^{1,\infty}(\mathbb{R}^n)$; in fact, each u_j is C^∞, and we have

$$\|u_j\|_{L^\infty} \le K_1, \quad \|\nabla u_j\|_{L^\infty} \le K_2. \tag{10.27}$$

Also $u_j \to u$ locally uniformly. The second estimate in (10.27) implies

$$|u_j(x) - u_j(y)| \le K_2|x-y|, \tag{10.28}$$

since $u_j(x) - u_j(y) = \int_0^1 (x-y)\cdot \nabla u(tx + (1-t)y)\, dt$. Thus in the limit $j \to \infty$, we get also $|u(x) - u(y)| \le K_2|x-y|$. This completes the proof.

We next show that, when $p \in [1,n)$, $H^{1,p}(\mathbb{R}^n)$ is contained in $L^q(\mathbb{R}^n)$ for some $q > p$. One technical tool which is useful for our estimates is the following generalized Hölder inequality.

Lemma 10.5. *If $p_j \in [1,\infty]$, $\sum p_j^{-1} = 1$, then*

$$\int_M |u_1 \cdots u_m|\, dx \le \|u_1\|_{L^{p_1}(M)} \cdots \|u_m\|_{L^{p_m}(M)}. \tag{10.29}$$

The proof follows by induction from the case $m = 2$, which is the usual Hölder inequality.

Proposition 10.6. *For $p \in [1,n)$,*

$$H^{1,p}(\mathbb{R}^n) \subset L^{np/(n-p)}(\mathbb{R}^n). \tag{10.30}$$

In fact, there is an estimate

$$\|u\|_{L^{np/(n-p)}} \le C\|\nabla u\|_{L^p} \tag{10.31}$$

for $u \in H^{1,p}(\mathbb{R}^n)$, with $C = C(p,n)$.

Proof. It suffices to establish (10.31) for $u \in C_0^\infty(\mathbb{R}^n)$. Clearly

$$|u(x)| \le \int_{-\infty}^{\infty} |\partial_j u|\, dy_j, \tag{10.32}$$

where the integrand, written more fully, is $|\partial_j u(x_1, \dots, y_j, \dots, x_n)|$. (Note that the right side of (10.32) is independent of x_j.) Hence

$$|u(x)|^{n/(n-1)} \le \prod_{j=1}^{n} \Bigl(\int_{-\infty}^{\infty} |\partial_j u|\, dy_j\Bigr)^{1/(n-1)}. \tag{10.33}$$

We can integrate (10.33) successively over each variable x_j, $j = 1, \dots, n$, and apply the generalized Hölder inequality (10.29) with $m = p_1 = \cdots = p_m = n-1$ after each integration. We get

$$\|u\|_{L^{n/(n-1)}} \le \Bigl\{\prod_{j=1}^{n} \int_{\mathbb{R}^n} |\partial_j u|\, dx\Bigr\}^{1/n} \le C\|\nabla u\|_{L^1}. \tag{10.34}$$

This establishes (10.31) in the case $p = 1$. We can apply this to $v = |u|^\gamma$, $\gamma > 1$, obtaining

$$\bigl\||u|^\gamma\bigr\|_{L^{n/(n-1)}} \le C\bigl\||u|^{\gamma-1}|\nabla u|\bigr\|_{L^1} \le C\bigl\||u|^{\gamma-1}\bigr\|_{L^{p'}}\bigl\|\nabla u\bigr\|_{L^p}. \tag{10.35}$$

For $p < n$, pick $\gamma = (n-1)p/(n-p)$. Then (10.35) gives (10.31) and the proposition is proved.

There are also Sobolev spaces $H^{k,p}(\mathbb{R}^n)$, for each $k \in \mathbb{Z}^+$. By definition $u \in H^{k,p}(\mathbb{R}^n)$ provided

$$\partial^\alpha u = f_\alpha \in L^p(\mathbb{R}^n), \quad \forall\, |\alpha| \le k, \tag{10.36}$$

where $\partial^\alpha = \partial_1^{\alpha_1} \cdots \partial_n^{\alpha_n}$, $|\alpha| = \alpha_1 + \cdots + \alpha_n$, and, as in (10.2), (10.36) means

$$(-1)^{|\alpha|} \int \frac{\partial^\alpha \varphi}{\partial x^\alpha} u\, dx = \int \varphi f_\alpha\, dx, \quad \forall\, \varphi \in C_0^\infty(\mathbb{R}^n). \tag{10.37}$$

Given $u \in H^{k,p}(\mathbb{R}^n)$, we can apply Proposition 10.6 to estimate the $L^{np/(n-p)}$-norm of $\partial^{k-1} u$ in terms of $\|\partial^k u\|_{L^p}$, where we use the notation

$$\partial^k u = \{\partial^\alpha u : |\alpha| = k\}, \quad \|\partial^k u\|_{L^p} = \sum_{|\alpha|=k} \|\partial^\alpha u\|_{L^p}, \tag{10.38}$$

and proceed inductively, obtaining the following corollary.

Proposition 10.7. *For $kp < n$,*

$$H^{k,p}(\mathbb{R}^n) \subset L^{np/(n-kp)}(\mathbb{R}^n). \tag{10.39}$$

The next result provides a generalization of Proposition 10.2.

Proposition 10.8. *We have*

$$H^{k,p}(\mathbb{R}^n) \subset C(\mathbb{R}^n) \cap L^\infty(\mathbb{R}^n) \ \ \textit{for} \ \ kp > n. \tag{10.40}$$

Proof. If $p > n$, we can apply Proposition 10.2. If $p = n$ and $k \ge 2$, since it suffices to obtain an L^∞ bound for $u \in H^{k,p}(\mathbb{R}^n)$ with support in the unit ball, just use $u \in H^{2,n-\varepsilon}(\mathbb{R}^n)$ and proceed to the next step of the argument.

If $p \in [1, n)$, it follows from Proposition 10.6 that

$$H^{k,p}(\mathbb{R}^n) \subset H^{k-1,p_1}(\mathbb{R}^n), \quad p_1 = \frac{np}{n-p}. \tag{10.41}$$

Thus the hypothesis $kp > n$ implies $(k-1)p_1 > kp > n$. Iterating this argument, we obtain $H^{k,p}(\mathbb{R}^n) \subset H^{\ell,q}(\mathbb{R}^n)$, for some $\ell \ge 1$ and $q > n$, and again we can apply Proposition 10.2.

Exercises

1. Write down the details for the proof of the identities in (10.4).

2. Verify the estimates in (10.14).
Hint. Write the first integral in (10.13) as $1/A$ times

$$\int_\Sigma \int_0^1 v_+(z)\cdot\nabla u(tz,\tfrac{1}{2}t)\,dt\,dz + \int_\Sigma \int_0^1 v_-(z)\cdot\nabla u(tz,1-\tfrac{1}{2}t)\,dt\,dz,$$

where $v_\pm(z) = (\pm z, 1/2)$. Then calculate an appropriate Jacobian determinant to obtain the second integral in (10.13).

3. Suppose $1 < p < \infty$. If $\tau_y f(x) = f(x-y)$, show that f belongs to $H^{1,p}(\mathbb{R}^n)$ if and only if $\tau_y f$ is a Lipschitz function of y with values in $L^p(\mathbb{R}^n)$, i.e.,

$$\|\tau_y f - \tau_z f\|_{L^p} \le C|y-z|. \tag{10.42}$$

Hint. Consider the proof of Proposition 10.4.
What happens in the case $p = 1$?

4. Show that $H^{n,1}(\mathbb{R}^n) \subset C(\mathbb{R}^n)\cap L^\infty(\mathbb{R}^n)$.
Hint. $u(x) = \int_{-\infty}^0 \cdots \int_{-\infty}^0 \partial_1\cdots\partial_n u(x+y)\,dy_1\cdots dy_n$.

5. If $p_j \in [1,\infty]$ and $u_j \in L^{p_j}$, show that $u_1u_2 \in L^r$ provided $1/r = 1/p_1 + 1/p_2$ and

$$\|u_1u_2\|_{L^r} \le \|u_1\|_{L^{p_1}}\|u_2\|_{L^{p_2}}. \tag{10.43}$$

Show that this implies (10.29).

6. Given $u \in L^2(\mathbb{R}^n)$, show that

$$u \in H^{k,2}(\mathbb{R}^n) \Longleftrightarrow \big(1+|\xi|\big)^k \hat{u} \in L^2(\mathbb{R}^n). \tag{10.44}$$

7. Let $f \in L^1(\mathbb{R})$, and set $g(x) = \int_{-\infty}^x f(y)\,dy$. Continuity of g follows from the Dominated Convergence Theorem. Show that

$$\partial_1 g = f. \tag{10.45}$$

Hint. Given $\varphi \in C_0^\infty(\mathbb{R})$, start with

$$\int \frac{d\varphi}{dx} g(x)\,dx = \int \int_{-\infty}^{x} \varphi'(x) f(y)\,dy\,dx, \tag{10.46}$$

and use Fubini's Theorem. Then use $\int_y^\infty \varphi'(x)\,dx = -\varphi(y)$.
Alternative. Write the left side of (10.46) as

$$\lim_{h\to 0} \frac{1}{h} \int \bigl[\varphi(x+h) - \varphi(x)\bigr] g(x)\,dx = -\lim_{h\to 0} \int \int_x^{x+h} f(y)\varphi(x)\,dy\,dx,$$

and use (4.64).

8. If $u \in H^{1,p}(\mathbb{R}^n)$ for some $p \in [1,\infty)$ and $\partial_j u = 0$ on a connected open set $U \subset \mathbb{R}^n$, for $1 \le j \le n$, show that u is (equal a.e. to a) constant on U.
Hint. Approximate u by (10.6), i.e., by $u_\nu = \varphi_\nu * u$, where $\varphi_\nu \in C_0^\infty(\mathbb{R}^n)$ has support in $\{|x| < 1/\nu\}$, $\int \varphi_\nu\,dx = 1$. Show that $\partial_j(\varphi_\nu * u) = 0$ on $U_\nu \subset\subset U$, where $U_\nu \nearrow U$ as $\nu \to \infty$.

   More generally, if $\partial_j u = f_j \in C(U)$, $1 \le j \le n$, show that u is equal a.e. to a function in $C^1(U)$.

9. In case $n = 1$, deduce from Exercises 7 and 8 that, if $u \in L^1_{\mathrm{loc}}(\mathbb{R})$,

$$\partial_1 u = f \in L^1(\mathbb{R}) \Longrightarrow u(x) = c + \int_{-\infty}^{x} f(y)\,dy, \quad \text{a.e. } x \in \mathbb{R}, \tag{10.47}$$

   for some constant c.

10. Let $g \in H^{2,1}(\mathbb{R})$, $I = [a,b]$, and $f = g\big|_I$. Show that the estimate (9.75) concerning the trapezoidal rule holds in this setting.

Chapter 11

Maximal Functions and A.E. Phenomena

We will be concerned with results that various sequences of functions converge almost everywhere. One example is the following.

Theorem 11.1. *Given* $f \in \mathcal{L}^1(\mathbb{R}^n, dx)$*, consider*

$$A_r f(x) = \frac{1}{m(B_r)} \int_{B_r(x)} f(y)\, dy, \quad r > 0. \tag{11.1}$$

Then

$$\lim_{r \to 0} A_r f(x) = f(x) \quad a.e. \tag{11.2}$$

Here and below, $m(S)$ denotes the Lebesgue measure of a set S, and $B_r(x) = \{y \in \mathbb{R}^n : |x - y| \le r\}$. From results in previous sections, we know that $A_r f \to f$ in L^1-norm, as $r \to 0$. Hence, for any fixed $f \in L^1(\mathbb{R}^n)$, there exists a sequence $r_\nu \to 0$ such that $A_{r_\nu} f(x) \to f(x)$ a.e. However, this does not imply (11.2).

A successful systematic approach to deriving both pointwise a.e. convergence results and additional interesting quantitative information was achieved by Hardy and Littlewood. We look at the "Hardy-Littlewood maximal function," defined by

$$M(f)(x) = \sup_{r>0} \frac{1}{m(B_r)} \int_{B_r(x)} |f(y)|\, dy. \tag{11.3}$$

Note that this depends only on the class of f in $L^1(\mathbb{R}^n)$, as does (11.1). The basic estimate on this maximal function is the following.

Theorem 11.2. *There is a constant $C = C(n)$ such that, for any $\lambda > 0$, $f \in L^1(\mathbb{R}^n)$, we have the estimate*

$$m(\{x \in \mathbb{R}^n : M(f)(x) > \lambda\}) \le \frac{C}{\lambda}\|f\|_{L^1}. \tag{11.4}$$

We also note that the estimate

$$m(\{x \in \mathbb{R}^n : |f(x)| > \lambda\}) \le \frac{1}{\lambda}\|f\|_{L^1}, \tag{11.5}$$

known as Tchebychev's inequality, follows by integrating the inequality $|f| \ge \lambda\chi_{S_\lambda}$, where S_λ is the set $\{x : |f(x)| > \lambda\}$.

We show how these estimates allow us to prove Theorem 11.1. Indeed, given $f \in \mathcal{L}^1(\mathbb{R}^n, dx)$ and $\varepsilon > 0$, pick $g \in C_0^\infty(\mathbb{R}^n)$ such that $\|f - g\|_{L^1} < \varepsilon$. Clearly $A_r g(x) \to g(x)$ uniformly as $r \to 0$, so if

$$E_\lambda = \{x \in \mathbb{R}^n : \limsup_{r\to 0} |A_r f(x) - f(x)| > \lambda\}, \tag{11.6}$$

then E_λ is unchanged if f is replaced by $f - g$. Now

$$\sup_{r>0} |A_r(f - g) - (f - g)| \le M(f - g) + |f - g|,$$

so

$$E_\lambda \subset \Big\{x : M(f - g)(x) > \frac{\lambda}{2}\Big\} \cup \Big\{x : |f(x) - g(x)| > \frac{\lambda}{2}\Big\}.$$

If we apply (11.4)–(11.5) with f replaced by $f - g$ and λ replaced by $\lambda/2$, we get

$$m(E_\lambda) \le \frac{C}{\lambda}\|f - g\|_{L^1} < \frac{C\varepsilon}{\lambda}.$$

Since this holds for all $\varepsilon > 0$, we deduce that

$$m(E_\lambda) = 0 \quad \text{for all} \quad \lambda > 0. \tag{11.7}$$

This is precisely equivalent to (11.2).

We now take up the proof of Theorem 11.2. Let

$$F_\lambda = \{x \in \mathbb{R}^n : Mf(x) > \lambda\}. \tag{11.8}$$

We remark that, for any $f \in L^1(\mathbb{R}^n)$ and any $\lambda > 0$, F_λ is *open*. Given $x \in F_\lambda$, pick $r = r_x$ such that $A_r|f|(x) > \lambda$, and let $B_x = B_{r_x}(x)$. Thus $\{B_x : x \in F_\lambda\}$ is a *covering* of F_λ by balls. We will be able to obtain the estimate (11.4) from the following "covering lemma," due to N. Wiener.

Lemma 11.3. *If $\mathcal{C} = \{B_\alpha : \alpha \in \mathfrak{A}\}$ is a collection of open balls in $\mathbb{R}^n$, with union U, and if $m_0 < m(U)$, then there is a finite collection of disjoint balls $B_j \in \mathcal{C}$, $1 \le j \le K$, such that*

$$\sum m(B_j) > 3^{-n} m_0. \tag{11.9}$$

We show how the lemma allows us to prove (11.4). In this case, let $\mathcal{C} = \{\overset{\circ}{B}_x : x \in F_\lambda\}$. Thus, if $m_0 < m(F_\lambda)$, there exist disjoint balls $B_j = \overset{\circ}{B}_{r_j}(x_j)$ such that $m(\cup B_j) > 3^{-n} m_0$. This implies

$$m_0 < 3^n \sum m(B_j) \le \frac{3^n}{\lambda} \sum \int_{B_j} |f(x)|\, dx \le \frac{3^n}{\lambda} \int |f(x)|\, dx, \tag{11.10}$$

for all $m_0 < m(F_\lambda)$, which yields (11.4), with $C = 3^n$.

We now turn to the proof of Lemma 11.3. We can pick a compact $K \subset U$ such that $m(K) > m_0$. Then the covering $\mathcal{C}$ yields a finite covering of K, say $A_1, \dots, A_N$. Let B_1 be the ball A_j of the largest radius. Throw out all A_ℓ which meet B_1, and let B_2 be the remaining ball of largest radius. Continue until $\{A_1, \dots, A_N\}$ is exhausted. One gets disjoint balls $B_1, \dots, B_K$ in $\mathcal{C}$. Now each A_j meets some B_ℓ, having the property that the radius of B_ℓ is $\ge$ the radius of A_j. Thus, if $\widehat{B}_j$ is the ball concentric with B_j, with three times the radius, we have

$$\bigcup_{j=1}^{K} \widehat{B}_j \supset \bigcup_{\ell=1}^{N} A_\ell \supset K.$$

This yields (11.9).

There are more elaborate covering lemmas, due to Vitali and to Besicovitch (amongst others), which arise in measure theory. In appendices to this chapter, we discuss these covering lemmas.

Having proved Theorem 11.1, we next establish a slightly stronger result.

Proposition 11.4. *Given $f \in \mathcal{L}^1(\mathbb{R}^n, dx)$, then*

$$\lim_{r\to 0} \frac{1}{m(B_r)} \int_{B_r(x)} |f(y) - f(x)|\, dy = 0, \quad \textit{for a.e. } x. \tag{11.11}$$

The proof of this goes just as the proof of Theorem 11.1, given the following maximal function estimate. Let

$$\widetilde{M}(f)(x) = \sup_{r>0} \frac{1}{m(B_r)} \int_{B_r(x)} |f(y) - f(x)|\, dy. \tag{11.12}$$

We claim $\widetilde{M}$ satisfies an estimate of the form (11.4). Indeed, we have

$$\widetilde{M}(f)(x) \leq M(f)(x) + |f(x)|, \tag{11.13}$$

so this is an easy consequence.

Clearly, when (11.11) holds, we also have (11.2). Given $f \in \mathcal{L}^1(\mathbb{R}^n, dx)$, a point $x \in \mathbb{R}^n$ for which (11.11) holds is called a *Lebesgue point* of f. Thus the content of Proposition 11.4 is that almost every $x \in \mathbb{R}^n$ is a Lebesgue point for any given $f \in \mathcal{L}^1(\mathbb{R}^n, dx)$.

In analogy with (11.12), define $\widetilde{M}_p(f)(x)$ by

$$\widetilde{M}_p(f)(x)^p = \sup_{r>0} \frac{1}{m(B_r)} \int\limits_{B_r(x)} |f(y) - f(x)|^p \, dy, \tag{11.14}$$

for $p \in [1, \infty)$. Equivalently,

$$\widetilde{M}_p(f)(x) = \sup_{r>0} \|f - f(x)\|_{L^p(B_r(x), dV_r)}, \tag{11.15}$$

where dV_r is $dx/m(B_r)$, normalized to be a probability measure on $B_r(x)$. Using Minkowski's inequality, we obtain

$$\widetilde{M}_p(f)(x) \leq M(|f|^p)(x)^{1/p} + |f(x)|. \tag{11.16}$$

Note that, if $f \in L^p(\mathbb{R}^n)$, the estimate (11.4) applied to $|f|^p$ yields

$$m\big(\{x \in \mathbb{R}^n : M(|f|^p)(x) > \lambda^p\}\big) \leq \frac{C}{\lambda^p} \|f\|_{L^p}^p. \tag{11.17}$$

Now, by (11.16), if $\widetilde{M}_p(f)(x) > \lambda$, then either $M(|f|^p)(x) > (\lambda/2)^p$ or $|f(x)|^p > (\lambda/2)^p$, so we have the maximal function estimate

$$m\big(\{x \in \mathbb{R}^n : \widetilde{M}_p(f)(x) > \lambda\}\big) \leq \frac{C}{\lambda^p} \|f\|_{L^p}^p. \tag{11.18}$$

Thus we have the following extension of Proposition 11.4.

Proposition 11.5. *Given $p \in [1, \infty)$, $f \in \mathcal{L}^p(\mathbb{R}^n, dx)$, then*

$$\lim_{r\to 0} \frac{1}{m(B_r)} \int\limits_{B_r(x)} |f(y) - f(x)|^p \, dy = 0, \quad \textit{for a.e. } x. \tag{11.19}$$

The proof is again like that of Theorem 11.1. One uses the denseness of $C_0^\infty(\mathbb{R}^n)$ in $L^p(\mathbb{R}^n)$, for $p \in [1, \infty)$. When the limit is 0 in (11.19), we call x an L^p-Lebesgue point for f.

We next establish some results on differentiability almost everywhere of certain classes of functions. A function f defined on an open set $\Omega \subset \mathbb{R}^n$, with values in $\mathbb{R}^m$, is differentiable at a point $x \in \Omega$ provided there is a linear transformation $A : \mathbb{R}^n \to \mathbb{R}^m$ such that

$$|f(x+y) - f(x) - Ay| = o(|y|), \text{ as } |y| \to 0, \tag{11.20}$$

i.e., the ratio of the left side of (11.20) to $|y|$ tends to 0 as $y \to 0$. If such A exists, it must be unique; we denote it by $Df(x)$. If $m > 1$, then f is differentiable at x if and only if each component of f is differentiable at x, as is easily verified.

Recall from Chapter 10 the Sobolev spaces $H^{1,p}(\mathbb{R}^n)$, consisting of $f \in L^p(\mathbb{R}^n)$ with weak derivative $\nabla f \in L^p(\mathbb{R}^n)$. We showed that

$$H^{1,p}(\mathbb{R}^n) \subset C(\mathbb{R}^n), \text{ provided } p > n \text{ or } p = n = 1. \tag{11.21}$$

We now establish the following.

Proposition 11.6. *If $f \in H^{1,p}(\mathbb{R}^n)$ and either $n < p < \infty$ or $p = n = 1$, then f is differentiable at almost every $x \in \mathbb{R}^n$. More precisely, if x is an L^p-Lebesgue point for ∇f, then f is differentiable at x and $Df(x) = \nabla f(x)$.*

Proof. Under the hypotheses on f, it follows from (10.20) that

$$\bigl|f(x+y) - f(x)\bigr| \le C|y| \Bigl(\frac{1}{m(B_r)} \int\limits_{B_r(x)} |\nabla f(z)|^p \, dz\Bigr)^{1/p}, \quad r = |y|. \tag{11.22}$$

Now suppose x is an L^p-Lebesgue point of (each component of) $\nabla f \in L^p(\mathbb{R}^n)$. If we replace $f(z)$ by $g(z) = f(z) - \nabla f(x) \cdot z$ in (11.22), we have $\nabla g = \nabla f - \nabla f(x)$, so

$$\begin{aligned} &\bigl|f(x+y) - f(x) - \nabla f(x) \cdot y\bigr| \\ &\le C|y| \Bigl(\frac{1}{m(B_r)} \int\limits_{B_r(x)} |\nabla f(z) - \nabla f(x)|^p \, dz\Bigr)^{1/p}, \end{aligned} \tag{11.23}$$

with $r = |y|$. If x is an L^p-Lebesgue point for ∇f, it follows that the right side of (11.23) is $o(|y|)$ as $|y| \to 0$, so the proposition is proved.

The following immediate consequence is known as Rademacher's Theorem.

Corollary 11.7. *If $\Omega \subset \mathbb{R}^n$ is open and $f : \Omega \to \mathbb{R}^m$ is Lipschitz, then f is differentiable almost everywhere.*

Proof. Given any $\varphi \in C_0^\infty(\Omega)$, $\varphi f \in H^{1,p}(\mathbb{R}^n)$ for all $p \le \infty$ and hence for some $p \in (n, \infty)$.

Corollary 11.7 establishes the result stated after Proposition 7.6. Therefore we have that if $\mathcal{O}$ and Ω are open in $\mathbb{R}^n$ and if $F : \mathcal{O} \to \Omega$ is a Lipschitz map which is bijective (hence a homeomorphism), then, whenever $u \in \mathcal{M}^+(\Omega)$,

$$\int_\Omega u(x)\,dx = \int_{\mathcal{O}} u\big(F(x)\big)\,J(x)\,dx, \quad J(x) = |\det DF(x)|, \tag{11.24}$$

a result which was established in Theorem 7.2 when F is a C^1 diffeomorphism. We will record a statement and separate proof of an important special case of this result.

Proposition 11.8. *Assume $\mathcal{O}$ and Ω are open in $\mathbb{R}^n$ and that $F : \mathcal{O} \to \Omega$ is a Lipschitz homeomorphism with a Lipschitz inverse $G : \Omega \to \mathcal{O}$. Then* (11.24) *holds, for all $u \in \mathcal{M}^+(\Omega)$.*

While this result is already contained in Proposition 7.6, augmented by Corollary 11.7, we include a proof which is simpler, in that it avoids the appeal to topological degree theory made in the proof of Proposition 7.6.

In fact, the simple part of the argument used in Proposition 7.6, via (7.45), shows that

$$\int_{\mathcal{O}} v(x)\,dx \ge \int_\Omega v\big(G(y)\big)\,K(y)\,dy, \quad K(y) = |\det DG(y)|, \tag{11.25}$$

as in (7.16). Now the chain rule implies that $J(x)K\big(F(x)\big) = 1$ a.e., and we obtain

$$\int_\Omega u(x)\,dx \le \int_{\mathcal{O}} u\big(F(x)\big)\,J(x)\,dx, \quad \forall\, u \in \mathcal{M}^+(\Omega). \tag{11.26}$$

As in the end of the proof of Theorem 7.2, we can now interchange the roles of Ω and $\mathcal{O}$ to get the reverse inequality, thus obtaining (11.24).

We next show that a Lipschitz function can be approximated in a rather strong sense by a C^1 function. This will be useful for results on Hausdorff measure of Lipschitz surfaces in Chapter 12.

Theorem 11.9. *Let $f \in \mathrm{Lip}(\mathbb{R}^n)$. Given $\varepsilon > 0$ and a closed ball $B \subset \mathbb{R}^n$, there exists a compact $K \subset B$ such that $m(B \setminus K) < \varepsilon$ and a $g \in C^1(\mathbb{R}^n)$ such that*

$$g = f \text{ on } K \quad \text{and} \quad \|g\|_{C^1} \le C(n)\|f\|_{\mathrm{Lip}}. \tag{11.27}$$

Proof. By Corollary 11.7, f is differentiable almost everywhere; we have $Df \in L^\infty(\mathbb{R}^n)$. We can apply Lusin's Theorem, given in Exercise 14 of Chapter 5, to see that there is a compact $K_1 \subset B$ such that $m(B \setminus K_1) < \varepsilon/2$ and $Df\big|_{K_1}$ is continuous. Now, for $x \in K_1$, $\delta > 0$, set

$$R(x, y) = f(y) - f(x) - Df(x)(y - x) \tag{11.28}$$

and

$$\eta_\delta(x) = \sup \{|R(x, y)|/|x - y| : 0 < |x - y| < \delta\}. \tag{11.29}$$

We know that $\eta_\delta(x) \to 0$ as $\delta \to 0$, for each $x \in K_1$. By Egoroff's Theorem, there exists compact $K \subset K_1$ such that $m(K_1 \setminus K) < \varepsilon/2$ and $\eta_\delta \to 0$ uniformly on K. Thus $f = f\big|_K$, $f^\# = Df\big|_K$ is a pair satisfying the hypotheses of Whitney's Extension Theorem; see Appendix C. That theorem produces a function $g \in C^1(\mathbb{R}^n)$ satisfying (11.27).

We return to a look at averaging operators of the form (11.1), but this time applied to finite (signed) measures in $\mathbb{R}^n$:

$$A_r\mu(x) = \frac{1}{m(B_r)} \int_{B_r(x)} d\mu = \frac{\mu(B_r(x))}{m(B_r(x))}. \tag{11.30}$$

Using the Lebesgue-Radon-Nikodym Theorem, we can write $\mu = fm + \nu$, with $f \in \mathcal{L}^1(\mathbb{R}^n, dx)$ and $\nu \perp m$. Thus we are left with the analysis of

$$A_r\nu(x) = \frac{\nu(B_r(x))}{m(B_r(x))}, \quad \nu \perp m. \tag{11.31}$$

We will establish

Proposition 11.10. *If ν is a finite measure on $\mathbb{R}^n$ which is purely singular with respect to Lebesgue measure, i.e., $\nu \perp m$, then*

$$\lim_{r \to 0} A_r\nu(x) = 0, \quad \textit{for } m\textit{-a.e. } x. \tag{11.32}$$

Proof. There is no loss in assuming ν is positive. Write $\mathbb{R}^n = X \cup X^c$, $\nu(X) = m(X^c) = 0$. Given $\lambda > 0$, set

$$F_\lambda = \{x \in X : \limsup_{r \to 0} |A_r\nu(x)| > \lambda\}. \tag{11.33}$$

We want to show that $m(F_\lambda) = 0$. We will use an argument not very different from the proof of Theorem 11.2.

Given $\varepsilon > 0$, there exists an open $U \supset X$ such that $\nu(U) < \varepsilon$. Now, for each $x \in F_\lambda$, there is a ball $B_x \subset U$, centered at x, such that $\nu(B_x) > \lambda m(B_x)$. Thus

$$F_\lambda \subset \bigcup_{x \in F_\lambda} B_x = V \subset U.$$

By Wiener's Covering Lemma, if $m_0 < m(V)$, there exist *disjoint* $B_{x_1}, \dots, B_{x_J}$ such that $\sum m(B_{x_j}) > 3^{-n} m_0$. Thus we have

$$3^{-n} m_0 < \frac{1}{\lambda} \sum \nu(B_{x_j}) \le \frac{1}{\lambda} \nu(V) \le \frac{1}{\lambda} \nu(U) \le \frac{\varepsilon}{\lambda}. \tag{11.34}$$

Hence

$$m(F_\lambda) \le m(V) \le \frac{3^n}{\lambda} \varepsilon, \tag{11.35}$$

for all $\varepsilon > 0$. This shows $m(F_\lambda) = 0$, and the proposition is proved.

The original work of Lebesgue on the one-dimensional case of Theorem 11.1 was done in order to extend the Fundamental Theorem of Calculus beyond the formulation given in Theorems 1.6–1.7 to the two results we now state. The following result extends Theorem 1.6.

Proposition 11.11. *If $f \in L^1(\mathbb{R})$, the function*

$$g(x) = \int_{-\infty}^{x} f(y)\, dy \tag{11.36}$$

is differentiable at almost every $x \in \mathbb{R}$, and

$$Dg(x) = f(x), \quad a.e.\ x \in \mathbb{R}. \tag{11.37}$$

Proof. It follows from (10.45) that the weak derivative of g exists and in fact $\partial_1 g = f$. Then (11.37) is the $n = 1$ case of Proposition 11.6.

There is also an extension of Theorem 1.7, which we state next. In the generalization, we replace the hypothesis that G be of class C^1 by the hypothesis that G be *absolutely continuous*. By definition, this means that, for every $\varepsilon > 0$, there is a $\delta > 0$ with the property that, for any finite collection of disjoint intervals $(a_1, b_1), \dots, (a_N, b_N)$,

$$\sum (b_j - a_j) < \delta \Longrightarrow \sum |G(b_j) - G(a_j)| < \varepsilon. \tag{11.38}$$

Clearly (by Theorem 1.7), if G is C^1, with $|G'(x)| \le K$, or, more generally, if G is Lipschitz, with Lipschitz constant K, then (11.38) holds, with $\delta = \varepsilon / K$.

Proposition 11.12. *Assume that $G : \mathbb{R} \to \mathbb{R}$ is absolutely continuous. Then G is differentiable almost everywhere, and $DG \in L^1(\mathbb{R})$. Furthermore, for $-\infty < a < b < \infty$,*

$$\int_a^b DG(x)\,dx = G(b) - G(a). \tag{11.39}$$

For the proof it is convenient to use material that will be developed in Chapter 13. As will be seen in Exercise 8 of Chapter 13, the hypothesis that G be absolutely continuous implies that $\partial_1 G = f \in L^1(\mathbb{R})$. Again, by Proposition 11.6, we have $DG = f$ a.e. Hence, by (10.47), for some constant c,

$$G(x) = c + \int_{-\infty}^{x} DG(y)\,dy.$$

Since both sides are continuous in x, this a.e. identity is valid everywhere. Then (11.39) is an immediate consequence.

Another part of Exercise 8 in Chapter 13 will be that, if $\partial_1 G = f \in L^1(\mathbb{R})$, then G is absolutely continuous.

11A. The Vitali Covering Lemma

There is a covering lemma that is a bit more elaborate than the Wiener Covering Lemma (Lemma 11.3), known as the Vitali Covering Lemma, which has widespread use in analysis, and we discuss it in this appendix. It is natural to divide the result into two parts, each of which is occasionally called the Vitali Covering Lemma. The first part does not mention measure.

Lemma 11A.1. *Let $\mathcal{C}$ be a collection of closed balls in $\mathbb{R}^n$ (with positive radius) such that $\operatorname{diam} B < C_0 < \infty$ for all $B \in \mathcal{C}$. Then there exists a countable family $\mathcal{F}$ of disjoint balls in $\mathcal{C}$ with the following property:*

(11A.1) *Every ball B in $\mathcal{C}$ meets a ball in $\mathcal{F}$ with at least half the radius of B,*

and hence

$$\bigcup_{B\in\mathcal{F}} \widehat{B} \supset \bigcup_{B\in\mathcal{C}} B, \tag{11A.2}$$

where $\widehat{B}$ is a ball concentric with B with five times the radius.

Proof. Set

$$\mathcal{C}_j = \{B \in \mathcal{C} : 2^{-j}C_0 \le \operatorname{diam} B < 2^{-j+1}C_0\}.$$

Let $\mathcal{F}_1$ be a maximal disjoint collection of balls in $\mathcal{C}_1$. Inductively, let $\mathcal{F}_k$ be a maximal disjoint set of balls in

$$\{B \in \mathcal{C}_k : B \text{ disjoint from all balls in } \mathcal{F}_1, \dots, \mathcal{F}_{k-1}\}.$$

Then set $\mathcal{F} = \bigcup \mathcal{F}_k$. For such $\mathcal{F}$, (11A.1) holds, and then (11A.2) follows.

The second part involves measure.

Lemma 11A.2. *Let A be a subset of $\mathbb{R}^n$, and let $\mathcal{C}$ be a collection of closed balls centered at points of A such that for each $a \in A$, $\varepsilon > 0$, there is a ball $B_r(a) \in \mathcal{C}$ of radius $r < \varepsilon$. Then there exists a family $\mathcal{F}$ of disjoint balls in $\mathcal{C}$ such that*

$$m\Big(A \setminus \bigcup_{B \in \mathcal{F}} B\Big) = 0. \tag{11A.3}$$

Proof. Without loss of generality, we can assume all balls in $\mathcal{C}$ have radius < 1. Apply Lemma 11A.1 and consider the resulting family $\mathcal{F}$. We claim (11A.3) holds for this family. It suffices to show that (11A.3) holds with A replaced by $A_K = A \cap B_K(0)$, for each $K < \infty$. In fact, let $\mathcal{F}_K = \{B \in \mathcal{F} : B \subset B_{K+2}(0)\}$. We will show that

$$m\Big(A_K \setminus \bigcup_{B \in \mathcal{F}_K} B\Big) = 0. \tag{11A.4}$$

Write $\mathcal{F}_K = \{B_j : j \in \mathbb{Z}^+\}$. Clearly $\sum_j m(B_j) < \infty$. Note that for any ball B, with 5-fold dilation $\widehat{B}$,

$$m(\widehat{B}) \le Cm(B), \tag{11A.5}$$

with $C = 5^n$, so $\sum_j m(\widehat{B}_j) < \infty$, and hence

$$\sum_{j>N} m(\widehat{B}_j) \to 0, \quad \text{as } N \to \infty. \tag{11A.6}$$

Hence it suffices to show that

$$A_K \setminus \bigcup_{j=1}^{N} B_j \subset \bigcup_{j>N} \widehat{B}_j. \tag{11A.7}$$

To see this, pick $a \in A_K \setminus \bigcup_{j=1}^N B_j$. Since the balls B_j are closed, we can find some small ball $B_r(a) \in \mathcal{C}$ that does not meet any of the balls B_j for $j \le N$. Now, by Lemma 11A.1, $B_r(a)$ meets some $B \in \mathcal{F}$ with radius at least $r/2$. Furthermore, such B must belong to $\mathcal{F}_K$, so $B = B_j$, for some $j > N$, and hence $B_r(a) \subset \widehat{B}_j$. Thus (11A.7) holds and Lemma 11A.2 is proven.

We now use Lemma 11A.2 to produce a proof of Theorem 11.1, slightly different from the one given at the beginning of Chapter 11. We start the same way: given $f \in \mathcal{L}^1(\mathbb{R}^n, dx)$ and $\varepsilon > 0$, pick $g \in C_0^\infty(\mathbb{R}^n)$ such that $\|f - g\|_{L^1} < \varepsilon$, and note that

$$E_\lambda = \{x \in \mathbb{R}^n : \limsup_{r\to 0} |A_r f(x) - f(x)| > \lambda\} \tag{11A.8}$$

is unchanged if f is replaced by $f - g$. We also have

$$\limsup_{r\to 0} |A_r(f-g) - (f-g)| \le M_0(f-g) + |f-g|, \tag{11A.9}$$

where we set

$$M_0 f(x) = \limsup_{r\to 0} \frac{1}{m(B_r)} \int_{B_r(x)} |f(y)|\, dy. \tag{11A.10}$$

We claim that for all $f \in L^1(\mathbb{R}^n)$,

$$m\Big(\{x \in \mathbb{R}^n : M_0 f(x) > \lambda\}\Big) \le \frac{1}{\lambda}\|f\|_{L^1}. \tag{11A.11}$$

Assuming this for the moment, we have from (11A.8)–(11A.9) that

$$E_\lambda \subset \Big\{x : M_0(f-g)(x) > \frac{\lambda}{2}\Big\} \cup \Big\{x : |f(x) - g(x)| > \frac{\lambda}{2}\Big\}, \tag{11A.12}$$

so that by (11A.11) and Tchebychev's inequality (11.5),

$$m(E_\lambda) \le \frac{4}{\lambda}\|f - g\|_{L^1} < \frac{4\varepsilon}{\lambda}, \tag{11A.13}$$

for all $\varepsilon > 0$; hence $m(E_\lambda) = 0$ for all $\lambda > 0$ and we have (11.2). This gives a second proof of Theorem 11.1, modulo the proof of (11A.11).

To prove (11A.11), we let $A = \{x \in \mathbb{R}^n : M_0 f(x) > \lambda\}$, and we let $\mathcal{C}$ consist of all balls of radius < 1 such that $\int_B |f(x)|\, dx > \lambda m(B)$. If $A \ne \emptyset$, the hypotheses of Lemma 11A.2 are satisfied, so we can take a family of disjoint balls $\mathcal{F} = \{B_j : j \in \mathbb{Z}^+\} \subset \mathcal{C}$ so that (11A.3) holds. We hence have

$$m(A) \le \sum_j m(B_j) < \frac{1}{\lambda} \sum_j \int_{B_j} |f(x)|\, dx \le \frac{1}{\lambda} \|f\|_{L^1}, \tag{11A.14}$$

so (11A.11) is proven.

REMARK 1. We do not elevate the estimate (11A.11) to the status of a theorem, parallel to Theorem 11.2, for the simple reason that once the dust clears and Theorem 11.1 is established, we see that $M_0 f(x) = |f(x)|$ almost everywhere, so (11A.11) carries no information beyond Tchebychev's inequality (11.5).

REMARK 2. The Vitali Covering Lemma can be extended in scope considerably. Indeed, Lemma 11A.1 holds when $\mathbb{R}^n$ is replaced by a general metric space. As for Lemma 11A.2, Lebesgue measure on $\mathbb{R}^n$ can be replaced by any locally finite Borel measure on $\mathbb{R}^n$ for which condition (11A.5), called the "doubling condition," holds, and again the setting can be extended to more general metric spaces. For details and further results and references, see the monograph [**Hei**] by J. Heinonen. We also mention the work of [**Che**], extending the scope of Rademacher's Theorem.

11B. The Besicovitch Covering Lemma

The Besicovitch Covering Lemma is a bit more elaborate than the Vitali Covering Lemma, but it permits a generalization of Theorem 11.1 in which Lebesgue measure is replaced by a general locally finite Borel measure on $\mathbb{R}^n$. As in Appendix 11A, the covering lemma comes in two parts, the first part not mentioning measure.

Lemma 11B.1. *There is an integer $K(n)$ with the following property. If $\mathcal{C}$ is a collection of closed balls in $\mathbb{R}^n$ with radii in $(0, R]$, for some $R < \infty$, and if A is the set of centers of balls in $\mathcal{C}$, then there exist subcollections $\mathcal{G}_k \subset \mathcal{C}$, $1 \le k \le K(n)$, such that each $\mathcal{G}_k$ is a countable collection of disjoint balls and*

$$A \subset \bigcup_{k=1}^{K(n)} \bigcup_{B \in \mathcal{G}_k} B. \tag{11B.1}$$

We postpone the proof of Lemma 11B.1. Now we show how it leads to its measure-theoretic counterpart.

Lemma 11B.2. *Let μ be a Borel measure on $\mathbb{R}^n$. Assume $A \subset \mathbb{R}^n$, $\mu^*(A) < \infty$, and $\mathcal{C}$ is a collection of closed balls (of positive radius), centered at points of A, such that*

$$\inf \{r : B_r(a) \in \mathcal{C}\} = 0, \quad \forall\, a \in A. \tag{11B.2}$$

Then there is a countable collection $\mathcal{F} \subset \mathcal{C}$, consisting of disjoint balls, such that

$$\mu^*\Bigl(A \setminus \bigcup_{B \in \mathcal{F}} B\Bigr) = 0. \tag{11B.3}$$

Proof. Without loss of generality, we can assume each ball in $\mathcal{C}$ has radius ≤ 1. Then Lemma 11B.1 applies and we have collections $\mathcal{G}_1, \dots, \mathcal{G}_{K(n)}$, each consisting of disjoint balls, such that the union of all these balls covers A. Hence there is one collection, say $\mathcal{G}_k$, and disjoint balls $B_1, \dots, B_L \in \mathcal{G}_k$ such that

$$\mu^*\Bigl(A \cap \bigcup_{j=1}^{L} B_j\Bigr) \ge \frac{\mu^*(A)}{K(n)+1}. \tag{11B.4}$$

Since $X = \bigcup_{j=1}^{L} B_j$ is μ^*-measurable, we have by (5.6) that $\mu^*(A) = \mu^*(A \cap X) + \mu^*(A \setminus X)$, so

$$\mu^*\Bigl(A \setminus \bigcup_{j=1}^{L} B_j\Bigr) \le \Bigl(1 - \frac{1}{K(n)+1}\Bigr)\mu^*(A). \tag{11B.5}$$

Now we can set $A_2 = A \setminus \bigcup_{j=1}^{L} B_j$ and take $\mathcal{C}_2$ to be the collection of balls in $\mathcal{C}$, centered at points in A_2, disjoint from $B_1, \dots, B_L$, which, by (11B.2), contains balls with center at each point of A_2. Then apply the argument above to $(A_2, \mathcal{C}_2)$, obtaining disjoint balls $B_{L+1}, \dots, B_{L_2} \in \mathcal{C}_2$ such that

$$\mu^*\Bigl(A \setminus \bigcup_{j=1}^{L_2} B_j\Bigr) \le \Bigl(1 - \frac{1}{K(n)+1}\Bigr)^2 \mu^*(A). \tag{11B.6}$$

Continuing in this fashion, we obtain $\mathcal{F} = \{B_j : j \ge 1\}$ such that (11B.3) holds.

We use Lemma 11B.2 to produce a maximal function estimate more general than (11A.12). To set it up, let μ be a (positive) locally finite Borel measure on $\mathbb{R}^n$. There is a maximal open set $U \subset \mathbb{R}^n$ such that $\mu(U) = 0$, namely the union of all balls $B_r(a)$ with $a \in \mathbb{Q}^n$ and $r \in (0, \infty) \cap \mathbb{Q}$ such that $\mu(B_r(a)) = 0$. Then $\operatorname{supp}\mu = \mathbb{R}^n \setminus U$ is the smallest closed set in $\mathbb{R}^n$ supporting μ. For $x \in \operatorname{supp}\mu$ and $f \in L^1(\mathbb{R}^n, \mu)$, set

$$M_\mu f(x) = \limsup_{r\to 0} \frac{1}{\mu(B_r(x))} \int\limits_{B_r(x)} |f|\, d\mu. \tag{11B.7}$$

This function coincides with (11A.10) when μ is Lebesgue measure. We have the following maximal function estimate.

Lemma 11B.3. *Given $f \in L^1(\mathbb{R}^n, \mu)$, $\lambda > 0$,*

$$\mu^*\big(\{x \in \operatorname{supp}\mu : M_\mu f(x) > \lambda\}\big) \le \frac{1}{\lambda}\|f\|_{L^1(\mathbb{R}^n,\mu)}. \tag{11B.8}$$

Proof. Let $A = \{x \in \operatorname{supp}\mu : M_\mu f(x) > \lambda\}$, and let $\mathcal{C}$ consist of all balls B of radius < 1, centered at some point of A, such that $\int_B |f|\, d\mu > \lambda\mu(B)$. If $A \neq \emptyset$, the hypotheses of Lemma 11B.2 hold, so we can take a family of disjoint balls $\mathcal{F} = \{B_j : j \in \mathbb{N}\} \subset \mathcal{C}$ such that (11B.3) holds. Hence

$$\mu^*(A) \le \sum_j \mu(B_j) < \frac{1}{\lambda}\sum_j \int\limits_{B_j} |f|\, d\mu \le \frac{1}{\lambda}\|f\|_{L^1(\mathbb{R}^n,\mu)}, \tag{11B.9}$$

proving (11B.8).

We are now ready for the following extension of Theorem 11.1, known as the Lebesgue-Besicovitch Differentiation Theorem.

Theorem 11B.4. *Let μ be a locally finite Borel measure on $\mathbb{R}^n$. Given $x \in \operatorname{supp}\mu$, $f \in L^1(\mathbb{R}^n, \mu)$, consider*

$$A_r f(x) = \frac{1}{\mu(B_r(x))} \int\limits_{B_r(x)} f\, d\mu, \quad r > 0. \tag{11B.10}$$

Then

$$\lim_{r\to 0} A_r f(x) = f(x), \quad \mu\text{-a.e.} \tag{11B.11}$$

Proof. Without loss of generality, we can assume $K = \operatorname{supp}\mu$ is compact. By Proposition 4.5, for each $\varepsilon > 0$ there exists $g \in C(K)$ such that

$$\|f - g\|_{L^1(\mathbb{R}^n\mu)} < \varepsilon.$$

Clearly we have $A_r g(x) \to g(x)$ for each $x \in K$, so for each $\lambda > 0$

$$E_\lambda = \{x \in \operatorname{supp} \mu : \limsup_{r\to 0} |A_r f(x) - f(x)| > \lambda\} \tag{11B.12}$$

is unchanged if f is replaced by $f - g$. We also have

$$\limsup_{r\to 0} |A_r(f-g) - (f-g)| \le M_\mu(f-g) + |f-g|. \tag{11B.13}$$

Hence, arguing as in (11A.11)–(11A.13), we have

$$\mu^*(E_\lambda) \le \frac{4\varepsilon}{\lambda}, \tag{11B.14}$$

for all $\varepsilon > 0$; hence $\mu^*(E_\lambda) = 0$, giving (11B.11).

We turn to the proof of Lemma 11B.1. We first produce the quantity $K(n)$.

Lemma 11B.5. *There exists $K_0(n)$ such that if B and $B_1, \dots, B_{K_0(n)}$ are closed unit balls in $\mathbb{R}^n$ and each $B \cap B_j \neq \emptyset$, then some B_j contains the center of B_k (for some $k \neq j$). In fact, one can take $K_0(n) = 4^n$.*

Proof. If there are unit balls $B, B_1, \dots, B_K$ such that each $B \cap B_j \neq \emptyset$ and no B_j contains the center of another B_k, then $(1/2)B_j$ are disjoint balls of radius $1/2$ contained in $2B$ (where rB denotes the ball of radius r concentric with B), so $\mathrm{vol}(2B) > K\ \mathrm{vol}((1/2)B)$, or $K < 4^n$.

We will use the following variant, whose proof can safely be left to the reader.

Lemma 11B.6. *There exists $K_1(n)$ such that if $r \in (0,\infty)$, B is a closed ball of radius r, and $B_1, \dots, B_{K_1(n)}$ are closed balls of radius $\ge (2/3)r$, and each $B \cap B_j \neq \emptyset$, then some B_j contains the center of B_k (for some $k \neq j$).*

We will prove Lemma 11B.1 with

$$K(n) = K_1(n) + 1. \tag{11B.15}$$

To start things off, let $R_0 = \sup\{\text{radius}\, B : B \in \mathcal{C}\}$, pick $B \in \mathcal{C}$ such that radius $B \ge (9/10)R_0$, and put B in $\mathcal{G}_1$. Throw away all other balls $B' \in \mathcal{C}$ whose centers are contained in B. We proceed from here by a process of transcendental induction, making use of Zorn's Lemma. We set this up.

We denote by $\mathfrak{K}$ the collection of partitions of $\mathcal{C}$ into $\{\mathcal{G}_1, \dots, \mathcal{G}_{K(n)}, \mathcal{T}, \mathcal{R}\}$ (some pieces of which might be empty) such that the following holds.

(11B.16)

(1) Each $\mathcal{G}_k$ consists of disjoint balls.

(2) If $B \in \mathcal{G}_k$ and $(9/10)B$ contains the center of $B' \neq B$, then $B' \in \mathcal{T}$.

(3) If $B \in \mathcal{G}_k$ and $B' \in \mathcal{R}$, then radius $B \geq (9/10)$ radius B'.

(4) If $B \in \mathcal{T}$, then some $B' \in \mathcal{G}_1 \cup \cdots \cup \mathcal{G}_{K(n)}$ contains the center of B.

In the first step given above, we have $\mathcal{G}_1 = \{B\}$, $\mathcal{G}_k = \emptyset$ for $2 \leq k \leq K(n)$, and $\mathcal{T}$ consists of the balls thrown away and $\mathcal{R}$ the remainder. If we show that there exists $\{\mathcal{G}_1, \dots, \mathcal{G}_{K(n)}, \mathcal{T}, \mathcal{R}\}$ in $\mathfrak{K}$ with $\mathcal{R} = \emptyset$, then Lemma 11B.1 is proven.

To accomplish this, we define a partial ordering on $\mathfrak{K}$ as follows. We say

(11B.17)
$$\begin{aligned} &\{\mathcal{G}_1, \dots, \mathcal{G}_{K(n)}, \mathcal{T}, \mathcal{R}\} \prec \{\mathcal{G}'_1, \dots, \mathcal{G}'_{K(n)}, \mathcal{T}', \mathcal{R}'\} \\ &\Longleftrightarrow \mathcal{G}_k \subset \mathcal{G}'_k,\ \forall k, \quad \mathcal{T} \subset \mathcal{T}', \quad \mathcal{R} \supset \mathcal{R}'. \end{aligned}$$

According to Zorn's Lemma, if each totally ordered family in $\mathfrak{K}$ has an upper bound, then $\mathfrak{K}$ has a maximal element. But if $\{\mathcal{G}_1^\alpha, \dots, \mathcal{G}_{K(n)}^\alpha, \mathcal{T}^\alpha, \mathcal{R}^\alpha\}$ is totally ordered, then

(11B.18)
$$\mathcal{G}_k = \bigcup_\alpha \mathcal{G}_k^\alpha, \quad \mathcal{T} = \bigcup_\alpha \mathcal{T}^\alpha, \quad \mathcal{R} = \bigcap_\alpha \mathcal{R}^\alpha$$

defines an upper bound. Hence $\mathfrak{K}$ has a maximal element, say

$$\{\mathcal{G}_1, \dots, \mathcal{G}_{K(n)}, \mathcal{T}, \mathcal{R}\}.$$

It remains to show that for any such maximal element, $\mathcal{R} = \emptyset$.

So let $\{\mathcal{G}_1, \dots, \mathcal{G}_{K(n)}, \mathcal{T}, \mathcal{R}\}$ be an element of $\mathfrak{K}$ and suppose $\mathcal{R} \neq \emptyset$. Let $R_1 = \sup\{\text{radius } B : B \in \mathcal{R}\}$ and pick $B \in \mathcal{R}$ of radius $\geq (9/10)R_1$. If B were not disjoint from all the balls in at least one family $\mathcal{G}_k$, we would have a contradiction to Lemma 11B.6. Let k_1 be the smallest $k \in \{1, \dots, K(n)\}$ with this disjointness property. Set $\mathcal{G}_{k_1}^\# = \mathcal{G}_{k_1} \cup \{B\}$, $\mathcal{G}_k^\# = \mathcal{G}_k$ for other k, form $\mathcal{T}^\#$ by throwing into $\mathcal{T}$ each ball from $\mathcal{R}$ whose center is contained in B (other than B itself), and let $\mathcal{R}^\#$ denote the remainder. Then $\{\mathcal{G}_1^\#, \dots, \mathcal{G}_{K(n)}^\#, \mathcal{T}^\#, \mathcal{R}^\#\}$ satisfies the four conditions in (11B.16) and

(11B.19)
$$\{\mathcal{G}_1, \dots, \mathcal{G}_{K(n)}, \mathcal{T}, \mathcal{R}\} \prec \{\mathcal{G}_1^\#, \dots, \mathcal{G}_{K(n)}^\#, \mathcal{T}^\#, \mathcal{R}^\#\}.$$

The sought-for property of any maximal element of $\mathfrak{K}$ is hence established, and Lemma 11B.1 is proven.

REMARK. See [**Dug**] for a discussion of Zorn's Lemma. In [**EG**] there is a proof of Lemma 11B.1 that does not use Zorn's Lemma.

Exercises

1. Let μ be a finite signed measure on $\mathbb{R}$. Set

$$f(x) = \int_{-\infty}^{x} d\mu = \int \chi_{(-\infty,x]} \, d\mu,$$

where the third quantity defines the first two.
(a) Show that f is bounded and right-continuous, i.e.,

$$\lim_{h\searrow 0} f(x+h) = f(x), \quad \forall\, x \in \mathbb{R}.$$

(b) Let the Lebesgue-Radon-Nikodym decomposition of μ be $\mu = gm + \nu$, where m is Lebesgue measure, $g \in L^1(\mathbb{R})$, and $\nu \perp m$. Show that f is differentiable at almost all $x \in \mathbb{R}$ and

$$f'(x) = g(x), \quad m\text{-a.e. } x.$$

(c) Show that $\partial_1 f = \mu$ in the weak sense, i.e.,

$$\int v \, d\mu = -\int \frac{dv}{dx} f \, dx, \quad \forall\, v \in C_0^\infty(\mathbb{R}).$$

Compare Exercise 7 in Chapter 10.

2. Give an example of a sequence $f_j \in L^1(\mathbb{R}^n)$ such that $f_j \to f$ in L^1-norm but $f_j(x)$ does *not* converge a.e. to $f(x)$.

3. Use the Vitali Covering Lemma, Lemma 11A.1, in place of Wiener's Covering Lemma, Lemma 11.3, to prove the Hardy-Littlewood estimate, Theorem 11.2.

4. Let χ_{B_r} denote the characteristic function of $B_r = \{x \in \mathbb{R}^n : |x| \le r\}$. Set

$$\psi_r(x) = \frac{1}{m(B_r)} \chi_{B_r}(x),$$

pick $g \in L^1(\mathbb{R}^+)$, $g \ge 0$, and set

$$\varphi(x) = \int_0^\infty \psi_s(x) g(s) \, ds, \quad \varphi_r(x) = r^{-n} \varphi\Big(\frac{x}{r}\Big).$$

Define the maximal function

$$\mathcal{M}^\varphi f(x) = \sup_{r>0} \int \varphi_r(x-y) |f(y)| \, dy.$$

Show that

$$\mathcal{M}^{\varphi} f(x) \leq \|g\|_{L^1} Mf(x),$$

where $Mf(x)$ is given by (11.3).

5. In the setting of Exercise 4, show that

$$f \in \mathcal{L}^1(\mathbb{R}^n, dx) \Longrightarrow \lim_{r\to 0} \varphi_r * f(x) = Af(x), \quad \text{a.e.},$$

where $A = \int_0^\infty g(s)\, ds$.

6. Let $\Phi : (0,\infty) \to \mathbb{R}^+$ be a monotone decreasing C^1 function with the property that

$$-s^n \Phi'(s) \in L^1(\mathbb{R}^+).$$

Show that Exercises 4–5 apply to

$$\varphi(x) = \Phi(|x|), \quad x \in \mathbb{R}^n.$$

7. Let $f \in L^1(\mathbb{R}^n)$ and let $u(t,x)$ be the solution to the heat equation, given by (9.58)–(9.60). Show that

$$\lim_{t\to 0} u(t,x) = f(x), \quad \text{a.e.}$$

8. Let ν_1 and ν_2 be locally finite Borel measures on $\mathbb{R}^n$, and assume $\nu_1 \perp \nu_2$. Show that

$$\lim_{r\to 0} \frac{\nu_1(B_r(x))}{\nu_2(B_r(x))} = 0, \quad \text{for } \nu_2\text{-a.e. } x.$$

Hint. Set $\mu = \nu_1 + \nu_2$, $\nu_j = f_j \mu$, and apply Theorem 11B.4 to $f = f_1$. Note that this result generalizes Proposition 11.10.

Chapter 12

Hausdorff's r-Dimensional Measures

Hausdorff r-dimensional measure $\mathcal{H}^r$ is defined on any separable metric space (X, d) and for any real $r \geq 0$. To define $\mathcal{H}^r$, let us first set for any $\delta > 0$, $S \subset X$,

$$h^*_{r,\delta}(S) = \inf \Big\{ \sum_{j\geq 1} (\text{diam } B_j)^r : S \subset \bigcup_{j\geq 1} B_j, \ \text{diam } B_j \leq \delta \Big\}, \tag{12.1}$$

with the convention that $\inf \emptyset = +\infty$. Here diam $B_j = \sup \{d(x,y) : x, y \in B_j\}$. By Proposition 5.1, each set function $h^*_{r,\delta}$ is an outer measure. As δ decreases, $h^*_{r,\delta}(S)$ increases. Set

$$h^*_r(S) = \lim_{\delta\to 0} h^*_{r,\delta}(S). \tag{12.2}$$

It is an exercise to deduce that h^*_r is an outer measure. The following allows us to use Caratheodory's results. (In fact, for $r \in \mathbb{N}$, this construction is due to Caratheodory.)

Lemma 12.1. *The set function h^*_r is a metric outer measure.*

Proof. If $A, B \subset X$ and $\inf \{d(x,y) : x \in A, y \in B\} = \varepsilon > 0$, then any set of diameter $\leq \delta < \varepsilon$ which intersects $A \cup B$ must in fact intersect only one of these sets, so it follows that

$$\delta < \varepsilon \Longrightarrow h^*_{r,\delta}(A \cup B) = h^*_{r,\delta}(A) + h^*_{r,\delta}(B),$$

which yields $h_r^*(A \cup B) = h_r^*(A) + h_r^*(B)$ in the limit.

By Proposition 5.8, every closed subset of X, hence every Borel subset of X, is h_r^*-measurable. For any h_r^*-measurable set A, we set

$$\mathcal{H}^r(A) = \gamma_r \, h_r^*(A), \quad \gamma_r = \frac{\pi^{r/2} 2^{-r}}{\Gamma(\frac{r}{2}+1)}. \tag{12.3}$$

As for the factor γ_r, note that when $r = n \in \mathbb{Z}^+$, then γ_n is equal to the Lebesgue measure of a ball of diameter 1 in $\mathbb{R}^n$. (See Exercise 6 in the first exercise set of Chapter 7.) This is set up so that $\mathcal{H}^n$ coincides with Lebesgue measure on $\mathbb{R}^n$ (which we denote $\mathcal{L}^n$ in this chapter). This fact is not trivial, but it will be proved below. For use in the next proposition, we mention that this identity is very easy in the case $n = 1$.

One can make a construction parallel to (12.1)–(12.3), in which the countable covering of S in (12.1) is required to consist of *balls* of diameter $\le \delta$. We denote the resulting measure by $\mathcal{H}_B^r$, which is clearly $\ge \mathcal{H}^r$. Now a set of diameter d_j might not be contained in a ball of diameter d_j (unless $X = \mathbb{R}$), but it is certainly contained in a ball of *radius* d_j, i.e., of diameter $2d_j$, so we always have

$$\mathcal{H}^r(A) \le \mathcal{H}_B^r(A) \le 2^r \mathcal{H}^r(A). \tag{12.4}$$

We begin the task of showing that $\mathcal{H}^n(A) = \mathcal{L}^n(A)$ for measurable $A \subset \mathbb{R}^n$. The first ingredient is known as the *isodiametric inequality.*

Proposition 12.2. *For any measurable* $S \subset \mathbb{R}^n$,

$$\mathcal{L}^n(S) \le \gamma_n \big(\operatorname{diam} S\big)^n. \tag{12.5}$$

In other words, $\mathcal{L}^n(S) \le \mathcal{L}^n(B)$ where B is a ball in $\mathbb{R}^n$ with the same diameter as S.

Proof. We will apply to S a sequence of transformations that do not change the Lebesgue measure and do not increase the diameter. Given $\omega \in \mathbb{R}^n$, $|\omega| = 1$, $y \in \mathbb{R}^n$, set

$$P_\omega = \{x \in \mathbb{R}^n : x \cdot \omega = 0\}, \quad L_y^\omega = \{y + t\omega : t \in \mathbb{R}\}, \tag{12.6}$$

and

$$\Sigma_\omega(S) = \bigcup \Big\{ y + t\omega : |t| \le \frac{1}{2} \mathcal{H}^1(S \cap L_y^\omega),\ y \in P_\omega,\ S \cap L_y^\omega \ne \emptyset \Big\}. \tag{12.7}$$

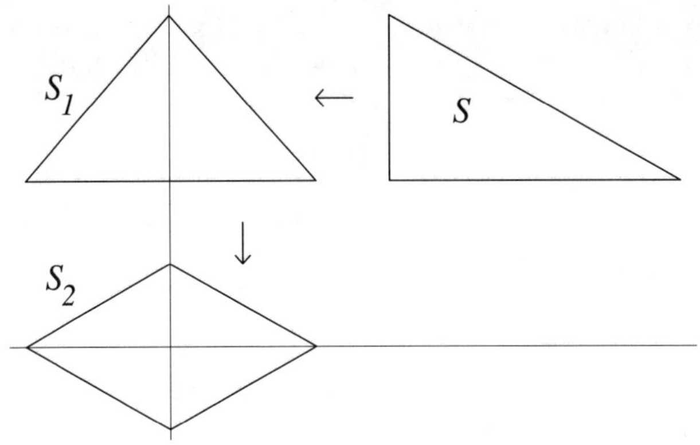

Figure 12.1

The set $\Sigma_\omega(S)$ is called the *Steiner symmetrization* of S about the hyperplane P_ω. See Figure 12.1. It is easy to verify that $\Sigma_\omega(S)$ is Lebesgue measurable if S is and that

$$\text{diam}\ \Sigma_\omega(S) \le \ \text{diam}\ S, \quad \mathcal{L}^n\big(\Sigma_\omega(S)\big) = \mathcal{L}^n(S), \tag{12.8}$$

the latter identity via Fubini's Theorem.

Now, if $\{e_j : 1 \le j \le n\}$ is the standard orthonormal basis of $\mathbb{R}^n$, we can pass from S to $S_1 = \Sigma_{e_1}(S),\ S_2 = \Sigma_{e_2}(S_1), \cdots, S_n = \Sigma_{e_n}(S_{n-1})$, obtaining $S_n \subset \mathbb{R}^n$ with $\mathcal{L}^n(S_n) = S_1$ and diam $S_n \le$ diam S. Furthermore, it is easily verified that S_n is symmetric about the origin, i.e., $x \in S_n \Rightarrow -x \in S_n$. In fact, each S_j is invariant with respect to reflection about $P_{e_1}, \dots, P_{e_j}$. Thus, if diam $S_n = 2\rho$, it follows that $S_n \subset B_\rho(0)$, so

$$\mathcal{L}^n(S) = \mathcal{L}^n(S_n) \le \mathcal{L}^n\big(B_\rho(0)\big), \tag{12.9}$$

as asserted.

In light of the estimate (12.5), we see that, for any covering of a measurable $A \subset \mathbb{R}^n$ of the form (12.1),

$$\sum_{j\ge1} \mathcal{L}^n(B_j) \le \gamma_n \sum_{j\ge1} \big(\text{diam}\ B_j\big)^n, \tag{12.10}$$

which immediately yields $\mathcal{L}^n(A) \le \gamma_n h^*_{n,\delta}(A)$, for each $\delta > 0$, and hence

$$\mathcal{L}^n(A) \le \mathcal{H}^n(A). \tag{12.11}$$

Before establishing that $\mathcal{L}^n(A) \ge \mathcal{H}^n(A)$, we note that a weaker estimate is quite simple. Namely, given $\varepsilon > 0$, we can cover A by a countable family of cubes whose total volumes sum to $\le \mathcal{L}^n(A) + \varepsilon$. Since for a cube $Q \subset \mathbb{R}^n$, $\mathcal{L}^n(Q) = \rho^n$ when $\operatorname{diam} Q = \rho\sqrt{n}$, we have easily that

$$\mathcal{H}^n(A) \le \kappa_n \mathcal{L}^n(A), \quad \kappa_n = \gamma_n \, n^{n/2}. \tag{12.12}$$

The proof of the sharper result will use Wiener's Covering Lemma.

Lemma 12.3. *If $U \subset \mathbb{R}^n$ is open, then $\mathcal{H}^n(U) \le \mathcal{L}^n(U)$.*

Proof. We can assume that $\mathcal{L}^n(U) < \infty$. Take $\delta > 0$, and write U as a union of a collection $\mathcal{C}_\delta = \{B_{j\delta}\}$ of open balls of diameter $\le \delta$. By Lemma 11.3, there is a finite collection of disjoint balls $B_j \in \mathcal{C}_\delta$, $1 \le j \le K$, such that $\sum_{j=1}^K \mathcal{L}^n(B_j) > 4^{-n}\mathcal{L}^n(U)$. If we apply this argument again to $U \setminus \bigcup_{j=1}^K \overline{B}_j$ and repeat this process a countable number of times, we obtain a countable collection of mutually disjoint balls $\{B_k\}$, all of diameter $\le \delta$, such that

$$U_1 = \bigcup_{j\ge 1} B_j, \quad \mathcal{L}^n(U \setminus U_1) = 0. \tag{12.13}$$

Then $\gamma_n h^*_{n,\delta}(U_1) \le \mathcal{L}^n(U)$.

Note that U_1 is open. Repeat this construction, with U replaced by U_1 and δ replaced by $\delta/2$, obtaining open U_2 such that

$$U_2 \subset U_1 \subset U, \quad \mathcal{L}^n(U_1 \setminus U_2) = 0, \quad \gamma_n h^*_{n,\delta/2}(U_2) \le \mathcal{L}^n(U_1). \tag{12.14}$$

Continue, obtaining U_ν, and let $V = \bigcap_{\nu\ge 1} U_\nu$. Then

$$V \subset U, \quad \mathcal{L}^n(U \setminus V) = 0, \tag{12.15}$$

and for all ν, $\gamma_n h^*_{n,\delta/2^\nu}(V) \le \gamma_n h^*_{n,\delta/2^\nu}(U_\nu) \le \mathcal{L}^n(U)$, so

$$\mathcal{H}^n(V) = \lim_{\nu\to\infty} \gamma_n h^*_{n,\delta/2^\nu}(V) \le \mathcal{L}^n(U). \tag{12.16}$$

On the other hand, by (12.12) we see that

$$\mathcal{L}^n(U \setminus V) = 0 \Rightarrow \mathcal{H}^n(U \setminus V) = 0 \Rightarrow \mathcal{H}^n(U) = \mathcal{H}^n(V), \tag{12.17}$$

so the lemma is proved.

It is now easy to prove

Proposition 12.4. *For any Borel set $A \subset \mathbb{R}^n$,*

$$\mathcal{L}^n(A) = \mathcal{H}^n(A). \tag{12.18}$$

Proof. If $K \subset \mathbb{R}^n$ is compact, put K in an open ball B. Then $\mathcal{L}^n$ and $\mathcal{H}^n$ coincide on B and on $B \setminus K$, so $\mathcal{L}^n(K) = \mathcal{H}^n(K)$. On the other hand, as a consequence of (5.60), we have that, for any Borel set $A \subset \mathbb{R}^n$,

$$\mu(A) = \sup \{\mu(K) : K \subset A,\ K \text{ compact}\}, \tag{12.19}$$

both for $\mu = \mathcal{L}^n$ and for $\mu = \mathcal{H}^n$, so we have (12.18).

Note that, without using a covering lemma argument, we could say the following. Both $\mathcal{H}^n$ and $\mathcal{L}^n$ are clearly translation-invariant measures on $\mathbb{R}^n$. Having the inequalities (12.11) and (12.12), we can apply Proposition 7.7 and deduce that, for all Borel sets $A \subset \mathbb{R}^n$, $\mathcal{L}^n(A) = c_n \mathcal{H}^n(A)$, for some constant $c_n \in [\kappa_n^{-1}, 1]$. The reader might ponder whether it can be shown that $c_n = 1$, without using a covering lemma argument.

It is useful to estimate the effect a change of metric has in changing the Hausdorff measure. We have the following result.

Proposition 12.5. *Let d_1 and d_2 be two metrics on a compact space X, with the property that, for some $a, b \in (0, \infty)$,*

$$a d_1(x,y) \le d_2(x,y) \le b d_1(x,y). \tag{12.20}$$

Alternatively, assume

$$d_2(x,y) = \varphi(x,y) d_1(x,y), \tag{12.21}$$

where $\varphi : X \times X \to \mathbb{R}^+$ is continuous and

$$a \le \varphi(x,x) \le b, \quad \forall\, x \in X. \tag{12.22}$$

In either case, if $\mathcal{H}_j^r$ denotes the r-dimensional Hausdorff measure on X arising from d_j, then, for each Borel set $S \subset X$,

$$a^r \mathcal{H}_1^r(S) \le \mathcal{H}_2^r(S) \le b^r \mathcal{H}_1^r(S). \tag{12.23}$$

Proof. We concentrate on the implication (12.21)–(12.22) $\Rightarrow$ (12.23), since the implication (12.20) $\Rightarrow$ (12.23) is easier.

Let $\vartheta_k(B)$ denote the diameter of a set $B \subset X$, for the metric d_k. Let $\{B_j\}$ be a cover of S by sets satisfying $\vartheta_1(B_j) \le \delta$. Then, under the hypothesis (12.21)–(12.22),

$$\vartheta_2(B_j) \le \bigl(1 + \varepsilon(\delta)\bigr) b\, \vartheta_1(B_j),$$

where $\varepsilon(\delta) \to 0$ as $\delta \to 0$. Now (with obvious notation), $h^*_{r,\delta,1}(S)$ is the infimum of sums $\sum_j \vartheta_1(B_j)^r$, while $h^*_{r,(1+\varepsilon)\delta,2}(S)$ is $\le$ the infimum of sums $\sum_j \vartheta_2(B_j)^r$, as $\{B_j\}$ runs over the same family of covers. Hence

$$h^*_{r,(1+\varepsilon)\delta,2}(S) \le (1+\varepsilon)^r b^r\, h^*_{r,\delta,1}(S). \tag{12.24}$$

Taking $\delta \to 0$, we get

$$h^*_{r,2}(S) \le b^r\, h^*_{r,1}(S), \tag{12.25}$$

which gives one of the inequalities in (12.23). The other follows by reversing the roles of d_1 and d_2.

Let us recall the measure V constructed in Chapter 7 on a Riemannian manifold M, assumed to be a C^1 manifold with a C^0 metric tensor. If S is a Borel subset of M, contained in a coordinate chart $\mathcal{O}$, in which the metric tensor has components g_{jk}, then

$$V(S) = \int_{\mathcal{O}} \chi_S \sqrt{g}\, dx, \quad g = \det(g_{jk}). \tag{12.26}$$

As shown in Chapter 7, this construction of V is independent of the choice of C^1 coordinate charts for M. Now, for such a Riemannian manifold, a C^1 curve $\gamma : [a,b] \to M$ has a well-defined length,

$$L(\gamma) = \int_a^b \langle \gamma'(s), \gamma'(s)\rangle^{1/2}\, ds. \tag{12.27}$$

The infimum of lengths of curves from x to y in M defines a distance function on M, making it into a metric space. The following is a useful extension of Proposition 12.4.

Proposition 12.6. *If M is a C^1 manifold of dimension n with a C^0 metric tensor, making M a metric space, then, for any Borel set $S \subset M$,*

$$V(S) = \mathcal{H}^n(S). \tag{12.28}$$

Proof. It suffices to prove (12.28) when $S = K$ is compact. Given $\varepsilon > 0$, cover K by a finite union of coordinate charts $\mathcal{O}_\nu$, so that, on each $\mathcal{O}_\nu$,

$$g_{jk}(x) = \delta_{jk} + r_{jk}(x), \quad |r_{jk}(x)| < \varepsilon. \tag{12.29}$$

Partition K into Borel subsets $K_\nu \subset \mathcal{O}_\nu$. Then

$$(12.30)\quad \begin{aligned} V(K) &= \sum V(K_\nu) = \sum_\nu \int_{\mathcal{O}_\nu} \chi_{K_\nu}(x)\,\sqrt{g_\nu}\,dx \\ &= \sum_\nu \int_{\mathcal{O}_\nu} \chi_{K_\nu}(x)\bigl(1+O(\varepsilon)\bigr)\,dx = \bigl(1+O(\varepsilon)\bigr)\sum_\nu \mathcal{L}^n(K_\nu). \end{aligned}$$

Now, on each $\mathcal{O}_\nu$, identified with an open set in $\mathbb{R}^n$, we have two metrics, arising from the metric tensors g_{jk} and δ_{jk}, defining two n-dimensional Hausdorff measures on $\mathcal{O}_\nu$. The first coincides with the Hausdorff measure appearing in (12.28), while the second coincides with Hausdorff measure on $\mathbb{R}^n$, which by Proposition 12.4 coincides with Lebesgue measure on $\mathbb{R}^n$ (which is the measure used in the integrals in (12.26) and (12.30)). Applying Proposition 12.5, we have

$$(12.31)\quad (1-C\varepsilon)\mathcal{L}^n(K_\nu) \le \mathcal{H}^n(K_\nu) \le (1+C\varepsilon)\mathcal{L}^n(K_\nu),$$

where $\mathcal{H}^n$ here comes from the metric on M. Therefore, summing on ν and using (12.30), we have

$$(1-C\varepsilon)V(K) \le \mathcal{H}^n(K) \le (1+C\varepsilon)V(K),$$

for all $\varepsilon > 0$. This implies (12.28).

Often one has an n-dimensional manifold M imbedded in an m-dimensional Riemannian manifold Ω, perhaps $\Omega = \mathbb{R}^m$. Give M the induced Riemannian metric. Then M has two distance functions, $d_M(p,q)$, from minimizing the lengths of curves in M from p to q, and $d_\Omega(p,q)$, from considering all curves in Ω from p to q. Clearly

$$(12.32)\quad p,q \in M \Longrightarrow d_M(p,q) \ge d_\Omega(p,q).$$

Each metric on M yields r-dimensional Hausdorff measures. We show they coincide.

Proposition 12.7. *Let Ω be a C^1 manifold with C^0 metric tensor, and let M be a C^1 imbedded submanifold of Ω, with the induced metric tensor, so M has two metrics, as in* (12.32). *If $\mathcal{H}^r_M$ and $\mathcal{H}^r_\Omega$ denote the respective r-dimensional Hausdorff measures, then, for any $r \in \mathbb{R}^+$,*

$$(12.33)\quad S \subset M \text{ Borel} \Longrightarrow \mathcal{H}^r_M(S) = \mathcal{H}^r_\Omega(S).$$

Proof. It suffices to note that

$$d_M(p,q) = \varphi(p,q)d_\Omega(p,q), \quad \varphi(p,p) = 1, \tag{12.34}$$

with $\varphi : M \times M \to \mathbb{R}^+$ continuous. Then we can apply Proposition 12.5.

Applying (7.26), we have immediately:

Corollary 12.8. *If $\mathcal{O} \subset \mathbb{R}^n$ is open and $u : \mathcal{O} \to \mathbb{R}^k$ is C^1, then, for the surface*

$$M = \big\{\big(x, u(x)\big) : x \in \mathcal{O}\big\} \subset \mathbb{R}^{n+k}, \tag{12.35}$$

we have, using Hausdorff measure on $\mathbb{R}^{n+k}$,

$$\mathcal{H}^n(M) = \int_{\mathcal{O}} \big(1 + |\nabla u(x)|^2\big)^{1/2}\, dx. \tag{12.36}$$

Using Theorem 11.9, we can extend the last result to Lipschitz mappings.

Proposition 12.9. *If $\mathcal{O} \subset \mathbb{R}^n$ is open and $v : \mathcal{O} \to \mathbb{R}^k$ is Lipschitz, then, for the surface*

$$M = \{(x, v(x)) : x \in \mathcal{O}\} \subset \mathbb{R}^{n+k}, \tag{12.37}$$

we have, using Hausdorff measure on $\mathbb{R}^{n+k}$,

$$\mathcal{H}^n(M) = \int_{\mathcal{O}} \big(1 + |\nabla v(x)|^2\big)^{1/2}\, dx. \tag{12.38}$$

Proof. It suffices to prove the result for bounded $\mathcal{O}$. Assume $|v(x)-v(y)| \le L|x-y|$ for $x, y \in \mathcal{O}$. Take $\varepsilon > 0$. By Theorem 11.9, there exists a C^1 map $u : \mathcal{O} \to \mathbb{R}^k$ such that $|\nabla u| \le CL$ and such that $u = v$ on a compact set $K \subset \mathcal{O}$ such that $\mathcal{L}^n(\mathcal{O} \setminus K) < \varepsilon$, and furthermore, $\nabla u = \nabla v$ on K. Thus we have a C^1 surface

$$\widetilde{M} = \big\{(x, u(x)) : x \in \mathcal{O}\big\} \subset \mathbb{R}^{n+k},$$

and, by Corollary 12.8,

$$\mathcal{H}^n(\widetilde{M}) = \int_{\mathcal{O}} \big(1 + |\nabla u(x)|^2\big)^{1/2}\, dx. \tag{12.39}$$

Note that the symmetric difference $M \triangle \widetilde{M} = (M \setminus \widetilde{M}) \cup (\widetilde{M} \setminus M)$ is contained in $S \cup \widetilde{S}$, where

$$S = \{(x, v(x)) : x \in \mathcal{O} \setminus K\}, \quad \widetilde{S} = \{(x, u(x)) : x \in \mathcal{O} \setminus K\}. \tag{12.40}$$

Now we have elementary estimates

$$\mathcal{H}^n(S) \le L\varepsilon, \quad \mathcal{H}^n(\widetilde{S}) \le CL\varepsilon, \tag{12.41}$$

so $|\mathcal{H}^n(M) - \mathcal{H}^n(\widetilde{M})| \le (C+1)L\varepsilon$. On the other hand, the right sides of (11.37) and (11.38) differ by a quantity that does not exceed

$$\int_{\mathcal{O}\setminus K} \left(1 + |\nabla v(x)|^2\right)^{1/2} dx + \int_{\mathcal{O}\setminus K} \left(1 + |\nabla u(x)|^2\right)^{1/2} dx \le C' L\varepsilon. \tag{12.42}$$

Hence the two sides of (12.37) differ by a quantity of magnitude $\le C'' L\varepsilon$. Since $\varepsilon > 0$ can be taken arbitrarily small, the proof is done.

Having Proposition 12.9, note that, if K is a compact subset of the open set $\mathcal{O} \subset \mathbb{R}^n$ and $v : \mathcal{O} \to \mathbb{R}^k$ is Lipschitz and if we set

$$S = \{(x, v(x)) : x \in K\}, \tag{12.43}$$

then, with M as in (12.37), we have a disjoint union

$$M = S \cup M^b, \quad M^b = \{(x, v(x)) : x \in \mathcal{O} \setminus K\}. \tag{12.44}$$

Proposition 12.9 applies to both M and M^b, so we get

$$\mathcal{H}^n(S) = \int_K \left(1 + |\nabla v(x)|^2\right)^{1/2} dx. \tag{12.45}$$

We can extend this result to the following:

Proposition 12.10. *If $E \subset \mathbb{R}^n$ is a Borel set and $w : E \to \mathbb{R}^k$ is Lipschitz, then ∇w is well defined as an element of $L^\infty(E)$, and, for the surface*

$$\Gamma = \{(x, w(x)) : x \in E\} \tag{12.46}$$

we have

$$\mathcal{H}^n(\Gamma) = \int_E \left(1 + |\nabla w|^2\right)^{1/2} dx. \tag{12.47}$$

Proof. We first establish this for $E = K$, compact. Now by Proposition C.3 (in Appendix C), there is a Lipschitz map $v : \mathbb{R}^n \to \mathbb{R}^k$ such that $v = w$ on K. We consequently deduce (12.47) from (12.45), provided

$$\nabla w = \nabla v\big|_K, \quad \mathcal{L}^n\text{-a.e.} \tag{12.48}$$

We claim that the right side of (12.48) defines an element of $L^\infty(K)$, independent of the choice of the Lipschitz extension v of w. To demonstrate this point, it suffices to establish

Lemma 12.11. *If $f \in \mathrm{Lip}(\mathbb{R}^n)$, $K \subset \mathbb{R}^n$ compact, and $f = 0$ on K, then $\nabla f = 0$, $\mathcal{L}^n$-a.e. on K.*

Proof. Let S be as in (12.43), with $v = f$. Then (12.45) holds. On the other hand, in this case $S = \{(x, 0) : x \in K\}$, so $\mathcal{H}^n(S) = \mathcal{L}^n(K)$. Hence

$$\int_K \left(1 + |\nabla f(x)|^2\right)^{1/2} dx = \mathcal{L}^n(K). \tag{12.49}$$

This implies that $\left(1 + |\nabla f(x)|^2\right)^{1/2} = 1$, $\mathcal{L}^n$-a.e. on K, which proves the lemma.

To pass from $E = K$, compact, to general Borel E in Proposition 12.10, it suffices to show that, if $F = E \setminus K$ is a Borel set for which $\mathcal{L}^n(F)$ is small, then the Hausdorff measure $\mathcal{H}^n\big(\{(x, w(x)) : x \in F\}\big)$ is small. We establish the following general result.

Proposition 12.12. *Let X and Y be locally compact metric spaces and $F : X \to Y$ a Lipschitz map satisfying $d_Y(F(p), F(q)) \le L d_X(p, q)$. If $S \subset X$ is a Borel set and $\mathcal{H}^r(S) < \infty$, then $F(S)$ is $\mathcal{H}^r$-measurable and*

$$\mathcal{H}^r\big(F(S)\big) \le L^r \mathcal{H}^r(S). \tag{12.50}$$

Proof. Since we can pick compact $S_0 \subset S$ such that $\mathcal{H}^r(S \setminus S_0) < \varepsilon$ and since $F(S_0)$ is compact in Y, it suffices to show that

$$h_r^*\big(F(S)\big) \le L^r h_r^*(S). \tag{12.51}$$

In fact, the hypotheses easily yield

$$h_{r,L\delta}^*\big(F(S)\big) \le L^r h_{r,\delta}^*(S). \tag{12.52}$$

Remark. If $S \subset X$ is a Borel set, $F(S)$ need not be a Borel set.

Now that Proposition 12.10 has been proved, it is desirable to consider the Hausdorff measure of images of sets in $\mathbb{R}^n$ under more general maps into $\mathbb{R}^m = \mathbb{R}^{n+k}$. For example, suppose $\mathcal{O} \subset \mathbb{R}^n$ is open and $u : \mathcal{O} \to \mathbb{R}^m$ is a C^1 map which is one-to-one and has injective derivative at each point $x \in \mathcal{O}$. Then u is said to be an imbedding (loosely speaking), and $M = u(\mathcal{O})$ acquires the structure of a Riemannian manifold, with metric tensor

$$g_{ij}(x) = Du(x)e_i \cdot Du(x)e_j = \sum_\ell \frac{\partial u_\ell}{\partial x_i}\frac{\partial u_\ell}{\partial x_j}. \tag{12.53}$$

In other words, we have the matrix identity

$$\big(g_{ij}\big) = Du(x)^t Du(x). \tag{12.54}$$

Hence the volume element of M is

$$dV = \sqrt{g}\, dx = \big(\det Du^t\, Du\big)^{1/2}\, dx. \tag{12.55}$$

The same argument used to establish Corollary 12.8 establishes the first part of the following.

Proposition 12.13. *If $\mathcal{O} \subset \mathbb{R}^n$ is open and $u : \mathcal{O} \to \mathbb{R}^m$ is a C^1 imbedding, then for $M = u(\mathcal{O})$ we have*

$$\mathcal{H}^n(M) = \int_{\mathcal{O}} J_u(x)\, dx, \tag{12.56}$$

where

$$J_u(x) = \big(\det Du(x)^t Du(x)\big)^{1/2}. \tag{12.57}$$

Furthermore, if $K \subset \mathcal{O}$ is a compact set and $S = u(K)$, then

$$\mathcal{H}^n(S) = \int_K J_u(x)\, dx. \tag{12.58}$$

The second part of the proposition follows just as in (12.43)–(12.45).

Suppose more generally that $u : \mathcal{O} \to \mathbb{R}^m$ is a Lipschitz map that is one-to-one and satisfies $\det\big(Du(x)^t Du(x)\big) > 0$ for $\mathcal{L}^n$-a.e. $x \in \mathcal{O}$. Assume $\mathcal{O}$ is bounded. Given $\varepsilon > 0$, there is a compact $K \subset \mathcal{O}$ such that $\mathcal{L}^n(\mathcal{O} \setminus K) < \varepsilon$ and $Du\big|_K$ is continuous, and also $\det\big(Du(x)^t Du(x)\big) \ge \delta > 0$ for $x \in K$. If v is a C^1 extension of $u\big|_K$, then $\det\big(Dv(x)^t Dv(x)\big) \ge \delta/2 > 0$ on a neighborhood U of K. We claim that also $v : U \to \mathbb{R}^m$ is one-to-one if U is a small enough neighborhood of K, and we pause to establish this.

Lemma 12.14. *Suppose $\mathcal{O} \subset \mathbb{R}^n$ is open, $v : \mathcal{O} \to \mathbb{R}^m$ is C^1, and $K \subset \mathcal{O}$ is compact. If v maps K one-to-one into $\mathbb{R}^m$ and $Dv(x)$ is injective for each $x \in K$, then v maps some neighborhood U of K one-to-one into $\mathbb{R}^m$.*

Proof. Let U_ν be a family of neighborhoods shrinking to K. If there are $x_\nu, y_\nu \in U_\nu$ such that $x_\nu \neq y_\nu$ and $v(x_\nu) = v(y_\nu)$, then x_ν and y_ν must have limit points $x, y \in K$. Furthermore $v(x) = v(y)$, so it is not possible that $x \neq y$. But if $x = y \in K$ and $Dv(x)$ is injective, then v must be an imbedding on a neighborhood of x, by the Implicit Function Theorem, and this contradicts the convergence of distinct x_ν and y_ν to x.

Therefore (12.57) applies to $S = v(K) = u(K)$. Letting $\varepsilon \to 0$ and arguing as in the proof of Proposition 12.9, we conclude that, with $M = u(\mathcal{O})$, (12.56) continues to hold. We have

Proposition 12.15. *The conclusions of Proposition 12.13 continue to hold as long as $u : \mathcal{O} \to \mathbb{R}^m$ is Lipschitz, one-to-one, and*

$$J_u(x) > 0, \tag{12.59}$$

for $\mathcal{L}^n$-a.e. $x \in \mathcal{O}$, where J_u is defined a.e. on $\mathcal{O}$ by (12.57).

It is now within reach to produce a more definitive result.

Theorem 12.16. *If $E \subset \mathbb{R}^n$ is a Borel set and $u : E \to \mathbb{R}^m$ is one-to-one and Lipschitz, then $\Gamma = u(E)$ is $\mathcal{H}^n$-measurable and*

$$\mathcal{H}^n(\Gamma) = \int_E J_u(x)\, dx. \tag{12.60}$$

Proof. Write $E = E_1 \cup E_2$, a disjoint union of Borel sets, such that $J_u > 0$ a.e. on E_1 and $J_u = 0$ a.e. on E_2. It suffices to show that (12.60) holds with E replaced by E_1 and Γ by $\Gamma_1 = u(E_1)$ and that, for $\Gamma_2 = u(E_2)$, $\mathcal{H}^n(\Gamma_2) = 0$.

Chopping E_1 into a countable number of pieces if necessary, we can assume $\mathcal{L}^n(E_1) < \infty$. Given $\varepsilon > 0$, pick compact $K_1 \subset E_1$ such that $\mathcal{L}^n(E_1 \setminus K_1) < \varepsilon$, $Du\big|_{K_1}$ is continuous, and $J_u \geq \delta > 0$ on K_1. Then the argument, involving Lemma 12.14, used to extend (12.54), works here to yield

$$\mathcal{H}^n\big(u(K_1)\big) = \int_{K_1} J_u(x)\, dx.$$

Taking $\varepsilon \to 0$ and using Proposition 12.12, we have the desired result for E_1.

To analyze u on E_2, we use the following trick. Given $\varepsilon > 0$, define $v_\varepsilon : E_2 \to \mathbb{R}^{m+n}$ by $v_\varepsilon(x) = \big(u(x), \varepsilon x\big)$. Then v_ε is Lipschitz and $J_{v_\varepsilon} = \varepsilon^n$ a.e. on E_2. By the first part of the argument, we have that

$$\mathcal{H}^n\big(v_\varepsilon(E_2)\big) = \varepsilon^n \mathcal{L}^n(E_2).$$

On the other hand, $\Gamma_2 = u(E_2)$ is the image of $v_\varepsilon(E_2)$ under a projection, and Proposition 12.12 implies $\mathcal{H}^n(\Gamma_2) \le \mathcal{H}^n\big(v_\varepsilon(E_2)\big)$. Hence $\mathcal{H}^n(\Gamma_2) \le \varepsilon^n \mathcal{L}^n(E_2)$ for all $\varepsilon > 0$, so $\mathcal{H}^n(\Gamma_2) = 0$. This proves the theorem.

We have concentrated on the Hausdorff measures $\mathcal{H}^r$ for $r = n \in \mathbb{Z}^+$. There are also sets for which it is interesting to consider $\mathcal{H}^r$ for $r \notin \mathbb{Z}$. Such sets are said to have nonintegral Hausdorff dimension, where the Hausdorff dimension of a nonempty subset S of a metric space X is

$$\begin{aligned} \text{Hdim } S &= \inf\{r \ge 0 : \mathcal{H}^r(S) = 0\} \\ &= \sup\{r \ge 0 : \mathcal{H}^r(S) > 0\}. \end{aligned} \tag{12.61}$$

Note that, if $0 \le r_1 < r_2 < \infty$, then

$$h^*_{r_1,\delta}(S) \ge \delta^{-(r_2-r_1)}\, h^*_{r_2,\delta}(S). \tag{12.62}$$

This implies that the two quantities on the right side of (12.61) are identical, and it also implies that

$$0 \le r < \text{ Hdim } S \Longrightarrow \mathcal{H}^r(S) = \infty. \tag{12.63}$$

Our first example of a set with fractional Hausdorff dimension is a subset of the unit interval $[0,1]$ known as the Cantor middle third set. It is constructed as follows. Let $K_0 = [0,1]$. Form K_1 by removing the open interval in the middle of K_0, of length $1/3$. Thus K_1 consists of two intervals, each of length $1/3$. Next, remove from each of the intervals making up K_1 the open interval in the middle, of length $1/3^2$. Continue this process. Thus, K_ν is a union of 2^ν disjoint closed intervals, each of length $3^{-\nu}$. We have

$$K_0 \supset K_1 \supset K_2 \supset K_3 \supset \cdots \searrow K. \tag{12.64}$$

Compactness implies $K = \bigcap_\nu K_\nu$ is nonempty; it is the Cantor middle third set.

Since K_ν is a cover of K by 2^ν intervals of length (i.e., diameter) $3^{-\nu}$, we have

$$h^*_{r,\delta}(K) \le \Big(\frac{2}{3^r}\Big)^\nu, \quad \delta = 3^{-\nu}. \tag{12.65}$$

The right side of (12.65) is independent of ν if $3^r = 2$ or, equivalently, if $r = (\log 2)/(\log 3)$. For such r we get $\mathcal{H}^r(K) \le \gamma_r$. We show that this is actually equality.

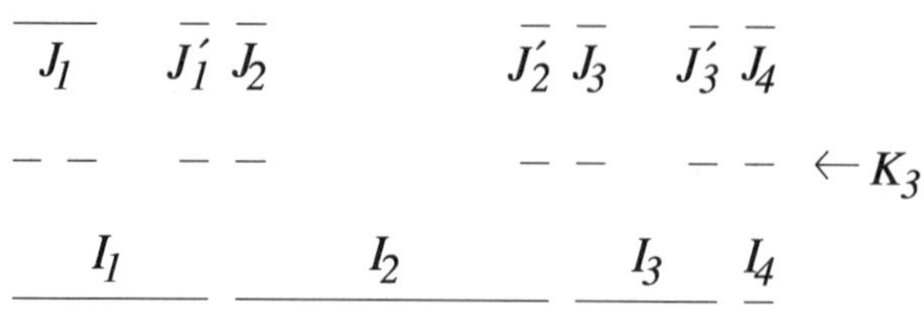

Figure 12.2

Proposition 12.17. *If K is the Cantor middle third set, then*

$$\mathcal{H}^r(K) = \gamma_r, \quad \textit{for} \ \ r = \frac{\log 2}{\log 3} \approx 0.6309. \tag{12.66}$$

Proof. We will show that, if $\{I_k\}$ is any collection of intervals covering K, then

$$\sum \ell(I_k)^r \ge 1, \tag{12.67}$$

when $r = (\log 2)/(\log 3)$. By expanding each interval slightly and using the compactness of K, it suffices to prove this for a *finite* cover of K by open intervals.

In such a case, $\{I_k\}$ is actually a cover of some K_ν by open intervals. Increasing ν if necessary, we can assume that each interval in K_ν is contained in some I_k. Then, possibly shrinking the intervals I_k and forming their closure, we arrange that each interval in K_ν has nonempty intersection with only one I_k and that all the endpoints of the I_k are also endpoints of intervals in K_ν. A typical case (with $\nu = 3$) is illustrated in Figure 12.2. One sees that, for each I_k in such a cover of K_ν, either

(a) I_k coincides with an interval in K_ν, or
(b) I_k contains an interval $L_k \subset \mathbb{R} \setminus K_\nu$ such that $\ell(L_k) \ge \ell(I_k)/3$.

In case (b), we can partition I_k into three intervals, say J_k, L_k, and J'_k, such that $J_k \cup J'_k$ covers $K_\nu \cap I_k$. Making use of concavity of the function $\psi(t) = t^r$, when $r \in (0,1)$, and the fact that $3^r = 2$, we have

$$\begin{aligned} \ell(I_k)^r &= \big[\ell(J_k) + \ell(L_k) + \ell(J'_k)\big]^r \ge \Big(\frac{3}{2}\big[\ell(J_k) + \ell(J'_k)\big]\Big)^r \\ &= 2\Big(\frac{1}{2}\ell(J_k) + \frac{1}{2}\ell(J'_k)\Big)^r \ge \ell(J_k)^r + \ell(J'_k)^r. \end{aligned}$$

Thus replacing I_k by the two subintervals J_k and J'_k does not increase the sum in (12.67). Iterating this argument, we conclude that the sum (12.67) is not increased if $\{I_k\}$ is replaced by the set of intervals making up K_ν. As (12.67) holds for this covering, the proof is done.

We can produce more general Cantor sets as follows. Given $\vartheta \in (0,1)$, we alter the construction of K, at each stage removing from each interval J making up K_ν the open interval in the middle whose length is $\vartheta\ell(J)$. This yields a Cantor set (i.e., a compact, totally disconnected subset of $[0,1]$) which we denote $K(\vartheta)$. The Cantor middle third set is $K(1/3)$. Note that, in this case, K_ν consists of 2^ν disjoint closed intervals, each of length $((1-\vartheta)/2)^\nu$. Thus, in place of (12.65), we have

$$h^*_{r,\delta}\big(K(\vartheta)\big) \le 2^\nu \Big(\frac{1-\vartheta}{2}\Big)^{r\nu}, \quad \delta = \Big(\frac{1-\vartheta}{2}\Big)^{\nu}.$$

In this case, the critical value of r is that for which $(2/(1-\vartheta))^r = 2$. The proof of Proposition 12.17 extends readily to yield

$$\mathcal{H}^r\big(K(\vartheta)\big) = \gamma_r, \quad r = \frac{\log 2}{\log b}, \quad b = \frac{2}{1-\vartheta}. \tag{12.68}$$

Note that, as ϑ increases from 0 to 1, r decreases from 1 to 0.

One interesting property the Cantor sets $K(\vartheta)$ have is self-similarity. We say that a Borel set $S \subset \mathbb{R}^n$ is self-similar, of similarity dimension

$$\operatorname{Sdim} S = \frac{\log k}{\log b}, \tag{12.69}$$

if, for some integer $k \ge 2$ and some real $b > 1$, there is a disjoint union $S = S_1 \cup \cdots \cup S_k$ such that, for each j, $bS_j = \{bx : x \in S_j\}$ is congruent to S.

It is easy to see that the Cantor set $K(\vartheta)$ is self-similar, of similarity dimension $r = (\log 2)/(\log b)$, where $b = 2/(1-\vartheta)$. We see that, for $K(\vartheta)$, the similarity dimension is equal to the Hausdorff dimension. The following general result loosely relates similarity dimension to Hausdorff dimension.

Proposition 12.18. *If $S \subset \mathbb{R}^n$ is self-similar, then*

$$r > \operatorname{Sdim} S \Longrightarrow \mathcal{H}^r(S) = 0, \tag{12.70}$$

provided $h^*_{r,\delta}(S) < \infty$ for some $\delta > 0$. Furthermore, for any $r \ge 0$,

$$0 < \mathcal{H}^r(S) < \infty \Longrightarrow r = \operatorname{Sdim} S. \tag{12.71}$$

Proof. If $\{B_j\}$ is a cover of S, then there is a another cover $\{B_{j\ell} : j \ge 1, 1 \le \ell \le k\}$, such that $B_{j\ell}$ is congruent to $b^{-1}B_j$. We have

$$\sum_{j,\ell} (\operatorname{diam} B_{j\ell})^r = kb^{-r} \sum_j (\operatorname{diam} B_j)^r.$$

This implies

$$h^*_{r,\delta/b}(S) \le (kb^{-r})h^*_{r,\delta}(S),$$

which yields (12.70), since $kb^{-r} < 1$ when $r >$ Sdim S.

We obtain (12.71) as follows. Since $\mathcal{H}^r(bS_j) = b^r\mathcal{H}^r(S_j)$, the hypothesis of self-similarity implies $kb^{-r}\mathcal{H}^r(S) = \mathcal{H}^r(S)$. Granted that $\mathcal{H}^r(S)$ is neither 0 nor ∞, this implies that $k = b^r$, or $r = (\log k)/(\log b)$, as asserted.

If $\mathcal{H}^r(S)$ does not lie in $(0,\infty)$ for any $r \ge 0$, we cannot apply (12.71), and the identity $\operatorname{Hdim} S = \operatorname{Sdim} S$ can fail. To give an example, let $\mathcal{Q} = \mathbb{Q} \cap [0,1)$. We can write $\mathcal{Q} = \mathcal{Q}_1 \cup \mathcal{Q}_2$ where $\mathcal{Q}_1 = \mathbb{Q} \cap [0,1/2)$ and $\mathcal{Q}_2 = \mathbb{Q} \cap [1/2,1)$. Consequently,

$$\operatorname{Sdim} \mathcal{Q} = 1, \text{ but } \operatorname{Hdim} \mathcal{Q} = 0. \tag{12.72}$$

In fact, the Hausdorff dimension of a countable set is always zero.

Here is another tool for estimating the Hausdorff dimension of a set.

Proposition 12.19. *Let $K \subset \mathbb{R}^n$ be compact, and take $a \in (0,\infty)$. Assume there is a positive Borel measure $\mu \ne 0$, supported on K, such that*

$$\iint \frac{d\mu(x)\,d\mu(y)}{|x-y|^a} < \infty. \tag{12.73}$$

Then $\mathcal{H}^a(K) = \infty$, so $\operatorname{Hdim} K \ge a$.

Proof. We will assume $\mathcal{H}^a(K) < \infty$ and show that the integral in (12.73) must be $+\infty$. To do this, consider the set

$$E = \{x \in K : \limsup_{r\to 0}\, r^{-a}\mu(B_r(x)) > 0\}. \tag{12.74}$$

Below we will show that if $\mathcal{H}^a(K) < \infty$, then

$$\mu(K \setminus E) = 0. \tag{12.75}$$

For now we assume this and proceed with the proof.

If $x \in E$, there is a sequence $r_j \searrow 0$ such that $\mu(B_{r_j}(x)) \ge Cr_j^a$, for some $C > 0$. We can assume $\mu(\{x\}) = 0$, since otherwise the integral in (12.73) is clearly infinite. Then there exists $q_j \in (0, r_j)$ such that

$$\mu(A_j) \ge \frac{1}{2}Cr_j^a, \quad A_j = B_{r_j}(x) \setminus B_{q_j}(x). \tag{12.76}$$

Passing to a subsequence, we can assume $r_{j+1} < q_j$, so the shells A_j are disjoint. Then

$$(12.77)\quad \begin{aligned} x \in E \Rightarrow \int_K \frac{d\mu(y)}{|x-y|^a} &\geq \sum_j \int_{A_j} \frac{d\mu(y)}{|x-y|^a} \\ &\geq \frac{C}{2} \sum_{j\geq 1} r_j^a r_j^{-a} \\ &= +\infty. \end{aligned}$$

By (12.75), $\mu(E) > 0$, so the integral in (12.73) is

$$(12.78)\quad \geq \int_E \int_K \frac{d\mu(y)\, d\mu(x)}{|x-y|^a} = +\infty.$$

This proves Proposition 12.19, modulo the proof of the next lemma.

Lemma 12.20. *Given a compact $K \subset \mathbb{R}^n$ with $\mathcal{H}^a(K) < \infty$ and a finite Borel measure μ supported on K,*

$$(12.79)\quad F = \{x \in K : \limsup_{r\to 0}\, r^{-a}\mu(B_r(x)) = 0\} \Longrightarrow \mu(F) = 0.$$

Proof. Given $\varepsilon, \rho > 0$, set

$$F_{\varepsilon\rho} = \{x \in K : \mu(B_r(x)) \leq \varepsilon r^a,\ \forall\, r \leq \rho\}.$$

If $\{U_j\}$ is a countable cover of $F_{\varepsilon\rho}$ by sets of diameter $\leq \delta \leq \rho$ and each U_j contains a point of $F_{\varepsilon\rho}$, then there exist balls B_j, centered in $F_{\varepsilon\rho}$, such that $B_j \supset U_j$ and $\operatorname{diam} B_j \leq 2 \operatorname{diam} U_j \leq 2\rho$. Then

$$(12.80)\quad \mu(F_{\varepsilon\rho}) \leq \sum_j \mu(B_j) < 2^{-a}\varepsilon \sum_j (\operatorname{diam} B_j)^a \leq \varepsilon \sum_j (\operatorname{diam} U_j)^a.$$

Hence $\mu(F_{\varepsilon\rho}) \leq \varepsilon \mathcal{H}^a_\delta(F_{\varepsilon\rho})$, so

$$(12.81)\quad \mu(F_{\varepsilon\rho}) \leq \varepsilon \mathcal{H}^a(F_{\varepsilon\rho}) \leq \varepsilon \mathcal{H}^a(K), \quad \forall\, \varepsilon, \rho > 0.$$

This gives (12.79).

Proposition 12.19 has the following counterpart.

Proposition 12.21. *Given a compact $K \subset \mathbb{R}^n$ and $a \in (0, \infty)$, if for each nonzero positive Borel measure μ supported in K one has*

$$\iint \frac{d\mu(x)\, d\mu(y)}{|x-y|^a} = +\infty, \tag{12.82}$$

then $\mathcal{H}^b(K) = 0$ for all $b > a$, so $\operatorname{Hdim} K \le a$.

We will not prove Proposition 12.21; see [**Fal**], p. 78.

In Chapter 16, we will use Proposition 12.19 to prove that for almost every Brownian path ω in $\mathbb{R}^n$, if $n \ge 2$, $T \in (0, \infty)$,

$$\operatorname{Hdim} \omega([0,T]) = 2. \tag{12.83}$$

More precisely, Proposition 12.19 will be used to prove $\operatorname{Hdim} \omega([0,T]) \ge 2$. The proof that $\operatorname{Hdim} \omega([0,T]) \le 2$ will not use Proposition 12.21, but rather the result of Exercise 9 below.

When the Hausdorff dimension of a set $S \subset \mathbb{R}^n$ is not equal to its topological dimension, following B. Mandelbrot, we call S a "fractal." The Cantor sets $K(\vartheta)$, being totally disconnected, have topological dimension zero, hence they are fractals. By (12.83), Brownian paths in $\mathbb{R}^n$ are also fractals, when $n \ge 2$. Many other examples of fractals, a number of which are self-similar, together with speculations on their role in the description of nature, can be found in [**Mdb**]. Further measure-theoretic results on fractals can be found in [**Fal**].

REMARK. More generally than r-dimensional Hausdorff measure, one can take any monotone, continuous function $\varphi : [0, \infty) \to [0, \infty)$, satisfying $\varphi(0) = 0$, $\varphi(t) > 0$ for $t > 0$, and set

$$\begin{aligned} h^*_{\varphi,\delta}(S) &= \inf \Big\{ \sum_{j \ge 1} \varphi(\operatorname{diam} B_j) : S \subset \bigcup_{j \ge 1} B_j,\ \operatorname{diam} B_j \le \delta \Big\}, \\ h^*_\varphi(S) &= \lim_{\delta \to 0} h^*_{\varphi,\delta}(S), \end{aligned} \tag{12.84}$$

obtaining a metric outer measure. Then

$$h^\varphi(S) = h^*_\varphi(S), \quad \text{for } S\ h^*_\varphi\text{-measurable}, \tag{12.85}$$

defines a measure. An example is

$$\varphi(t) = t^2 \Big(\log \frac{1}{t}\Big) \Big(\log \log \log \frac{1}{t}\Big), \tag{12.86}$$

for $t << 1$. It has been shown that $0 < h^\varphi(\omega([0,T])) < \infty$ for almost all Brownian paths in $\mathbb{R}^2$; cf. [**Ray**], [**TS**].

Exercises

1. Suppose μ_j^* is an increasing sequence of outer measures on a set X and $\mu_j^*(S) \nearrow \mu^*(S)$ for all $S \subset X$. Show that μ^* is an outer measure. Hence deduce that h_r^* in (12.2) is an outer measure.

2. Let d_j be a sequence of metrics on a compact metrizable space X such that

$$ad_0(x,y) \le d_j(x,y) \le bd_0(x,y), \tag{12.87}$$

for some $a, b \in (0,\infty)$, all j, and, for a metric d on X,

$$d_j(x,y) \to d(x,y) \text{ as } j \to \infty, \tag{12.88}$$

for each $x, y \in X$. Let these metrics determine Hausdorff r-dimensional measures $\mathcal{H}^r_{d_j}$, $\mathcal{H}^r_d$. Show that, for any Borel set $S \subset X$,

$$\limsup_{j\to\infty} \mathcal{H}^r_{d_j}(S) \le \mathcal{H}^r_d(S).$$

In particular, if (12.88) is sharpened to

$$d_j(x,y) \searrow d(x,y), \tag{12.89}$$

then

$$\mathcal{H}^r_{d_j}(S) \searrow \mathcal{H}^r_d(S). \tag{12.90}$$

Hint. With $\lambda = b/a$, show that, for any $\varepsilon > 0$ and $\delta > 0$,

$$h^*_{r,\lambda\delta,d_j}(S) \le h^*_{r,\delta,d}(S) + \varepsilon, \text{ for } j \text{ large.}$$

3. Give a detailed demonstration of the first inequality in (12.8), stating that a Steiner symmetrization of a set $S \subset \mathbb{R}^n$ does not increase diameter.

4. Let $F : \mathbb{R}^n \to \mathbb{R}^n$ be a C^1 diffeomorphism. Show that $\mathcal{H}^k$ and $F_*\mathcal{H}^k$, defined by $F_*\mathcal{H}^k(S) = \mathcal{H}^k(F^{-1}(S))$, are mutually absolutely continuous, but if $0 < k < n$, there does not exist a measurable function g such that $F_*\mathcal{H}^k = g\mathcal{H}^k$, unless $DF(x)$ is everywhere a scalar multiple of an isometry. How does this relate to the Radon-Nikodym Theorem?

5. Give an elementary proof of Lemma 12.11 when $f \in C^1(\mathbb{R}^n)$.
Hint. If $x_0 \in K$ and $\nabla f(x_0) \neq 0$, use the Implicit Function Theorem to describe the zero set of f near x_0.

6. With $\mathcal{H}_B^r$ as in (12.4), show that, for Borel sets $A \subset \mathbb{R}^n$, $\mathcal{H}_B^n(A) = \mathcal{H}^n(A) = \mathcal{L}^n(A)$.
Hint. Show that the proof of Lemma 12.3 actually yields $\mathcal{H}_B^n(U) \leq \mathcal{L}^n(U)$, for open $U \subset \mathbb{R}^n$.
Note. The isodiametric inequality is not needed in the argument (otherwise parallel to that giving (12.11)) that $\mathcal{L}^n(A) \leq \mathcal{H}_B^n(A)$.

Exercises 7–8 deal with exterior normals. Let $\Omega \subset \mathbb{R}^n$ be an open set, with closure $\overline{\Omega}$, and let $p \in \partial\Omega$. We say a unit vector N is a unit exterior normal to $\partial\Omega$ at p provided

$$\lim_{r \searrow 0} \frac{m(\{x \in \Omega : (x-p)\cdot N > 0,\ |x-p| < r\})}{m(B_r)} = 0 \tag{12.91}$$

and

$$\lim_{r \searrow 0} \frac{m(\{x \in \Omega^c : (x-p)\cdot N < 0,\ |x-p| < r\})}{m(B_r)} = 0. \tag{12.92}$$

7. Show that at each $p \in \partial\Omega$ there is *at most one* exterior unit normal.

8. Assume $u : \mathbb{R}^{n-1} \to \mathbb{R}$ is Lipschitz, and let

$$\Omega = \{(x', x_n) \in \mathbb{R}^n : x_n < u(x')\}. \tag{12.93}$$

Show that, whenever u is differentiable at $p' \in \mathbb{R}^{n-1}$, then

$$N = (1 + |\nabla u(p')|^2)^{-1/2}\,(-\nabla u(p'), 1) \tag{12.94}$$

is the unit exterior normal to $\partial\Omega$ at $p = (p', u(p'))$. Deduce that in this situation the normal N, defined by (12.91)–(12.92), is given $\mathcal{H}^{n-1}$-a.e. on $\partial\Omega$ by (12.94).

9. Let $I \subset \mathbb{R}$ be a compact interval. Assume $F : I \to \mathbb{R}^n$ is Hölder continuous of exponent $s \in (0,1)$. Show that $\mathcal{H}^r(F(I)) < \infty$, with $r = 1/s$.
10. Give an example of a function $F : [0,1] \to \mathbb{R}^2$, Hölder continuous of exponent $1/2$, whose image is the square $[0,1] \times [0,1]$.

11. Give an example of a continuous function $F : [0,1] \to \mathbb{R}^2$ that is one-to-one and whose image has Hausdorff dimension 2.
12. Verify that the set functions h_φ^* defined by (12.84) are metric outer measures and that the set functions h^φ defined by (12.85) are measures, under the stated hypotheses on φ.
13. Find a compact set $K \subset \mathbb{R}$ such that $\mathcal{H}^r(K) = 0$ for all $r > 0$ but $h^\varphi(K) > 0$, where h^φ is defined as in (12.84)–(12.85), with

$$\varphi(t) = \left(\log \frac{1}{t}\right)^{-1}, \quad \text{for } t \leq \frac{1}{2}.$$

Chapter 13

Radon Measures

Recall that if X is a compact metric space, $C(X)$, the space of continuous (real-valued) functions on X, is a Banach space with the norm

$$\|f\| = \sup_{x \in X} |f(x)|. \tag{13.1}$$

We want to identify the dual of $C(X)$ with the space of (finite) signed Borel measures on X, also known as the space of Radon measures on X.

Before identifying the dual of $C(X)$, we will first identify the set of *positive* linear functionals on $C(X)$. By definition, a linear functional

$$\alpha : C(X) \longrightarrow \mathbb{R} \tag{13.2}$$

is positive provided

$$f \in C(X),\ f \geq 0 \Longrightarrow \alpha(f) \geq 0. \tag{13.3}$$

Clearly, if μ is a (positive) finite Borel measure on X, then

$$\alpha(f) = \int f \, d\mu \tag{13.4}$$

is a positive linear functional. We will establish the converse, that every positive linear functional on $C(X)$ is of the form (13.4).

It is easy to see that every positive linear functional α on $C(X)$ is bounded. In fact, applying α to $f(x) - a$ and to $b - f(x)$, we see that, when a and b are real numbers

$$f \in C(X),\ a \leq f \leq b \Longrightarrow a\alpha(1) \leq \alpha(f) \leq b\alpha(1), \tag{13.5}$$

so

$$|\alpha(f)| \le A\|f\|, \quad A = \alpha(1). \tag{13.6}$$

To begin the construction of μ, we construct a set function μ_0 on the collection $\mathcal{O}$ of open subsets of X by

$$\mu_0(U) = \sup\{\alpha(f) : f \prec U\}, \tag{13.7}$$

where we say

$$f \prec U \Longleftrightarrow f \in C(X),\ 0 \le f \le 1, \text{ and } \operatorname{supp} f \subset U. \tag{13.8}$$

Here, supp f is the closure of $\{x : f(x) \ne 0\}$. Clearly μ_0 is monotone. Then we set, for any $E \subset X$,

$$\mu^*(E) = \inf\{\mu_0(U) : E \subset U \in \mathcal{O}\}. \tag{13.9}$$

Of course, $\mu^*(U) = \mu_0(U)$ when U is open.

Lemma 13.1. *The set function μ^* is an outer measure.*

Proof. By Proposition 5.1, it suffices to show that

$$U_j \in \mathcal{O},\ U = \bigcup_{j\ge 1} U_j \Longrightarrow \mu_0(U) \le \sum_{j\ge 1} \mu_0(U_j), \tag{13.10}$$

so that we have an analogue of (5.5). Suppose $f \prec U$. We need to show that $\alpha(f) \le \sum \mu_0(U_j)$. Now, since supp $f = K$ is compact, we have $K \subset U_1 \cup \cdots \cup U_\ell$ for some finite ℓ. We claim there are $g_j \prec U_j$, $1 \le j \le \ell$, such that $\sum g_j = 1$ on K. Granted this, we can set $f_j = f g_j$. Then $f_j \prec U_j$, so $\alpha(f_j) \le \mu_0(U_j)$. Hence

$$\alpha(f) = \sum \alpha(f_j) \le \sum \mu_0(U_j), \tag{13.11}$$

as desired. Thus Lemma 13.1 will be proved once we have

Lemma 13.2. *If $K \subset X$ is compact, U_j are open, and $K \subset U_1 \cup \cdots \cup U_\ell = V$, then there exist $g_j \prec U_j$ such that $\sum g_j = 1$ on K.*

Proof. Set $U_{\ell+1} = X \setminus K$. Then $\{U_j : 1 \le j \le \ell+1\}$ is an open cover of X. Let $\{g_j : 1 \le j \le \ell+1\}$ be a partition of unity subordinate to this cover. (See Exercise 9 at the end of this chapter.) Then $\{g_j : 1 \le j \le \ell\}$ has the desired properties.

Now that we know μ^* is an outer measure, we prepare to apply Caratheodory's Theorem.

Lemma 13.3. *The outer measure μ^* is a metric outer measure.*

Proof. Let $S_j \subset X$, and assume

$$\text{dist}(S_1, S_2) \geq 4\varepsilon > 0. \tag{13.12}$$

Take $\delta > 0$. Given $U \supset S = S_1 \cup S_2$, U open, such that $\mu_0(U) \leq \mu^*(S) + \delta$, set

$$U_j = U \cap \{x \in X : \text{dist}(x, S_j) < \varepsilon\}.$$

It follows that

$$S_j \subset U_j, \quad U_1 \cup U_2 \subset U, \quad U_1 \cap U_2 = \emptyset. \tag{13.13}$$

Now, whenever $f \prec U_1 \cup U_2$, we have $f = f_1 + f_2$ with $f_j = f\big|_{U_j} \prec U_j$, which implies

$$\mu_0(U_1) + \mu_0(U_2) = \mu_0(U_1 \cup U_2).$$

Hence

$$\mu^*(S_1) + \mu^*(S_2) \leq \mu_0(U_1) + \mu_0(U_2) = \mu_0(U_1 \cup U_2) \leq \mu_0(U). \tag{13.14}$$

Thus $\mu^*(S) \geq \mu^*(S_1) + \mu^*(S_2) - \delta$, for all $\delta > 0$, which, together with subadditivity of μ^*, yields the desired identity $\mu^*(S) = \mu^*(S_1) + \mu^*(S_2)$.

It follows from Proposition 5.8 that every closed set in X is μ^*-measurable. Hence, by Theorem 5.2, every Borel set in X is μ^*-measurable, and the restriction of μ^* to $\mathfrak{B}(X)$, which we denote μ, is a measure.

We make a few useful comments about μ. First, in addition to (13.7), we have

$$\mu(U) = \sup\{\alpha(f) : f \preccurlyeq U\}, \tag{13.15}$$

for U open, where

$$f \preccurlyeq U \Longleftrightarrow f \in C(X),\ 0 \leq f \leq 1,\ f = 0 \text{ on } X \setminus U. \tag{13.16}$$

To see that (13.7) and (13.15) coincide, take $f \preccurlyeq U$. Then set $f_j = \xi_j(f)$, where $\xi_j(s)$ is defined by

$$\xi_j(s) = \begin{cases} 0 & \text{for } 0 \leq s \leq \dfrac{1}{j}, \\[2mm] 2s - \dfrac{2}{j} & \text{for } \dfrac{1}{j} \leq s \leq \dfrac{2}{j}, \\[2mm] s & \text{for } s \geq \dfrac{2}{j}. \end{cases}$$

It follows that $f_j \prec U$ and $f_j \nearrow f$ (uniformly); hence $\alpha(f_j) \nearrow \alpha(f)$.

Using the identity (13.15), we can establish the following.

Lemma 13.4. *If $K \subset X$ is compact, then*

$$\mu(K) = \inf \{\alpha(f) : f \in C(X),\ f \geq \chi_K\}. \tag{13.17}$$

Proof. Denote the right side of (13.17) by $\mu_1(K)$. It suffices to take the inf of $\alpha(f)$ over $f \succcurlyeq K$, where

$$f \succcurlyeq K \Longleftrightarrow f \in C(X),\ 0 \leq f \leq 1,\ f = 1 \text{ on } K. \tag{13.18}$$

Comparing this with (13.16), we see that $f \succcurlyeq K \Leftrightarrow 1 - f \preccurlyeq X \setminus K$, so

$$\begin{aligned} \mu_1(K) &= \inf \{\alpha(1) - \alpha(g) : g \preccurlyeq X \setminus K\} \\ &= \mu(X) - \mu(X \setminus K). \end{aligned} \tag{13.19}$$

On the other hand, since K is μ^*-measurable, $\mu(X \setminus K) + \mu(K) = \mu(X)$, so the identity (13.17) is proved.

We are now ready to prove

Theorem 13.5. *If X is a compact metric space and α is a positive linear functional on $C(X)$, then there exists a unique finite, positive Borel measure μ such that*

$$\alpha(f) = \int f \, d\mu \tag{13.20}$$

for all $f \in C(X)$.

Proof. We have constructed a positive Borel measure μ, which is finite since (13.7) implies $\mu(X) = \alpha(1)$. We next show that (13.20) holds. It suffices to check this when $f : X \to [0, 1]$.

In such a case, take $N \in \mathbb{Z}^+$ and define

$$\begin{aligned} \varphi_j(x) &= \min\,(f(x), jN^{-1}), \qquad & 0 \leq j \leq N, \\ f_j(x) &= \varphi_j(x) - \varphi_{j-1}(x), \qquad & 1 \leq j \leq N. \end{aligned} \tag{13.21}$$

We have $\sum f_j = f$ and

$$\frac{1}{N}\chi_{K_j} \leq f_j \leq \frac{1}{N}\chi_{K_{j-1}}, \quad K_j = \Big\{x \in X : f(x) \geq \frac{j}{N}\Big\}. \tag{13.22}$$

Hence

$$\frac{1}{N}\mu(K_j) \leq \int f_j \, d\mu \leq \frac{1}{N}\mu(K_{j-1}). \tag{13.23}$$

We claim that also

$$\frac{1}{N}\mu(K_j) \le \alpha(f_j) \le \frac{1}{N}\mu(K_{j-1}). \tag{13.24}$$

To see this, first note that if $K_{j-1} \subset U$ is open, then $Nf_j \prec U$, so $N\alpha(f_j) \le \mu_0(U)$. This implies the second inequality of (13.24). On the other hand, $Nf_j \succ K_j$, so Lemma 13.4 gives the first inequality of (13.24).

Summing (13.23) and (13.24), we have

$$\begin{aligned} \frac{1}{N}\sum_{j=1}^{N}\mu(K_j) \le \int f\,d\mu \le \frac{1}{N}\sum_{j=0}^{N-1}\mu(K_j), \\ \frac{1}{N}\sum_{j=1}^{N}\mu(K_j) \le \alpha(f) \le \frac{1}{N}\sum_{j=0}^{N-1}\mu(K_j). \end{aligned} \tag{13.25}$$

Hence

$$\left|\alpha(f) - \int f\,d\mu\right| \le \frac{1}{N}\mu(K_0) - \frac{1}{N}\mu(K_N) \le \frac{1}{N}\mu(X). \tag{13.26}$$

Letting $N \to \infty$, we have (13.20).

Only the uniqueness of μ remains to be proved. To see this, let λ be a positive Borel measure on X such that

$$\alpha(f) = \int f\,d\lambda \tag{13.27}$$

for all $f \in C(X)$. Let $K \subset X$ be compact, and apply this to

$$f_\nu(x) = \bigl(1 + \nu\,\operatorname{dist}(x, K)\bigr)^{-1}; \quad f_\nu \searrow \chi_K. \tag{13.28}$$

By the Monotone Convergence Theorem, we have $\int f_\nu\,d\mu \searrow \int \chi_K\,d\mu$ and $\int f_\nu\,d\lambda \searrow \int \chi_K\,d\lambda$, so $\alpha(f_\nu) \searrow \mu(K)$ and $\alpha(f_\nu) \searrow \lambda(K)$. Hence $\mu(K) = \lambda(K)$ for all compact K. Now, by (5.60), for every positive Borel measure λ on a compact metric space X, we have

$$E \in \mathfrak{B}(X) \Longrightarrow \lambda(E) = \sup\,\{\lambda(K) : K \subset E, K \text{ compact}\}. \tag{13.29}$$

This proves uniqueness.

Generally, if X is a compact Hausdorff space and λ is a finite (positive) measure on $\mathfrak{B}(X)$, λ is said to be *regular* if and only if (13.29) holds. The implication of Exercises 10–13 of Chapter 5 is that every finite Borel measure is regular when X is a compact *metric* space. If X is compact but not metrizable, a finite measure on $\mathfrak{B}(X)$ need not be regular. The generalization of Theorem 13.5 to this case is that, given a positive linear functional α on $C(X)$, there is a unique finite regular Borel measure μ such that (13.20) holds. (Note that if λ is any finite measure on $\mathfrak{B}(X)$, then (13.27) defines a positive linear functional on $C(X)$, which then gives rise to a regular Borel measure.) For this more general case, the construction of μ^* is the same as was done above in (13.7)–(13.9), but the proof that μ^* yields a regular measure on $\mathfrak{B}(X)$ is a little more elaborate than the proof given above for compact metric spaces. Treatments can be found in [**Fol**] and [**Ru**].

We want to extend Theorem 13.5 to the case of a general bounded linear functional

$$\omega : C(X) \to \mathbb{R}. \tag{13.30}$$

We start with an analogue of the Hahn decomposition.

Lemma 13.6. *If ω is a bounded (real) linear functional on $C(X)$, then there are positive linear functionals $\alpha^\pm$ on $C(X)$ such that*

$$\omega = \alpha^+ - \alpha^-. \tag{13.31}$$

Proof. We first define α^+ on

$$C^+(X) = \{f \in C(X) : f \ge 0\}. \tag{13.32}$$

For $f \in C^+(X)$, set

$$\alpha^+(f) = \sup\{\omega(g) : g \in C^+(X),\ 0 \le g \le f\}. \tag{13.33}$$

The hypothesis that ω is bounded implies $|\omega(g)| \le K\|g\| \le K\|f\|$ when $0 \le g \le f$, so

$$0 \le \alpha^+(f) \le K\|f\|, \text{ for } f \in C^+(X), \tag{13.34}$$

where $K = \|\omega\|$. Clearly, for $c \in \mathbb{R}$,

$$f \in C^+(X),\ c > 0 \Longrightarrow \alpha^+(cf) = c\alpha^+(f). \tag{13.35}$$

Now, suppose $f_1, f_2 \in C^+(X)$. If $g \in C^+(X)$ and $0 \le g \le f_1 + f_2$, we can write $g = g_1 + g_2$ with $g \in C^+(X)$ and $0 \le g_j \le f_j$. Just take $g_1 = \min(g, f_1)$. Hence

$$f_1, f_2 \in C^+(X) \Longrightarrow \alpha^+(f_1 + f_2) = \alpha^+(f_1) + \alpha^+(f_2). \tag{13.36}$$

We claim that α^+ has an extension to a linear functional on $C(X)$, which would necessarily be positive. In fact, given $f \in C(X)$, write

$$f = f_1 - f_2, \quad f_j \in C^+(X), \quad \alpha^+(f) = \alpha^+(f_1) - \alpha^+(f_2). \tag{13.37}$$

A given $f \in C(X)$ has many such representations as $f = f_1 - f_2$; showing that $\alpha^+(f)$ is independent of such a representation and defines a linear functional on $C(X)$ is a simple application of (13.36); compare the proof of Proposition 3.7. Note that, if we take $f_1 = f^+ = \max(f, 0)$ and $f_2 = f^- = \max(-f, 0)$, we see that

$$|\alpha^+(f)| \le K \max\big(\|f^+\|, \|f^-\|\big) \le K\|f\|. \tag{13.38}$$

Finally we set $\alpha^- = \alpha^+ - \omega$. It remains only to show that $f \in C^+(X) \Rightarrow \alpha^-(f) \ge 0$, i.e., $\alpha^+(f) \ge \omega(f)$. But that is immediate from the definition (13.33), so the lemma is proved.

We can combine Lemma 13.6 and Theorem 13.5 to prove the following, known as the Riesz Representation Theorem.

Theorem 13.7. *If X is a compact metric space and ω is a bounded (real) linear functional on $C(X)$, then there is a unique finite signed measure ρ on $\mathfrak{B}(X)$ such that*

$$\omega(f) = \int_X f \, d\rho, \tag{13.39}$$

for all $f \in C(X)$. Furthermore,

$$\|\rho\| = |\rho|(X) = \|\omega\|, \tag{13.40}$$

so there is an isometric isomorphism

$$C(X)' \approx \mathfrak{M}(X). \tag{13.41}$$

Here, $\mathfrak{M}(X)$ denotes the linear space of finite signed measures on $\mathfrak{B}(X)$, with norm given by the first identity in (13.40). This is also known as the space of finite Radon measures on X.

For the proof, write $\omega = \alpha^+ - \alpha^-$, as in (13.31), take finite positive measures $\mu^\pm$ on $\mathfrak{B}(X)$ so that $\alpha^\pm(f) = \int f \, d\mu^\pm$, and set $\rho = \mu^+ - \mu^-$. Thus we have the identity (13.39).

We need to prove (13.40). Let $\rho = \rho^+ - \rho^-$ be the Hahn decomposition of ρ, so $\rho^+ \perp \rho^-$ and

$$\omega(f) = \int f\, d\rho^+ - \int f\, d\rho^-, \tag{13.42}$$

for all $f \in C(X)$. Consequently

$$|\omega(f)| \le \|f\|\rho^+(X) + \|f\|\rho^-(X) = \|\rho\| \cdot \|f\|, \tag{13.43}$$

so we have $\|\omega\| \le \|\rho\|$. To prove the reverse inequality, let $\delta > 0$. Suppose $\rho^\pm$ are supported on disjoint Borel sets $X_\pm$. Let $K_\pm$ be compact sets in $X_\pm$ such that $\rho^\pm(K_\pm) \ge \rho^\pm(X) - \delta$. We have $K_+ \cap K_- = \emptyset$, so say $\operatorname{dist}(K_+, K_-) = \varepsilon > 0$. Let $U_\pm = \{x : \operatorname{dist}(x, K_\pm) < \varepsilon/4\}$, so $U_+ \cap U_- = \emptyset$. Using a simple variant of (13.28), we can construct $\varphi_\nu^\pm \prec U_\pm$ such that $\varphi_\nu^\pm \searrow \chi_{K_\pm}$. Hence, as $\nu \to \infty$,

$$\omega(\varphi_\nu^+ - \varphi_\nu^-) \to \rho^+(K_+) + \rho^-(K_-) \ge \|\rho\| - 2\delta, \tag{13.44}$$

while $\|\varphi_\nu^+ - \varphi_\nu^-\| = 1$. This proves the reverse inequality, $\|\rho\| \le \|\omega\|$, and establishes (13.40). This inequality also establishes uniqueness, so Theorem 13.7 is proved.

Though it is not needed to prove Theorem 13.7, we mention that $\mu^+ - \mu^-$ *is* the Hahn decomposition of ρ (provided α^+ is given by (13.33)). The proof is an exercise.

Applying Alaoglu's Theorem (Proposition 9.2 and Corollary 9.3) to $V = C(X)$, we have the following compactness result:

Proposition 13.8. *If X is a compact metric space, the closed unit ball in $\mathfrak{M}(X)$ is a compact metrizable space, with the weak* topology.*

By the definition of weak* topology given in Chapter 9, we see that a sequence μ_ν in $\mathfrak{M}(X)$ converges weak* to μ if and only if

$$\int f\, d\mu_\nu \longrightarrow \int f\, d\mu, \tag{13.45}$$

for each $f \in C(X)$. This topology on $\mathfrak{M}(X)$ is also called the *vague* topology; one says $\mu_\nu \to \mu$ vaguely provided (13.45) holds for all $f \in C(X)$.

There is an important subset of $\mathfrak{M}(X)$, the set of probability measures. An element μ of $\mathfrak{M}(X)$ is called a probability measure if and only if μ is positive and $\mu(X) = 1$. We use the notation

$$\operatorname{Prob}(X) = \{\mu \in \mathfrak{M}(X) : \mu \ge 0,\ \mu(X) = 1\}. \tag{13.46}$$

It is easy to see that

$$\operatorname{Prob}(X) = \{\mu \in \mathfrak{M}(X) : \|\mu\| \le 1,\ \mu(X) = 1\}. \tag{13.47}$$

Hence $\operatorname{Prob}(X)$ is a subset of the unit ball of $\mathfrak{M}(X)$ which is *closed* in the weak* topology. This has the following useful consequence.

Corollary 13.9. *If X is a compact metric space, the set* $\text{Prob}(X)$ *of probability measures on X is compact (and metrizable) in the weak* topology.*

It is often useful to consider Borel measures on a *locally compact* space Y, i.e., a Hausdorff space with the property that any $y \in Y$ has a compact neighborhood. In such a case, there is a Banach space $C_*(Y)$, the space of continuous functions on Y that "vanish at infinity." We say a continuous function $u : Y \to \mathbb{R}$ belongs to $C_*(Y)$ if, for any $\delta > 0$, there exists a compact $K \subset Y$ such that $|u(y)| < \delta$ for $y \in Y \setminus K$. We use the sup norm on $C_*(Y)$. We can construct the "one point compactification" $\widehat{Y} = Y \cup \{\infty\}$, declaring a set $U \subset \widehat{Y}$ to be open if either $U \subset Y$ is open or $\infty \in U$ and $\widehat{Y} \setminus U = K$ is a compact subset of Y. It readily follows that $\widehat{Y}$ is a compact Hausdorff space. Also, $C_*(Y)$ is naturally isomorphic to a closed linear subspace of $C(\widehat{Y})$:

$$C_*(Y) \approx \{u \in C(\widehat{Y}) : u(\infty) = 0\}. \tag{13.48}$$

Furthermore, given any $f \in C(\widehat{Y})$, if we set $a = f(\infty)$, then $f = g + a,\ g(\infty) = 0$, so

$$C(\widehat{Y}) \approx C_*(Y) \oplus \mathbb{R}, \tag{13.49}$$

and, for the duals, we have

$$C(\widehat{Y})' \approx C_*(Y)' \oplus \mathbb{R}. \tag{13.50}$$

In case Y has the additional properties of being metrizable and σ-compact, the one-point compactification $\widehat{Y}$ is also metrizable; cf. Exercise 15. Hence we can appeal to Theorem 13.7 to identify $C(\widehat{Y})'$ with $\mathfrak{M}(\widehat{Y})$, the space of finite signed measures on $\mathfrak{B}(\widehat{Y})$. (Even without these additional hypotheses, $C(\widehat{Y})'$ can be identified with the space of signed finite regular Borel measures on $\widehat{Y}$, though one must go to another source, such as [**Fol**] or [**Ru**], for a proof.) In the decomposition (13.50), we see that the last factor on the right consists of multiples of δ_∞, the measure defined by

$$\delta_\infty(S) = 1 \ \text{ if } \ \infty \in S, \quad 0 \ \text{ otherwise.}$$

Consequently we have the identification

$$C_*(Y)' \approx \{\mu \in \mathfrak{M}(\widehat{Y}) : \mu(\{\infty\}) = 0\} \approx \mathfrak{M}(Y), \tag{13.51}$$

where $\mathfrak{M}(Y)$ is the space of finite signed measures on $\mathfrak{B}(Y)$.

Exercises

In Exercises 1–3, X is a compact metric space and $F : X \to X$ is a continuous map. As in (7.32), we set

$$F_*\mu(S) = \mu_{F^{-1}}(S) = \mu\big(F^{-1}(S)\big), \tag{13.52}$$

for a Borel measure μ on X. We set $F^n = F \circ \cdots \circ F$ (n factors) and similarly define $F_*^n \mu$.

1. Define $T : C(X) \to C(X)$ by $Tu(x) = u(F(x))$. Show that, under the identification (13.41), the adjoint $T' : C(X)' \to C(X)'$ is given by (13.52).

2. Assume μ is a probability measure on X. Set

$$\mu_j = F_*^j \mu, \quad \nu_n = \frac{1}{n+1} \sum_{j=0}^{n} \mu_j. \tag{13.53}$$

   Show that $\nu_n \in \operatorname{Prob}(X)$ and

$$F_*\nu_n = \nu_n + \frac{1}{n+1}\big(\mu_{n+1} - \mu_0\big).$$

3. Suppose $\nu_{n_j} \to \nu$ in the weak* topology on $\operatorname{Prob}(X)$, as $j \to \infty$. Show that

$$F_*\nu = \nu. \tag{13.54}$$

   We say ν is an *invariant* measure for F.
   Hint. Given $u \in C(X)$, one has

$$\int u \circ F \, d\nu_{n_j} = \int u \, dF_*\nu_{n_j} = \int u \, d\nu_{n_j} + \frac{1}{n_j + 1}\Big(\int u \, d\mu_{n_j+1} - \int u \, d\mu_0\Big).$$

4. Let $f \in L^\infty(\mathbb{R})$. Show that the following are equivalent:
   (a) $h^{-1}(\tau_h f - f)$ is bounded in $L^1(\mathbb{R})$, for $h \in (0, 1]$.
   (b) $\partial_1 f = \mu$ for some finite signed measure μ on $\mathbb{R}$.
   (c) $f(x) = \int_{-\infty}^{x} d\mu$ a.e. on $\mathbb{R}$.

5. Let $f_0(x)$ denote the right side of (c) above, defined as in Exercise 1 of Chapter 11. Show that f_0 is the unique right continuous function on $\mathbb{R}$ equal to f a.e.

6. Let $f : \mathbb{R} \to \mathbb{R}$ be bounded and right-continuous. Show that f has the properties (a)–(c) of Exercise 4 if and only if there exists $C < \infty$ such that, for any finite set of real numbers $x_0 < x_1 < \cdots < x_\ell$,

$$\sum_{j=1}^{\ell} |f(x_j) - f(x_{j-1})| \le C. \tag{13.55}$$

Hint. To prove one implication, given $\nu \in \mathbb{Z}^+$, set $f_\nu(x) = f(\nu)$ if $x \ge \nu$, $f_\nu(x) = f(-\nu)$ if $x \le -\nu$, and

$$f_\nu(x) = f(-\nu + 2^{-\nu}j) \text{ if } -\nu + 2^{-\nu}j \le x < -\nu + 2^{-\nu}(j+1), \quad 0 \le j < 2\nu 2^\nu.$$

Show that $f_\nu \to f$ and $\{\partial_1 f_\nu : \nu \in \mathbb{Z}^+\}$ is a bounded set of measures on $\mathbb{R}^+$. Consider weak* limits. One says that f has *bounded variation* on $\mathbb{R}$ if this property holds, and one writes $f \in \mathrm{BV}(\mathbb{R})$.

7. Let $f \in L^\infty(\mathbb{R})$. Show that f is equal a.e. to an element of $\mathrm{BV}(\mathbb{R})$ if and only if you can write $f = g_1 - g_2$ a.e., with $g_j \in L^\infty(\mathbb{R})$ monotone $\nearrow$. In particular, if $g \in L^\infty(\mathbb{R})$ is monotone $\nearrow$, then $\partial_1 g$ is a positive measure. Reconsider Exercise 1 of Chapter 5, on Lebesgue-Stieltjes measures, in this light.
Hint. Given $f \in \mathrm{BV}(\mathbb{R})$, apply the Hahn decomposition to the signed measure μ arising in Exerise 4.

8. Let $f : \mathbb{R} \to \mathbb{R}$ be bounded and continuous. One says f is *absolutely continuous* provided that, for every $\varepsilon > 0$, there is a $\delta > 0$ with the property that, for any finite collection of disjoint intervals, $(a_1, b_1), \dots, (a_N, b_N)$,

$$\sum (b_j - a_j) < \delta \Longrightarrow \sum |f(b_j) - f(a_j)| < \varepsilon. \tag{13.56}$$

Show that the following are equivalent:
(a) f is absolutely continuous.
(b) $\partial_1 f = g \in L^1(\mathbb{R})$.
(c) $f(x) = \int_{-\infty}^{x} g(y)\, dy, \quad g \in L^1(\mathbb{R})$.
Hint. If (a) holds, first show that f has bounded variation and use Exercises 5–6 to get $\partial_1 f = \mu$. Then show μ is absolutely continuous with respect to Lebesgue measure. Compare Exercises 7–9 in Chapter

10 and also Exercise 1 in Chapter 11.

9. Let $\{U_1, \dots, U_{\ell+1}\}$ be an open cover of a compact metric space X. Show that there exist open sets V_j, $j = 1, \dots, \ell+1$, covering X, such that $\overline{V}_j \subset U_j$. Let
$$h_j(x) = \operatorname{dist}(x, X \setminus V_j).$$
Show that $h_j \in C(X)$, supp $H_j \subset \overline{V}_j \subset U_j$, and that $h = \sum_{j=1}^{\ell+1} h_j \in C(X)$ is > 0 on X. Deduce that
$$g_j(x) = h_j(x)/h(x)$$
form a partition of unity of X, subordinate to the cover $\{U_1, \dots, U_{\ell+1}\}$.
Hint. Set $H_j(x) = \operatorname{dist}(x, X \setminus U_j)$. Show that $H = \sum H_j > 0$ on X, $H \in C(X)$, hence $H \geq a > 0$. Set $V_j = \{x \in U_j : H_j(x) > a/(\ell+2)\}$.

10. The countable infinite product $Z = \prod_{j \geq 1} \{0, 1\}$ is compact, with the product topology, and metrizable (cf. Appendix A). Let $\mathcal{A} \subset C(Z)$ consist of continuous functions depending on only finitely many variables, so an element $f \in \mathcal{A}$ has the form $f(x) = f(x_1, \dots, x_k)$, for some $k \in \mathbb{Z}^+$. For such an f, set
$$\varphi(f) = 2^{-k} \sum_{x_j \in \{0,1\}, 1 \leq j \leq k} f(x_1, \dots, x_k).$$
Show that φ has a unique extension to a positive linear functional α on $C(Z)$. Show that, for $f \in C(Z)$, $\alpha(f) = \int_Z f \, d\mu$, where μ is the product measure on Z, discussed in Exercises 7–10 of Chapter 6.

11. Generalize Exercise 10 to other countable products of compact metric spaces, carrying positive Radon measures of total mass one.

12. Let $R = I_1 \times \cdots \times I_n$ be a compact cell in $\mathbb{R}^n$, with $I_\nu = [a_\nu, b_\nu]$. Consider the positive linear functional on $C(R)$ given by $f \mapsto I(f)$, where $I(f)$ is the Riemann integral of f, discussed in Chapter 1 (for $n = 1$) and in the second exercise set at the end of Chapter 7 (for $n > 1$). Show that the measure on R produced by Theorem 13.5 coincides with Lebesgue measure (on the Borel subsets of R), as constructed in Chapter 2 (for $n = 1$) and in Chapter 7 (for $n > 1$).

13. Let X be a compact Hausdorff space. Show that if $C(X)$ is separable, then X is metrizable.
Hint. Define $\varphi : X \to C(X)'$ by $\varphi(y)(f) = f(y)$ and use Corollary 9.3.

14. Let Y be a locally compact Hausdorff space. Assume Y is σ-compact, i.e., there is a countable family of compact $K_j \subset Y$ such that $Y = \bigcup_j K_j$. Show that there is a sequence $f_k \in C_*(Y)$ such that each f_k has compact support and for each $y \in Y$, some $f_k(y) \neq 0$. (Also arrange $0 \le f_k \le 1$.) Note that then

$$f = \sum_{k\ge 1} 2^{-k} f_k \in C_*(Y)$$

is > 0 at each point of Y and that $U_k = \{y \in Y : f(y) > 1/k\}$ satisfies

$$U_k \text{ open}, \quad \overline{U}_k \text{ compact}, \quad U_k \nearrow Y.$$

15. Let Y be a locally compact metrizable space. Assume Y is σ-compact. Show that $C_*(Y)$ is separable. Deduce that the one-point compactification $\widehat{Y}$ is metrizable.
Hint. With $U_k \subset Y$ as in Exercise 14, show that $V_k = \{f \in C_*(Y) : \operatorname{supp} f \subset \overline{U}_k\}$ is separable and $\bigcup_k V_k$ is dense in $C_*(Y)$.

In Exercises 16–17, take $C(X)$ to be the space of *complex-valued* continuous functions on the compact metric space X. Let ρ be a complex Borel measure on X, of the form $\rho = \nu_1 + i\nu_2$ (as in (8.19)), where ν_j are finite signed measures, with associated positive measures $|\nu_j|$ as in (8.10). Set $|\nu| = |\nu_1| + |\nu_2|$ and apply the Radon-Nikodym theorem to $\nu_j << |\nu|$ to obtain

$$\rho = f\,|\nu|.$$

16. Show that

$$\|\rho\| = \sup\Bigl\{\sum_{k\ge 0} |\rho(S_k)| : S_k \in \mathfrak{B}(X) \text{ disjoint}\Bigr\}$$

defines a norm on the set $\mathfrak{M}_{\mathbb{C}}(X)$ of complex Borel measures on X. Show that

$$\|\rho\| = \int |f|\, d|\nu|.$$

17. Show that the dual $C(X)'$ of $C(X)$ is isomorphic to $\mathfrak{M}_{\mathbb{C}}(X)$, with norm given in Exercise 16.

Chapter 14

Ergodic Theory

Throughout this chapter we assume $(X, \mathfrak{F}, \mu)$ is a probability space, i.e., a measure space with $\mu(X) = 1$. Ergodic theory studies properties of measure-preserving mappings $\varphi : X \to X$. That is, we assume

$$S \in \mathfrak{F} \Longrightarrow \varphi^{-1}(S) \in \mathfrak{F} \text{ and } \mu(\varphi^{-1}(S)) = \mu(S). \tag{14.1}$$

The map φ defines a linear map T on functions:

$$Tf(x) = f(\varphi(x)). \tag{14.2}$$

If (14.1) holds, then, given $f \in L^1(X, \mu)$,

$$\int_X f(\varphi(x))\, d\mu = \int_X f(x)\, d\mu. \tag{14.3}$$

Hence $T : L^p(X, \mu) \to L^p(X, \mu)$ is an isometry for each $p \in [1, \infty]$. A central object of study in ergodic theory is the sequence of means:

$$A_m f(x) = \frac{1}{m} \sum_{k=0}^{m-1} T^k f(x). \tag{14.4}$$

In particular, one considers whether $A_m f$ tends to a limit, as $m \to \infty$, and whether that limit is a constant, namely $c = \int_X f\, d\mu$.

The first basic result of this nature, due to J. von Neumann, deals with $f \in L^2(X, \mu)$. Actually it has a Hilbert space setting. Recall that if a linear operator $T : H \to H$ on a Hilbert space H is an isometry, then $T^*T = I$. The abstract result uses the following simple lemma.

Lemma 14.1. *If $T : H \to H$ is a linear isometry on a Hilbert space H, then there is an orthogonal direct sum*

$$H = K \oplus \overline{R}, \tag{14.5}$$

where

$$K = \operatorname{Ker}(I - T^*) = \operatorname{Ker}(I - T), \quad R = \operatorname{Range}(I - T), \tag{14.6}$$

and $\overline{R}$ is the closure of R.

Proof. First, note that

$$\begin{aligned} R^\perp &= \{v \in H : (v, (I - T)w) = 0, \ \forall\, w \in H\} \\ &= \{v \in H : ((I - T^*)v, w) = 0, \ \forall\, w \in H\} \\ &= \operatorname{Ker}(I - T^*). \end{aligned}$$

Now (14.5) follows by (4.29)–(4.30) and the rest of the paragraph there, which implies $\overline{R} = K^\perp$, with $K = \operatorname{Ker}(I - T^*)$.

It remains to show that $\operatorname{Ker}(I - T^*) = \operatorname{Ker}(I - T)$. Since $T^*T = I, I - T^* = -T^*(I - T)$, so clearly $\operatorname{Ker}(I - T) \subset \operatorname{Ker}(I - T^*)$. For the reverse inclusion, note that $T^*T = I \Rightarrow (TT^*)^2 = TT^*$, so $Q = TT^*$ is the orthogonal projection of H onto the range of T. (Cf. Exercises 16–17 of Chapter 9.) Now $T^*u = u \Rightarrow Qu = Tu$, but then $\|Qu\| = \|Tu\| = \|u\|$, so $Qu = u$ and hence $Tu = u$, giving the converse.

Here is the abstract Mean Ergodic Theorem.

Proposition 14.2. *In the setting of Lemma 14.1, for each $f \in H$,*

$$A_m f = \frac{1}{m} \sum_{k=0}^{m-1} T^k f \longrightarrow Pf, \tag{14.7}$$

in H-norm, where P is the orthogonal projection of H onto K.

Proof. Clearly $A_m f \equiv f$ if $f \in K$. If $f = (I - T)v \in R$, then

$$\frac{1}{m} \sum_{k=0}^{m-1} T^k f = \frac{1}{m}(v - T^m v) \to 0, \quad \text{as } m \to \infty, \tag{14.8}$$

and since the operator norm $\|A_m\| \le 1$ for each m, this convergence holds on $\overline{R}$. Now (14.7) follows from (14.5).

Proposition 14.2 immediately applies to (14.4) when $f \in L^2(X, \mu)$. We next establish a more general result.

Proposition 14.3. *Let P denote the orthogonal projection of $L^2(X,\mu)$ onto* Ker $(I-T)$. *Then, for $p \in [1,2]$, P extends to a continuous projection on $L^p(X,\mu)$, and*

$$f \in L^p(X,\mu) \Longrightarrow A_m f \to Pf \tag{14.9}$$

in L^p-norm, as $m \to \infty$.

Proof. Note that the L^p-operator norm $\|A_m\|_{\mathcal{L}(L^p)} \le 1$ for each m, and since $\|g\|_{L^p} \le \|g\|_{L^2}$ for $p \in [1,2]$, we have (14.9) in L^p-norm for each f in the dense subspace $L^2(X,\mu)$ of $L^p(X,\mu)$. Now, given $f \in L^p(X,\mu)$, $\varepsilon > 0$, pick $g \in L^2(X,\mu)$ such that $\|f-g\|_{L^p} < \varepsilon$. Then

$$\|A_n f - A_m f\|_{L^p} \le \|A_n g - A_m g\|_{L^2} + \|A_n(f-g)\|_{L^p} + \|A_m(f-g)\|_{L^p}. \tag{14.10}$$

Hence

$$\limsup_{m,n\to\infty} \|A_n f - A_m f\|_{L^p} \le 2\varepsilon, \quad \forall\, \varepsilon > 0. \tag{14.11}$$

This implies the sequence $(A_n f)$ is Cauchy in $L^p(X,\mu)$, for each $f \in L^p(X,\mu)$. Hence it has a limit; call it Qf. Clearly Qf is linear in f, $\|Qf\|_{L^p} \le \|f\|_{L^p}$, and $Qf = Pf$ for $f \in L^2(X,\mu)$. Hence Q is the unique continuous extension of P from $L^2(X,\mu)$ to $L^p(X,\mu)$ (so we change its name to P). Note that $P^2 = P$ on $L^p(X,\mu)$, since it holds on the dense linear subspace $L^2(X,\mu)$. Proposition 14.3 is proven.

REMARK. Note that $P = P^*$. It follows that $P : L^p(X,\mu) \to L^p(X,\mu)$ for all $p \in [1,\infty]$. We will show in Proposition 14.7 that (14.9) holds in L^p-norm for $p < \infty$. The subject of mean ergodic theorems has been considerably extended and abstracted by K. Yosida, S. Kakutani, W. Eberlein, and others. An account can be found in [**Kr**].

Such mean ergodic theorems were complemented by pointwise convergence results on $A_m f(x)$, first by G. Birkhoff. This can be done via estimates of Yosida and Kakutani on the maximal function

$$A^{\#} f(x) = \sup_{m\ge 1} A_m f(x) = \sup_{n \ge 1} A_n^{\#} f(x), \tag{14.12}$$

where

$$A_n^{\#} f(x) = \sup_{1 \le m \le n} A_m f(x). \tag{14.13}$$

We follow a clean route to such maximal function estimates given in [**Gar**].

Lemma 14.4. *With A_m given by (14.2)–(14.4) and $f \in L^1(X,\mu)$, set*

$$E_n = \{x \in X : A_n^\# f(x) \ge 0\}. \tag{14.14}$$

Then

$$\int_{E_n} f \, d\mu \ge 0. \tag{14.15}$$

Proof. For notational convenience, set

$$S_k f = k A_k f = f + Tf + \cdots + T^{k-1} f, \quad M_k f = k A_k^\# f = \sup_{1 \le \ell \le k} S_\ell f.$$

For $k \in \{1, \dots, n\}$, $(M_n f)^+ \ge S_k f$, and hence (because T is positivity preserving)

$$f + T(M_n f)^+ \ge f + T S_k f = S_{k+1} f.$$

Hence

$$f \ge S_k f - T(M_n f)^+, \quad \text{for} \ \ 1 \le k \le n,$$

this holding for $k \ge 2$ by the argument above, and trivially for $k = 1$. Taking the max over $k \in \{1, \dots, n\}$ yields

$$f \ge M_n f - T(M_n f)^+. \tag{14.16}$$

Integrating (14.16) over E_n yields

$$\begin{aligned} \int_{E_n} f \, d\mu &\ge \int_{E_n} (M_n f - T(M_n f)^+) \, d\mu \\ &= \int_{E_n} ((M_n f)^+ - T(M_n f)^+) \, d\mu \\ &= \int_X (M_n f)^+ \, d\mu - \int_{E_n} T(M_n f)^+ \, d\mu \\ &\ge \int_X (M_n f)^+ \, d\mu - \int_X T(M_n f)^+ \, d\mu = 0, \end{aligned} \tag{14.17}$$

the first and second identities on the right because $M_n f \ge 0$ precisely on E_n, the last inequality because $T(M_n f)^+ \ge 0$, and the last identity by (14.3). This proves the lemma.

Lemma 14.4 leads to the following maximal function estimate.

Proposition 14.5. *In the setting of Lemma 14.4, one has, for each* $\lambda > 0$,

$$\mu(\{x \in X : A_n^{\#} f(x) \geq \lambda\}) \leq \frac{1}{\lambda} \|f\|_{L^1}. \tag{14.18}$$

Proof. If we set $E_{n\lambda} = \{x \in X : A_n^{\#} f(x) \geq \lambda\} = \{x \in X : A_n^{\#}(f(x) - \lambda) \geq 0\}$, then Lemma 14.4 yields

$$\int_{E_{n\lambda}} (f - \lambda)\, d\mu \geq 0. \tag{14.19}$$

Thus

$$\|f\|_{L^1} \geq \int_{E_{n\lambda}} f\, d\mu \geq \lambda\, \mu(E_{n\lambda}), \tag{14.20}$$

as asserted in (14.18).

Note that

$$E_{n\lambda} \nearrow \{x \in X : A^{\#} f(x) \geq \lambda\} = E_\lambda, \tag{14.21}$$

so we have $\mu(E_\lambda) \leq \|f\|_{L^1}/\lambda$. Now we introduce the maximal function

$$\mathcal{A}^{\#} f(x) = \sup_{m \geq 1} |A_m f(x)| \leq A^{\#} |f|(x). \tag{14.22}$$

We have

$$\mu(\{x \in X : \mathcal{A}^{\#} f(x) \geq \lambda\}) \leq \frac{1}{\lambda} \|f\|_{L^1}. \tag{14.23}$$

We are now ready for Birkhoff's Pointwise Ergodic Theorem.

Theorem 14.6. *If* T *and* A_m *are given by* (14.2)–(14.4), *where* φ *is a measure-preserving map, then, given* $f \in L^1(X, \mu)$,

$$\lim_{m \to \infty} A_m f(x) = Pf(x), \quad \mu\text{-a.e.} \tag{14.24}$$

Proof. Given $f \in L^1(X, \mu)$, $\varepsilon > 0$, let us pick $f_1 \in L^2(X, \mu)$ such that $\|f - f_1\|_{L^1} \leq \varepsilon/2$. Then use Lemma 14.1, with $H = L^2(X, \mu)$, to produce

$$g \in \operatorname{Ker}(I - T), \quad h = (I - T)v, \quad \|f_1 - (g + h)\|_{L^2} \leq \frac{\varepsilon}{2}. \tag{14.25}$$

Here $v \in L^2(X,\mu)$. It follows that

$$\|f - (g+h)\|_{L^1} \le \varepsilon, \tag{14.26}$$

and we have

$$\begin{aligned} A_m f &= A_m g + A_m h + A_m(f-g-h) \\ &= g + \frac{1}{m}(v - T^m v) + A_m(f-g-h). \end{aligned} \tag{14.27}$$

Clearly $v(x)/m \to 0$, μ-a.e., as $m \to \infty$. Also

$$\int_X \sum_{m\ge 1} \Bigl|\frac{1}{m} T^m v(x)\Bigr|^2 \, d\mu = \|v\|_{L^2}^2 \sum_{m\ge 1} \frac{1}{m^2} < \infty, \tag{14.28}$$

which implies $T^m v(x)/m \to 0$, μ-a.e., as $m \to \infty$. We deduce that for each $\lambda > 0$,

(14.29)

$$\begin{aligned} &\mu\bigl(\{x \in X : \limsup A_m f(x) - \liminf A_m f(x) > \lambda\}\bigr) \\ &= \mu\bigl(\{x \in X : \limsup A_m(f-g-h) - \liminf A_m(f-g-h) > \lambda\}\bigr) \\ &\le \mu\Bigl(\Bigl\{x \in X : \mathcal{A}^\#(f-g-h) > \frac{\lambda}{2}\Bigr\}\Bigr) \\ &\le \frac{2}{\lambda}\|f-g-h\|_{L^1} \\ &\le \frac{2\varepsilon}{\lambda}. \end{aligned}$$

Since ε can be taken arbitrarily small, this implies that $A_m f(x)$ converges as $m \to \infty$, μ-a.e. We already know it converges to $Pf(x)$ in L^1-norm, so (14.24) follows.

We can use the maximal function estimate (14.23) to extend Proposition 14.3, as follows. First, there is the obvious estimate

$$\|\mathcal{A}^\# f\|_{L^\infty} \le \|f\|_{L^\infty}. \tag{14.30}$$

Now the Marcinkiewicz Interpolation Theorem (see Appendix D) applied to (14.23) and (14.30) yields

$$\|\mathcal{A}^\# f\|_{L^p} \le C_p \|f\|_{L^p}, \quad 1 < p < \infty. \tag{14.31}$$

Using this, we prove the following.

Proposition 14.7. *In the setting of Proposition 14.3, we have, for all $p \in [1,\infty)$,*

$$f \in L^p(X,\mu) \Longrightarrow A_m f \to Pf, \quad \textit{in } L^p\textit{-norm}, \tag{14.32}$$

as $m \to \infty$.

Proof. Take $p \in (1,\infty)$. Given $f \in L^p(X,\mu)$, we have

$$|A_m f(x)| \le \mathcal{A}^\# f(x), \quad \mathcal{A}^\# f \in L^p(X,\mu).$$

Since the convergence (14.24) holds pointwise μ-a.e., (14.32) follows from the Dominated Convergence Theorem. That just leaves $p = 1$, for which we rely on Proposition 14.3.

REMARK. Since $P^* = P$, it follows from Proposition 14.7 that

$$f \in L^p(X,\mu) \Longrightarrow A_m^* f \to Pf,$$

weak* in $L^p(X,\mu)$, for $p \in (1,\infty]$. More general ergodic theorems, such as can be found in [**Kr**], imply one has convergence in L^p-norm (and μ-a.e.), for $p \in [1,\infty)$. Of course if φ is invertible, then such a result is a simple application of the results given above, with φ replaced by φ^{-1}.

Having discussed the convergence of $A_m f$, we turn to the second question raised after (14.4), namely whether the limit must be constant. So far we see that the set of limits coincides with $\operatorname{Ker}(I - T)$, i.e., with the set of invariant functions, where we say $f \in L^p(X,\mu)$ is invariant if and only if

$$f(x) = f(\varphi(x)), \quad \mu\text{-a.e.} \tag{14.33}$$

We note that the following conditions are equivalent:

$$\begin{aligned} &\text{(a)} \qquad f \in L^1(X,\mu) \ \text{ invariant } \Rightarrow f \text{ constant } (\mu\text{-a.e.}),\\ &\text{(b)} \qquad f \in L^2(X,\mu) \ \text{ invariant } \Rightarrow f \text{ constant } (\mu\text{-a.e.}),\\ &\text{(c)} \qquad S \in \mathfrak{F} \ \text{ invariant } \Rightarrow \mu(S) = 0 \ \text{ or } \ \mu(S) = 1. \end{aligned} \tag{14.34}$$

Here we say $S \in \mathfrak{F}$ is invariant if and only if

$$\mu(\varphi^{-1}(S) \triangle S) = 0, \tag{14.35}$$

where $A \triangle B = (A \setminus B) \cup (B \setminus A)$. Note that if $S \in \mathfrak{F}$ satisfies (14.35), then

$$\widetilde{S} = \bigcap_{j \ge 0} \bigcup_{k \ge j} \varphi^{-k}(S) \Longrightarrow \varphi^{-1}(\widetilde{S}) = \widetilde{S} \ \text{ and } \ \mu(\widetilde{S} \triangle S) = 0. \tag{14.36}$$

To see the equivalence in (14.34), note that if $f \in L^1(X,\mu)$ is invariant, then all the sets $S_\lambda = \{x \in X : f(x) > \lambda\}$ are invariant, so (c)$\Rightarrow$(a). Meanwhile clearly (a)$\Rightarrow$(b)$\Rightarrow$(c). A measure-preserving map $\varphi : X \to X$ satisfying (14.34) is said to be *ergodic*.

Theorem 14.6 and Proposition 14.7 have the following corollary.

Proposition 14.8. *If $\varphi : X \to X$ is ergodic and $f \in L^p(X,\mu)$, $p \in [1,\infty)$, then*

$$A_m f \longrightarrow \int_X f\,d\mu, \quad \text{in } L^p\text{-norm and } \mu\text{-a.e.} \tag{14.37}$$

We now consider some examples of ergodic maps. First take the unit circle, $X = S^1 \approx \mathbb{R}/(2\pi\mathbb{Z})$, with measure $d\mu = d\theta/2\pi$. Take $e^{i\alpha} \in S^1$ and define

$$R_\alpha : S^1 \longrightarrow S^1, \quad R_\alpha(e^{i\theta}) = e^{i(\theta+\alpha)}. \tag{14.38}$$

Proposition 14.9. *The map R_α is ergodic if and only if $\alpha/2\pi$ is irrational.*

Proof. We compare the Fourier coefficients $\hat{f}(k) = \int f(\theta)e^{-ik\theta}\,d\mu = (f, e_k)$ with those of Tf. We have

$$\widehat{Tf}(k) = (Tf, e_k) = (f, T^{-1}e_k) = e^{ik\alpha}\hat{f}(k).$$

Thus

$$Tf = f,\ \hat{f}(k) \neq 0 \Longrightarrow e^{ik\alpha} = 1. \tag{14.39}$$

But $e^{ik\alpha} = 1$ for some nonzero $k \in \mathbb{Z}$ if and only if $\alpha/2\pi$ is rational.

In the next example, let $(X, \mathfrak{F}, \mu)$ be a probability space, and form the two-sided infinite product

$$\Omega = \prod_{k=-\infty}^{\infty} X, \tag{14.40}$$

which comes equipped with a σ-algebra $\mathcal{O}$ and a product measure ω, via the construction given at the end of Chapter 6. There is a map on Ω called the two-sided shift:

$$\Sigma : \Omega \to \Omega, \quad \Sigma(x)_k = x_{k+1}, \quad x = (\dots, x_{-1}, x_0, x_1, \dots). \tag{14.41}$$

Proposition 14.10. *The two-sided shift* (14.41) *is ergodic.*

Proof. We make use of the following orthonormal set. Let $\{u_j : j \in \mathbb{Z}^+\}$ be an orthonormal basis of $L^2(X,\mu)$, with $u_0 = 1$. Let $\mathcal{A}$ be the set of elements of $\prod_{k=-\infty}^{\infty} \mathbb{Z}^+$ of the form $\alpha = (\dots, \alpha_{-1}, \alpha_0, \alpha_1, \dots)$ such that $\alpha_k \neq 0$ for only finitely many k. Set

$$v_\alpha(x) = \prod_{k=-\infty}^{\infty} u_{\alpha_k}(x_k), \quad \alpha \in \mathcal{A}, \tag{14.42}$$

and note that for each $\alpha \in \mathcal{A}$ only finitely many factors in this product are not $\equiv 1$. We have the following:

$$\{v_\alpha : \alpha \in \mathcal{A}\} \text{ is an orthonormal basis of } L^2(\Omega, \omega). \tag{14.43}$$

(Cf. Exercise 13 of Chapter 6.) Note that if $Tf(x) = f(\Sigma(x))$,

$$Tv_\alpha = v_{\sigma(\alpha)}, \quad \sigma(\alpha)_k = \alpha_{k-1}. \tag{14.44}$$

Now assume $f \in L^2(\Omega, \omega)$ is invariant. Then

$$\hat{f}(\alpha) = (f, v_\alpha) = (Tf, Tv_\alpha) = \hat{f}(\sigma(\alpha)), \tag{14.45}$$

for each $\alpha \in \mathcal{A}$. Iterating this gives $\hat{f}(\alpha) = \hat{f}(\sigma^\ell(\alpha))$ for each $\ell \in \mathbb{Z}^+$. Since

$$\|f\|_{L^2}^2 = \sum_{\alpha \in \mathcal{A}} |\hat{f}(\alpha)|^2 < \infty, \tag{14.46}$$

and $\{\sigma^\ell(\alpha) : \ell \in \mathbb{Z}^+\}$ is an infinite set except for $\alpha = 0 = (\dots, 0, 0, 0, \dots)$, we deduce that $\hat{f}(\alpha) = 0$ for nonzero $\alpha \in \mathcal{A}$, and hence f must be constant.

A variant of the construction above yields the one-sided shift, on

$$\Omega_0 = \prod_{k=0}^{\infty} X, \tag{14.47}$$

with σ-algebra $\mathcal{O}_0$ and product measure ω_0 constructed in the same fashion. As in (14.41), one sets

$$\Sigma_0 : \Omega_0 \to \Omega_0, \quad \Sigma_0(x)_k = x_{k+1}, \quad x = (x_0, x_1, x_2, \dots). \tag{14.48}$$

The following result has essentially the same proof as Proposition 14.10.

Proposition 14.11. *The one-sided shift* (14.48) *is ergodic.*

Another proof of Proposition 14.11 goes as follows. Suppose that $f \in L^2(\Omega_0, \omega_0)$ is invariant, so $f(x_1, x_2, x_3, \dots) = f(x_{k+1}, x_{k+2}, x_{k+3}, \dots)$. Multiplying both sides by $g(x_1, \dots, x_k)$ and integrating, we have

$$(f, g)_{L^2} = (f, 1)_{L^2}(1, g)_{L^2}$$

for each $g \in L^2(\Omega_0, \omega_0)$ of the form $g = g(x_1, \dots, x_k)$, for any $k < \infty$. Since the set of such g is dense in $L^2(\Omega_0, \omega_0)$, we have this identity for all $g \in L^2(\Omega_0, \omega_0)$, and this implies that f is constant.

The concept of ergodicity defined above extends to a semigroup of measure-preserving transformations, i.e., a collection $\mathcal{S}$ of maps on X satisfying (14.1) for each $\varphi \in \mathcal{S}$ and such that

$$\varphi, \psi \in \mathcal{S} \Longrightarrow \varphi \circ \psi \in \mathcal{S}. \tag{14.49}$$

In such a case, one says a function $f \in L^p(X, \mu)$ is invariant provided (14.33) holds for each $\varphi \in \mathcal{S}$, one says $S \in \mathfrak{F}$ is invariant provided (14.35) holds for all $\varphi \in \mathcal{S}$, and one says the action of $\mathcal{S}$ on $(X, \mathfrak{F}, \mu)$ is ergodic provided the (equivalent) conditions in (14.34) hold. The study so far in this chapter has dealt with $\mathcal{S} = \{\varphi^k : k \in \mathbb{Z}^+\}$. Now we will consider one example of the action of a semigroup (actually a group) not isomorphic to $\mathbb{Z}^+$ (nor to $\mathbb{Z}$). This will lead to a result complementary to Proposition 14.10.

Let $\mathcal{S}_\infty$ denote the group of bijective maps $\sigma : \mathbb{Z} \to \mathbb{Z}$ with the property that $\sigma(k) = k$ for all but finitely many k. Let $(X, \mathfrak{F}, \mu)$ be a probability space and let $\Omega = \prod_{k=-\infty}^{\infty} X$, as in (14.40), with the product measure ω. The group $\mathcal{S}_\infty$ acts on Ω by

$$\varphi_\sigma : \Omega \longrightarrow \Omega, \quad \varphi_\sigma(x)_k = x_{\sigma(k)}, \tag{14.50}$$

where $x = (\dots, x_{-1}, x_0, x_1, \dots) \in \Omega$, $\sigma \in \mathcal{S}_\infty$. The following result is called the Hewitt-Savage 01 Law.

Proposition 14.12. *The action of $\mathcal{S}_\infty$ on Ω defined by* (14.50) *is ergodic.*

Proof. Let $\{v_\alpha : \alpha \in \mathcal{A}\}$ be the orthonormal basis of $L^2(\Omega, \omega)$ given by (14.42)–(14.43). Note that if $T_\sigma f(x) = f(\varphi_\sigma(x))$, then

$$T_\sigma v_\alpha = v_{\sigma^\# \alpha}, \quad (\sigma^\# \alpha)_k = \alpha_{\sigma^{-1}(k)}. \tag{14.51}$$

Now if $f \in L^2(\Omega, \omega)$ is invariant under the action of $\mathcal{S}_\infty$, then, parallel to (14.45), we have

$$\hat{f}(\alpha) = (f, v_\alpha) = (T_\sigma f, T_\sigma v_\alpha) = \hat{f}(\sigma^\# \alpha), \quad \forall\, \alpha \in \mathcal{A},\ \sigma \in \mathcal{S}_\infty. \tag{14.52}$$

Since $\|f\|_{L^2}^2 = \sum_\alpha |\hat{f}(\alpha)|^2 < \infty$ and $\{\sigma^\# \alpha : \sigma \in \mathcal{S}_\infty\}$ is an infinite set, for each nonzero $\alpha \in \mathcal{A}$, it follows that $\hat{f}(\alpha) = 0$ for nonzero $\alpha \in \mathcal{A}$, and hence f must be constant.

The same proof establishes the following result, which contains both Proposition 14.10 and Proposition 14.12. As above, $\mathcal{A}$ is the set defined in the beginning of the proof of Proposition 14.10.

Proposition 14.13. *Let $\mathcal{G}$ be a group of bijective maps on $\mathbb{Z}$ with the property that*

$$\{\sigma^{\#}\alpha : \sigma \in \mathcal{G}\} \text{ is an infinite set, for each nonzero } \alpha \in \mathcal{A}, \tag{14.53}$$

where $\sigma^{\#}\alpha$ is given by (14.51). Then the action of $\mathcal{G}$ on Ω, given by (14.50), is ergodic.

See Exercises 10–14 for a Mean Ergodic Theorem that applies in the setting of Proposition 14.12. Other ergodic theorems that apply to semigroups of transformations can be found in [**Kr**].

Exercises

1. Let $\mathbb{T}^n = S^1 \times \cdots \times S^1 \subset \mathbb{C}^n$, where $S^1 = \{z \in \mathbb{C} : |z| = 1\}$. Given $(e^{-\alpha_1}, \dots, e^{i\alpha_n}) \in \mathbb{T}^n$, define

$$R_\alpha : \mathbb{T}^n \to \mathbb{T}^n, \quad R_\alpha(e^{i\theta_1}, \dots, e^{i\theta_n}) = (e^{i(\theta_1+\alpha_1)}, \dots, e^{i(\theta_n+\alpha_n)}).$$

Give necessary and sufficient conditions that R_α be ergodic.
Hint. Adapt the argument used to prove Proposition 14.9.

2. Define $\varphi : S^1 \to S^1$ by $\varphi(z) = z^2$. Show that φ is ergodic.
Hint. Examine the Fourier series of an invariant function.

3. A measure-preserving map φ on $(X, \mathfrak{F}, \mu)$ is said to be *mixing* provided

$$\mu(\varphi^{-k}(E) \cap F) \to \mu(E)\mu(F), \quad \text{as } k \to \infty, \tag{14.54}$$

for each $E, F \in \mathfrak{F}$. Show that φ is mixing if and only if $Tf(x) = f(\varphi(x))$ has the property

$$(T^k f, g)_{L^2} \longrightarrow (f, 1)_{L^2}(1, g)_{L^2}, \quad \text{as } k \to \infty, \tag{14.55}$$

for all $f, g \in L^2(X, \mu)$.

4. Show that a mixing transformation is ergodic.
Hint. Show that

$$(A_k f, g) = \frac{1}{k}\sum_{j=0}^{k-1}(T^j f, g) \to (f, 1)(1, g). \tag{14.56}$$

Deduce that P in (14.7) is the orthogonal projection of $L^2(X, \mu)$ onto the space of constant functions. Alternatively, just apply (14.45) in case $Tf = f$.

5. Show that the map $\varphi : S^1 \to S^1$ in Exercise 2 is mixing.

6. Show that the two-sided and one-sided shifts Σ and Σ_0, given in (14.41) and (14.48), are mixing.
Hint. Verify (14.55) when f and g are elements of the orthonormal basis $\{v_\alpha\}$ described in (14.42). Alternatively, verify (14.55) when f and g are functions of x_j for $|j| \le M$.

7. Show that the maps R_α in (14.38) and in Exercise 1 are not mixing.

8. We assert that the ergodic transformation $\varphi : S^1 \to S^1$ in Exercise 2 is "equivalent" to the one-sided shift (14.48) for $X = \{0, 1\}$, with $\mu(\{0\}) = \mu(\{1\}) = 1/2$. Justify this.
Hint. Regard an element of Ω_0 as giving the binary expansion of a number $x \in [0, 1)$.

9. Let φ be an ergodic measure-preserving map on a probability space $(X, \mathfrak{F}, \mu)$, and take T as in (14.2). Show that

$$f \in \mathcal{M}^+(X),\ \int_X f\, d\mu = +\infty \Longrightarrow \lim_{k\to\infty} \frac{1}{k} \sum_{j=1}^k T^j f(x) = +\infty,\ \ \mu\text{-a.e.}$$

Exercises 10–14 extend the Mean Ergodic Theorem to the following setting. Let $\mathcal{S}$ be a countably infinite semigroup, represented by a family of isometries on a Hilbert space H, so we have $\{T_\alpha : \alpha \in \mathcal{S}\}$, satisfying $T_\alpha : H \to H$, $T_\alpha^* T_\alpha = I$, $T_\alpha T_\beta = T_{\alpha\beta}$, for $\alpha, \beta \in \mathcal{S}$. Let $M_k \subset \mathcal{S}$ be a sequence of finite subsets of $\mathcal{S}$, of cardinality $\# M_k$. Assume that for each fixed $\gamma \in \mathcal{S}$,

$$\lim_{k\to\infty} \frac{\#(M_k \triangle M_k\gamma)}{\# M_k} = 0, \tag{14.57}$$

where $M_k\gamma = \{\alpha\gamma : \alpha \in M_k\}$ and $M_k \triangle M_k\gamma$ is the symmetric difference. Set

$$S_k f = \frac{1}{\# M_k} \sum_{\alpha \in M_k} T_\alpha f, \quad f \in H. \tag{14.58}$$

10. Show that there is an orthogonal direct sum decomposition

$$H = K \oplus \overline{R}, \tag{14.59}$$

where

$$K = \{f \in H : T_\alpha f = f, \forall\, \alpha \in \mathcal{S}\},$$
$$R = \bigoplus_{\alpha \in \mathcal{S}} \text{Range}\,(I - T_\alpha).$$

Hint. Show that $R^\perp = \bigcap_{\alpha \in \mathcal{S}} \text{Ker}\,(I - T_\alpha^*)$ and that $\text{Ker}\,(I - T_\alpha^*) = \text{Ker}\,(I - T_\alpha)$.

11. Show that $f \in K \Rightarrow S_k f \equiv f$.

12. Show that

$$\begin{aligned} f = (I - T_\gamma)v \Rightarrow S_k f &= \frac{1}{\# M_k} \sum_{\alpha \in M_k} (T_\alpha v - T_{\alpha\gamma} v) \\ &= \frac{1}{\# M_k} \sum_{\alpha \in M_k \triangle M_k \gamma} (\pm T_\alpha v). \end{aligned} \tag{14.60}$$

Use hypothesis (14.57) to deduce that $S_k f \to 0$ as $k \to \infty$.

13. Now establish the following mean ergodic theorem, namely, under the hypothesis (14.57),

$$f \in H \Longrightarrow S_k f \to Pf, \tag{14.61}$$

in H-norm, where P is the orthogonal projection of H onto K.

14. In case $\mathcal{S} = \mathcal{S}_\infty$ is the group arising in Proposition 14.12, with action on $H = L^2(\Omega, \omega)$ given by (14.51), if we set

$$M_k = \{\sigma \in \mathcal{S}_\infty : \sigma(\ell) = \ell \text{ for } |\ell| > k\}, \tag{14.62}$$

show that hypothesis (14.57) holds, and hence the conclusion (14.61) holds. In this case, $Pf = \int_\Omega f\, d\omega$, by Proposition 14.12.

Chapter 15

Probability Spaces and Random Variables

We have already introduced the notion of a probability space, namely a measure space $(X, \mathfrak{F}, \mu)$ with the property that $\mu(X) = 1$. Here we look further at some basic notions and results of probability theory.

First, we give some terminology. A set $S \in \mathfrak{F}$ is called an *event*, and $\mu(S)$ is called the probability of the event, often denoted $P(S)$. The image to have in mind is that one picks a point $x \in X$ at random, and $\mu(S)$ is the probability that $x \in S$. A measurable function $f : X \to \mathbb{R}$ is called a (real) *random variable*. (One also speaks of a measurable map $F : (X, \mathfrak{F}) \to (Y, \mathfrak{G})$ as a Y-valued random variable, given a measurable space $(Y, \mathfrak{G})$.) If f is integrable, one sets

$$E(f) = \int_X f \, d\mu, \tag{15.1}$$

called the *expectation* of f, or the *mean* of f. One defines the *variance* of f as

$$\operatorname{Var}(f) = \int_X |f - a|^2 \, d\mu, \quad a = E(f). \tag{15.2}$$

The random variable f has finite variance if and only if $f \in L^2(X, \mu)$.

A random variable $f : X \to \mathbb{R}$ induces a probability measure ν_f on $\mathbb{R}$, called the probability distribution of f:

$$\nu_f(S) = \mu(f^{-1}(S)), \quad S \in \mathfrak{B}(\mathbb{R}), \tag{15.3}$$

where $\mathfrak{B}(\mathbb{R})$ denotes the σ-algebra of Borel sets in $\mathbb{R}$. It is clear that $f \in L^1(X, \mu)$ if and only if $\int |x|\, d\nu_f(x) < \infty$ and that

$$E(f) = \int_{\mathbb{R}} x\, d\nu_f(x). \tag{15.4}$$

We also have

$$\mathrm{Var}(f) = \int_{\mathbb{R}} (x-a)^2\, d\nu_f(x), \quad a = E(f). \tag{15.5}$$

To illustrate some of these concepts, consider coin tosses. We start with

$$X = \{h, t\}, \quad \mu(\{h\}) = \mu(\{t\}) = \frac{1}{2}. \tag{15.6}$$

The event $\{h\}$ is that the coin comes up heads, and $\{t\}$ gives tails. We also form

$$X_k = \prod_{j=1}^{k} X, \quad \mu_k = \mu \times \cdots \times \mu, \tag{15.7}$$

representing the set of possible outcomes of k successive coin tosses. If $H(k, \ell)$ is the event that there are exactly ℓ heads among the k coin tosses, its probability is

$$\mu_k(H(k, \ell)) = 2^{-k} \binom{k}{\ell}. \tag{15.8}$$

If $N_k : X_k \to \mathbb{R}$ yields the number of heads that come up in k tosses, i.e., $N_k(x)$ is the number of h's that occur in $x = (x_1, \dots, x_k) \in X_k$, then

$$E(N_k) = \frac{k}{2}. \tag{15.9}$$

The measure ν_{N_k} on $\mathbb{R}$ is supported on the set $\{\ell \in \mathbb{Z}^+ : 0 \le \ell \le k\}$, and

$$\nu_{N_k}(\{\ell\}) = \mu_k(H(k, \ell)) = 2^{-k} \binom{k}{\ell}. \tag{15.10}$$

A central area of probability theory is the study of the large k behavior of events in spaces such as (X_k, μ_k), particularly various limits as $k \to \infty$. This leads naturally to a consideration of the infinite product space

$$Z = \prod_{j=1}^{\infty} X, \tag{15.11}$$

with σ-algebra $\mathcal{Z}$ and product measure ν, constructed at the end of Chapter 6. In case (X, μ) is given by (15.6), one can consider $\widetilde{N}_k : Z \to \mathbb{R}$, the number of heads that come up in the first k throws; $\widetilde{N}_k = N_k \circ \pi_k$, where $\pi_k : Z \to X_k$ is the natural projection. It is a fundamental fact of probability theory that for almost all infinite sequences of random coin tosses (i.e., with probability 1) the fraction $k^{-1}\widetilde{N}_k$ of them that come up heads tends to $1/2$ as $k \to \infty$. This is a special case of the "law of large numbers," several versions of which will be established below.

Note that $k^{-1}\widetilde{N}_k$ has the form

$$\frac{1}{k}\sum_{j=1}^{k} f_j, \tag{15.12}$$

where $f_j : Z \to \mathbb{R}$ has the form

$$f_j(x) = f(x_j), \quad f : X \to \mathbb{R}, \quad x = (x_1, x_2, x_3, \dots) \in Z. \tag{15.13}$$

The random variables f_j have the important properties of being *independent* and *identically distributed.*

We define these last two terms in general, for random variables on a probability space $(X, \mathfrak{F}, \mu)$ that need not have a product structure. Say we have random variables $f_1, \dots, f_k : X \to \mathbb{R}$. Extending (15.3), we have a measure ν_{F_k} on $\mathbb{R}^k$ called the joint probability distribution:

$$\nu_{F_k}(S) = \mu(F_k^{-1}(S)), \quad S \in \mathfrak{B}(\mathbb{R}^k), \quad F_k = (f_1, \dots, f_k) : X \to \mathbb{R}^k. \tag{15.14}$$

We say

$$f_1, \dots, f_k \ \text{ are independent } \iff \nu_{F_k} = \nu_{f_1} \times \cdots \times \nu_{f_k}. \tag{15.15}$$

We also say

$$f_i \ \text{ and } \ f_j \ \text{ are identically distributed } \iff \nu_{f_i} = \nu_{f_j}. \tag{15.16}$$

If $\{f_j : j \in \mathbb{N}\}$ is an infinite sequence of random variables on X, we say they are independent if and only if each finite subset is. Equivalently, we can form

$$F = (f_1, f_2, f_3, \dots) : X \to \mathbb{R}^\infty = \prod_{j \ge 1} \mathbb{R}, \tag{15.17}$$

set

$$\nu_F(S) = \mu(F^{-1}(S)), \quad S \in \mathfrak{B}(\mathbb{R}^\infty), \tag{15.18}$$

and then independence is equivalent to

$$\nu_F = \prod_{j \ge 1} \nu_{f_j}. \tag{15.19}$$

It is an easy exercise to show that the random variables in (15.13) are independent and identically distributed.

Here is a simple consequence of independence.

Lemma 15.1. *If $f_1, f_2 \in L^2(X, \mu)$ are independent, then*

$$E(f_1 f_2) = E(f_1)E(f_2). \tag{15.20}$$

Proof. We have $x^2, y^2 \in L^1(\mathbb{R}^2, \nu_{f_1,f_2})$, and hence $xy \in L^1(\mathbb{R}^2, \nu_{f_1,f_2})$. Then

$$\begin{aligned} E(f_1 f_2) &= \int_{\mathbb{R}^2} xy \, d\nu_{f_1,f_2}(x,y) \\ &= \int_{\mathbb{R}^2} xy \, d\nu_{f_1}(x) \, d\nu_{f_2}(y) \\ &= E(f_1)E(f_2). \end{aligned} \tag{15.21}$$

The following is a version of the *weak law of large numbers*. (See the exercises for a more general version.)

Proposition 15.2. *Let $\{f_j : j \in \mathbb{N}\}$ be independent, identically distributed random variables on $(X, \mathfrak{F}, \mu)$. Assume f_j have finite variance, and set $a = E(f_j)$. Then*

$$\frac{1}{k} \sum_{j=1}^{k} f_j \longrightarrow a, \quad \textit{in } L^2\textit{-norm}, \tag{15.22}$$

and hence in measure, as $k \to \infty$.

Proof. Using Lemma 15.1, we have

$$\Bigl\| \frac{1}{k} \sum_{j=1}^{k} f_j - a \Bigr\|_{L^2}^2 = \frac{1}{k^2} \sum_{j=1}^{k} \|f_j - a\|_{L^2}^2 = \frac{b^2}{k}, \tag{15.23}$$

since $(f_j - a, f_\ell - a)_{L^2} = E(f_j - a)E(f_\ell - a) = 0,\ j \neq \ell$, and $\|f_j - a\|_{L^2}^2 = b^2$ is independent of j. Clearly (15.23) implies (15.22). Convergence in measure then follows by Tchebychev's inequality.

The strong law of large numbers produces pointwise a.e. convergence and relaxes the L^2-hypothesis made in Proposition 15.2. Before proceeding to the general case, we first treat the product case (15.11)–(15.13).

Proposition 15.3. *Let $(Z, \mathscr{Z}, \nu)$ be a product of a countable number of factors of a probability space $(X, \mathfrak{F}, \mu)$, as in (15.11). Assume $p \in [1, \infty)$, let $f \in L^p(X, \mu)$, and define $f_j \in L^p(Z, \nu)$, as in (15.13). Set $a = E(f)$. Then*

$$\frac{1}{k}\sum_{j=1}^{k} f_j \to a, \quad \text{in } L^p\text{-norm and } \nu\text{-a.e.}, \tag{15.24}$$

as $k \to \infty$.

Proof. This follows from ergodic theorems established in Chapter 14. In fact, note that $f_j = T^{j-1} f_1$, where $Tg(x) = f(\varphi(x))$, for

$$\varphi : Z \to Z, \quad \varphi(x_1, x_2, x_3, \dots) = (x_2, x_3, x_4, \dots). \tag{15.25}$$

By Proposition 14.11, φ is ergodic. We see that

$$\frac{1}{k}\sum_{j=1}^{k} f_j = \frac{1}{k}\sum_{j=0}^{k-1} T^j f_1 = A_k f_1, \tag{15.26}$$

as in (14.4). The convergence $A_k f_1 \to a$ asserted in (15.24) now follows from Proposition 14.8.

We now establish the following strong law of large numbers.

Theorem 15.4. *Let $(X, \mathfrak{F}, \mu)$ be a probability space, and let $\{f_j : j \in \mathbb{N}\}$ be independent, identically distributed random variables in $L^p(X, \mu)$, with $p \in [1, \infty)$. Set $a = E(f_j)$. Then*

$$\frac{1}{k}\sum_{j=1}^{k} f_j \to a, \quad \text{in } L^p\text{-norm and } \mu\text{-a.e.}, \tag{15.27}$$

as $k \to \infty$.

Proof. Our strategy is to reduce this to Proposition 15.3. We have a map $F : X \to \mathbb{R}^\infty$ as in (15.17), yielding a measure ν_F on $\mathbb{R}^\infty$, as in (15.18), which is actually a product measure, as in (15.19). We have coordinate functions

$$\xi_j : \mathbb{R}^\infty \longrightarrow \mathbb{R}, \quad \xi_j(x_1, x_2, x_3, \dots) = x_j, \tag{15.28}$$

and

$$f_j = \xi_j \circ F. \tag{15.29}$$

Note that $\xi_j \in L^p(\mathbb{R}^\infty, \nu_F)$ and

$$\int_{\mathbb{R}^\infty} \xi_j \, d\nu_F = \int_{\mathbb{R}} x_j \, d\nu_{f_j} = a. \tag{15.30}$$

Now Proposition 15.3 implies

$$\frac{1}{k} \sum_{j=1}^{k} \xi_j \to a, \quad \text{in } L^p\text{-norm and } \nu_F\text{-a.e.}, \tag{15.31}$$

on $(\mathbb{R}^\infty, \nu_F)$, as $k \to \infty$. Since (15.29) holds and F is measure-preserving, (15.27) follows.

Note that if f_j are independent random variables on X and $F_k = (f_1, \dots, f_k) : X \to \mathbb{R}^k$, then

$$\begin{aligned} \int_X G(f_1, \dots, f_k) \, d\mu &= \int_{\mathbb{R}^k} G(x_1, \dots, x_k) \, d\nu_{F_k} \\ &= \int_{\mathbb{R}^k} G(x_1, \dots, x_k) \, d\nu_{f_1}(x_1) \cdots d\nu_{f_k}(x_k). \end{aligned} \tag{15.32}$$

In particular, we have for the sum

$$S_k = \sum_{j=1}^{k} f_k \tag{15.33}$$

and a Borel set $B \subset \mathbb{R}$ that

$$\begin{aligned} \nu_{S_k}(B) &= \int_X \chi_B(f_1 + \cdots + f_k) \, d\mu \\ &= \int_{\mathbb{R}^k} \chi_B(x_1 + \cdots + x_k) \, d\nu_{f_1}(x_1) \cdots d\nu_{f_k}(x_k). \end{aligned} \tag{15.34}$$

Recalling the definition (9.60) of convolution of measures, we see that

$$\nu_{S_k} = \nu_{f_1} * \cdots * \nu_{f_k}. \tag{15.35}$$

Given a random variable $f : X \to \mathbb{R}$, the function

$$\chi_f(\xi) = E(e^{-i\xi f}) = \int_{\mathbb{R}} e^{-ix\xi} \, d\nu_f(x) = \sqrt{2\pi} \, \hat{\nu}_f(\xi) \tag{15.36}$$

is called the *characteristic function* of f (not to be confused with the characteristic function of a set). Combining (15.35) and (9.64), we see that if $\{f_j\}$ are independent and S_k is given by (15.33), then

$$\chi_{S_k}(\xi) = \chi_{f_1}(\xi) \cdots \chi_{f_k}(\xi). \tag{15.37}$$

There is a special class of probability distributions on $\mathbb{R}$ called *Gaussian.* They have the form

$$d\gamma_a^\sigma(x) = \frac{1}{\sqrt{2\pi\sigma}} e^{-(x-a)^2/2\sigma}\, dx. \tag{15.38}$$

That this is a probability distribution follows from Exercise 1 of Chapter 7. One computes

$$\int x\, d\gamma_a^\sigma = a, \quad \int (x-a)^2 d\gamma_a^\sigma = \sigma. \tag{15.39}$$

The distribution (15.38) is also called *normal,* with mean a and variance σ. (Frequently one sees σ^2 in place of σ in these formulas.) A random variable f on $(X, \mathfrak{F}, \mu)$ is said to be Gaussian if ν_f is Gaussian. The computation (9.43)–(9.48) shows that

$$\sqrt{2\pi}\, \hat{\gamma}_a^\sigma(\xi) = e^{-\sigma\xi^2/2 - ia\xi}. \tag{15.40}$$

Hence $f : X \to \mathbb{R}$ is Gaussian with mean a and variance σ if and only if

$$\chi_f(\xi) = e^{-\sigma\xi^2/2 - ia\xi}. \tag{15.41}$$

We also see that

$$\gamma_a^\sigma * \gamma_b^\tau = \gamma_{a+b}^{\sigma+\tau} \tag{15.42}$$

and that if f_j are independent Gaussian random variables on X, the sum $S_k = f_1 + \cdots + f_k$ is also Gaussian.

Gaussian distributions are often approximated by distributions of the sum of a large number of independent randon variables, suitably rescaled. Theorems to this effect are called Central Limit Theorems. We present one here.

Let $\{f_j : j \in \mathbb{N}\}$ be independent, identically distributed random variables on a probability space $(X, \mathfrak{F}, \mu)$, with

$$E(f_j) = a, \quad E((f_j - a)^2) = \sigma < \infty. \tag{15.43}$$

The appropriate rescaling of $f_1 + \cdots + f_k$ is suggested by the computation (15.23). We have

$$g_k = \frac{1}{\sqrt{k}} \sum_{j=1}^{k} (f_j - a) \Longrightarrow \|g_k\|_{L^2}^2 \equiv \sigma. \tag{15.44}$$

Note that if ν_1 is the probability distribution of $f_j - a$, then for any Borel set $B \subset \mathbb{R}$,

$$\nu_{g_k}(B) = \nu_k(\sqrt{k}B), \quad \nu_k = \nu_1 * \cdots * \nu_1 \quad (k \text{ factors}). \tag{15.45}$$

We have

$$\int x^2 \, d\nu_1 = \sigma, \quad \int x \, d\nu_1 = 0. \tag{15.46}$$

We are prepared to prove the following version of the Central Limit Theorem.

Proposition 15.5. *If $\{f_j : j \in \mathbb{N}\}$ are independent, identically distributed random variables on $(X, \mathfrak{F}, \mu)$, satisfying* (15.43)*, and g_k is given by* (15.44)*, then*

$$\nu_{g_k} \to \gamma_0^\sigma, \quad \textit{weak}^* \ \textit{in} \ \mathfrak{M}(\mathbb{R}) = C_*(\mathbb{R})'. \tag{15.47}$$

Proof. By (15.45) we have

$$\chi_{g_k}(\xi) = \chi(k^{-1/2}\xi)^k, \tag{15.48}$$

where $\chi(\xi) = \chi_{f_1 - a}(\xi) = \sqrt{2\pi}\hat{\nu}_1(\xi)$. By (15.46) we have $\chi \in C^2(\mathbb{R})$, $\chi'(0) = 0$, and $\chi''(0) = -\sigma$. Hence

$$\chi(\xi) = 1 - \frac{\sigma}{2}\xi^2 + r(\xi), \quad r(\xi) = o(\xi^2) \text{ as } \xi \to 0. \tag{15.49}$$

Equivalently,

$$\chi(\xi) = e^{-\sigma\xi^2/2 + \rho(\xi)}, \quad \rho(\xi) = o(\xi^2). \tag{15.50}$$

Hence

$$\chi_{g_k}(\xi) = e^{-\sigma\xi^2/2 + \rho_k(\xi)}, \tag{15.51}$$

where

$$\rho_k(\xi) = k\rho(k^{-1/2}\xi) \to 0 \ \text{ as } \ k \to \infty, \quad \forall \xi \in \mathbb{R}. \tag{15.52}$$

In other words,

$$\lim_{k\to\infty} \hat{\nu}_{g_k}(\xi) = \hat{\gamma}_0^\sigma(\xi), \quad \forall\, \xi \in \mathbb{R}. \tag{15.53}$$

Note that the functions in (15.53) are uniformly bounded by $\sqrt{2\pi}$. Making use of (15.53), the Fourier transform identity (9.58), and the Dominated Convergence Theorem, we obtain, for each $v \in \mathcal{S}(\mathbb{R})$,

$$\begin{aligned} \int v\, d\nu_{g_k} &= \int \tilde{v}(\xi)\hat{\nu}_{g_k}(\xi)\, d\xi \\ &\to \int \tilde{v}(\xi)\hat{\gamma}_0^\sigma(\xi)\, d\xi \\ &= \int v\, d\gamma_0^\sigma. \end{aligned} \tag{15.54}$$

Since $\mathcal{S}(\mathbb{R})$ is dense in $C_*(\mathbb{R})$ and all these measures are probability measures, this implies the asserted weak* convergence in (15.47).

Chapter 16 is devoted to the construction and study of a very important probability measure, known as Wiener measure, on the space of continuous paths in $\mathbb{R}^n$. There are many naturally occurring Gaussian random variables on this space.

We return to the strong law of large numbers and generalize Theorem 15.4 to a setting in which the f_j need not be independent. A sequence $\{f_j : j \in \mathbb{N}\}$ of real-valued random variables on $(X, \mathfrak{F}, \mu)$, giving a map $F : X \to R^\infty$ as in (15.17), is called a *stationary process* provided the probability measure ν_F on $\mathbb{R}^\infty$ given by (15.18) is invariant under the shift map

$$\theta : \mathbb{R}^\infty \longrightarrow \mathbb{R}^\infty, \quad \theta(x_1, x_2, x_3, \dots) = (x_2, x_3, x_4, \dots). \tag{15.55}$$

An equivalent condition is that for each $k, n \in \mathbb{N}$, the n-tuples $\{f_1, \dots, f_n\}$ and $\{f_k, \dots, f_{k+n-1}\}$ are identically distributed $\mathbb{R}^n$-valued random variables. Clearly a sequence of independent, identically distributed random variables is stationary, but there are many other stationary processes. (See the exercises.)

To see what happens to the averages $k^{-1}\sum_{j=1}^k f_j$ when one has a stationary process, we can follow the proof of Theorem 15.4. This time, an application of Theorem 14.6 and Proposition 14.7 to the action of θ on $(\mathbb{R}^\infty, \nu_F)$ gives

$$\frac{1}{k}\sum_{j=1}^k \xi_j \to P\xi_1, \quad \nu_F\text{-a.e. and in } L^p\text{-norm}, \tag{15.56}$$

provided

$$\xi_1 \in L^p(\mathbb{R}^\infty, \nu_F), \quad \text{i.e.,} \quad f_1 \in L^p(X, \mu), \quad p \in [1, \infty). \tag{15.57}$$

Here the map $P : L^2(\mathbb{R}^\infty, \nu_F) \to L^2(\mathbb{R}^\infty, \nu_F)$ is the orthogonal projection of $L^2(\mathbb{R}^\infty, \nu_F)$ onto the subspace consisting of θ-invariant functions, which, by Proposition 14.3, extends uniquely to a continuous projection on $L^p(\mathbb{R}^\infty, \nu_F)$ for each $p \in [1, \infty]$. Since $F : (X, \mathfrak{F}, \mu) \to (\mathbb{R}^\infty, \mathfrak{B}(\mathbb{R}^\infty), \nu_F)$ is measure preserving, the result (15.56) yields the following.

Proposition 15.6. *Let $\{f_j : j \in \mathbb{N}\}$ be a stationary process, consisting of $f_j \in L^p(X, \mu)$, with $p \in [1, \infty)$. Then*

$$\frac{1}{k} \sum_{j=1}^{k} f_j \to (P\xi_1) \circ F, \quad \mu\text{-a.e. and in } L^p\text{-norm}. \tag{15.58}$$

The right side of (15.58) can be written as a *conditional expectation*. See Exercise 12 in Chapter 17.

Exercises

1. The second computation in (15.39) is equivalent to the identity

$$\int_{-\infty}^{\infty} x^2 e^{-x^2} \, dx = \frac{\sqrt{\pi}}{2}.$$

Verify this identity.
Hint. Differentiate the identity

$$\int_{-\infty}^{\infty} e^{-sx^2} \, dx = \sqrt{\pi} s^{-1/2}.$$

2. Given a probability space $(X, \mathfrak{F}, \mu)$ and $A_j \in \mathfrak{F}$, we say the sets A_j, $1 \le j \le K$, are independent if and only if their characteristic functions χ_{A_j} are independent, as defined in (15.15). Show that such a collection of sets is independent if and only if, for any distinct $i_1, \dots, i_j$ in $\{1, \dots, K\}$,

$$\mu(A_{i_1} \cap \cdots \cap A_{i_j}) = \mu(A_{i_1}) \cdots \mu(A_{i_j}). \tag{15.59}$$

3. Let f_1, f_2 be random variables on $(X, \mathfrak{F}, \mu)$. Show that f_1 and f_2 are independent if and only if

$$E(e^{-i(\xi_1 f_1 + \xi_2 f_2)}) = E(e^{-i\xi_1 f_1}) E(e^{-i\xi_2 f_2}), \quad \forall\, \xi_j \in \mathbb{R}. \tag{15.60}$$

Extend this to a criterion for independence of $f_1, \dots, f_k$.
Hint. Write the left side of (15.60) as

$$\iint e^{-i(\xi_1 x_1 + \xi_2 x_2)}\, d\nu_{f_1,f_2}(x_1, x_2)$$

and the right side as a similar Fourier transform, using $d\nu_{f_1} \times d\nu_{f_2}$.

4. Demonstrate the following partial converse to Lemma 15.1.

Lemma 15.7. *Let f_1 and f_2 be random variables on (X, μ) such that $\xi_1 f_1 + \xi_2 f_2$ is Gaussian, of mean zero, for each $(\xi_1, \xi_2) \in \mathbb{R}^2$. Then*

$$E(f_1 f_2) = 0 \Longrightarrow f_1 \;\; \text{and} \;\; f_2 \;\; \text{are independent.}$$

More generally, if $\sum_1^k \xi_j f_j$ are all Gaussian and if $f_1, \dots, f_k$ are mutually orthogonal in $L^2(X, \mu)$, then $f_1, \dots, f_k$ are independent.

Hint. Use

$$E(e^{-i(\xi_1 f_1 + \xi_2 f_2)}) = e^{-\|\xi_1 f_1 + \xi_2 f_2\|^2/2},$$

which follows from (15.41).

Exercises 5–6 deal with results known as the Borel-Cantelli Lemmas. If $(X, \mathfrak{F}, \mu)$ is a probability space and $A_k \in \mathfrak{F}$, we set

$$A = \limsup_{k \to \infty} A_k = \bigcap_{\ell=1}^{\infty} \bigcup_{k=\ell}^{\infty} A_k,$$

the set of points $x \in X$ contained in infinitely many of the sets A_k. Equivalently,

$$\chi_A = \limsup \chi_{A_k}.$$

5. (First Borel-Cantelli Lemma) Show that

$$\sum_{k \ge 1} \mu(A_k) < \infty \Longrightarrow \mu(A) = 0.$$

Hint. $\mu(A) \le \mu\big(\bigcup_{k \ge \ell} A_k\big) \le \sum_{k \ge \ell} \mu(A_k)$.

6. (Second Borel-Cantelli Lemma) Assume $\{A_k : k \ge 1\}$ are independent events. Show that

$$\sum_{k \ge 1} \mu(A_k) = \infty \Longrightarrow \mu(A) = 1.$$

Hint. $X \setminus A = \bigcup_{\ell\ge1}\bigcap_{k\ge\ell} A_k^c$, so to prove $\mu(A) = 1$, we need to show $\mu(\bigcap_{k\ge\ell} A_k^c) = 0$, for each ℓ. Now independence implies

$$\mu\Big(\bigcap_{k=\ell}^{L} A_k^c\Big) = \prod_{k=\ell}^{L}(1-\mu(A_k)),$$

which tends to 0 as $L \to \infty$ provided $\sum \mu(A_k) = \infty$.

7. If x and y are chosen at random on $[0,1]$ (with Lebesgue measure), compute the probability distribution of $x-y$ and of $(x-y)^2$. Equivalently, compute the probability distribution of f and f^2, where $Q = [0,1]\times[0,1]$, with Lebesgue measure, and $f : Q \to \mathbb{R}$ is given by $f(x,y) = x - y$.

8. As in Exercise 7, set $Q = [0,1]\times[0,1]$. If $x = (x_1,x_2)$ and $y = (y_1,y_2)$ are chosen at random on Q, compute the probability distribution of $|x-y|^2$ and of $|x-y|$.
Hint. The random variables $(x_1-y_1)^2$ and $(x_2-y_2)^2$ are independent.

9. Suppose $\{f_j : j \in \mathbb{N}\}$ are independent random variables on $(X,\mathfrak{F},\mu)$, satisfying $E(f_j) = a$ and

$$\|f_j - a\|_{L^2}^2 = \sigma_j, \quad \lim_{k\to\infty}\frac{1}{k^2}\sum_{j=1}^{k}\sigma_j = 0. \tag{15.61}$$

Show that the conclusion (15.22) of Proposition 15.2 holds.

10. Suppose $\{f_j : j \in \mathbb{N}\}$ is a stationary process on $(X,\mathfrak{F},\mu)$. Let $G : \mathbb{R}^\infty \to \mathbb{R}$ be $\mathfrak{B}(\mathbb{R}^\infty)$-measurable, and set

$$g_j = G(f_j, f_{j+1}, f_{j+2}, \dots).$$

Show that $\{g_j : j \in \mathbb{N}\}$ is a stationary process on $(X,\mathfrak{F},\mu)$.

In Exercises 11–12, X is a compact metric space, $\mathfrak{F}$ the σ-algebra of Borel sets, μ a probability measure on $\mathfrak{F}$, and $\varphi : X \to X$ a continuous, measure-preserving map with the property that for each $p \in X$, $\varphi^{-1}(p)$ consists of exactly d points, where $d \ge 2$ is some fixed integer. (An example is $X = S^1$, $\varphi(z) = z^d$.) Set

$$\Omega = \prod_{k\ge1} X, \quad Z = \{(x_k) \in \Omega : \varphi(x_{k+1}) = x_k\}.$$

Note that Z is a closed subset of Ω.

11. Show that there is a unique probability measure ν on Ω with the property that for $x = (x_1, x_2, x_3, \dots) \in \Omega$, $A \in \mathfrak{F}$, the probability that $x_1 \in A$ is $\mu(A)$, and given $x_1 = p_1, \dots, x_k = p_k$, then $x_{k+1} \in \varphi^{-1}(p_k)$, and x_{k+1} has probability $1/d$ of being any one of these pre-image points.

    More formally, construct a probability measure ν on Ω such that if $A_j \in \mathfrak{F}$,

$$\nu(A_1 \times \cdots \times A_k) = \int\limits_{A_1} \cdots \int\limits_{A_k} d\gamma_{x_{k-1}}(x_k) \cdots d\gamma_{x_1}(x_2)\, d\mu(x_1),$$

where, given $p \in X$, we set

$$\gamma_p = \frac{1}{d} \sum_{q \in \varphi^{-1}(p)} \delta_q,$$

δ_q denoting the point mass concentrated at q. Equivalently, if $f \in C(\Omega)$ has the form $f = f(x_1, \dots, x_k)$, then

$$\int\limits_{\Omega} f\, d\nu = \int \cdots \int f(x_1, \dots, x_k)\, d\gamma_{x_{k-1}}(x_k) \cdots d\gamma_{x_1}(x_2)\, d\mu(x_1).$$

Such ν is supported on Z.
Hint. The construction of ν can be done in a fashion parallel to, but simpler than, the construction of Wiener measure made at the beginning of Chapter 16. One might read down to (16.13) and return to this problem.

12. Define $f_j : Z \to \mathbb{R}$ by $f_j(x) = x_j$. Show that $\{f_j : j \in \mathbb{N}\}$ is a stationary process on (Z, ν), as constructed in Exercise 11.

13. In the course of proving Proposition 15.5, it was shown that if ν_k and γ are probability measures on $\mathbb{R}$ and $\hat{\nu}_k(\xi) \to \hat{\gamma}(\xi)$ for each $\xi \in \mathbb{R}$, then $\nu_k \to \gamma$, weak* in $C_*(\mathbb{R})'$. Prove the converse:

    Assertion. *If ν_k and γ are probability measures and $\nu_k \to \gamma$ weak* in $C_*(\mathbb{R})'$, then $\hat{\nu}_k(\xi) \to \hat{\gamma}(\xi)$ for each $\xi \in \mathbb{R}$.*

    Hint. Show that for each $\varepsilon > 0$ there exist $R, N \in (0, \infty)$ such that $\nu_k(\mathbb{R} \setminus [-R, R]) < \varepsilon$ for all $k \geq N$.

14. Produce a counterexample to the assertion in Exercise 13 when ν_k are probability measures but γ is not.

15. Establish the following counterpart to Proposition 15.2 and Theorem 15.4. Let $\{f_j : j \in \mathbb{N}\}$ be independent, identically distributed random variables on a probability space $(X, \mathfrak{F}, \mu)$. Assume $f_j \geq 0$ and $\int f_j \, d\mu = +\infty$. Show that, as $k \to \infty$,

$$\frac{1}{k} \sum_{j=1}^{k} f_j \longrightarrow +\infty, \quad \mu\text{-a.e.}$$

16. Given $y \in \mathbb{R}$, $t > 0$, show that

$$\chi_{t,y}(\xi) = e^{-t(1-e^{-iy\xi})}$$

is the characteristic function of the probability distribution

$$\nu_{t,y} = \sum_{k=0}^{\infty} \frac{t^k}{k!} e^{-t} \, \delta_{ky}.$$

That is, $\chi_{t,y}(\xi) = \int e^{-ix\xi} \, d\nu_{t,y}(x)$. These probability distributions are called Poisson distributions. Recalling how (9.64) leads to (15.37), show that $\nu_{s,y} * \nu_{t,y} = \nu_{s+t,y}$.

17. Suppose $\psi(\xi)$ has the property that for each $t > 0$, $e^{-t\psi(\xi)}$ is the characteristic function of a probability distribution ν_t. (One says ν_t is infinitely divisible and that ψ generates a Lévy process.) Show that if $\varphi(\xi)$ also has this property, so does $a\psi(\xi) + b\varphi(\xi)$, given $a, b \in \mathbb{R}^+$.

18. Show that whenever μ is a positive Borel measure on $\mathbb{R} \setminus 0$ such that $\int_{\mathbb{R}} (|y^2| \wedge 1) \, d\mu < \infty$, then, given $A \geq 0$, $b \in \mathbb{R}$,

$$\psi(\xi) = A\xi^2 + ib\xi + \int \big(1 - e^{-iy\xi} - iy\xi\chi_I(y)\big) \, d\mu(y)$$

has the property exposed in Exercise 17, i.e., ψ generates a Lévy process. Here, $\chi_I = 1$ on $I = [-1, 1]$, 0 elsewhere.
Hint. Apply Exercise 17 and a limiting argument. For material on Lévy processes, see [**Sat**].

Wiener Measure and Brownian Motion

Diffusion of particles is a product of their apparently random motion. The density $u(t, x)$ of diffusing particles satisfies the "diffusion equation"

$$\frac{\partial u}{\partial t} = \Delta u. \tag{16.1}$$

If the initial condition $u(0, x) = f(x)$ for $x \in \mathbb{R}^n$ is given, Fourier analysis, as described in (9.69)–(9.71), can be used to provide the solution

$$\begin{aligned} u(t, x) &= (2\pi)^{-n/2} \int \hat{f}(\xi) e^{-t|\xi|^2} e^{ix\cdot\xi}\, d\xi \\ &= \int p(t, x, y) f(y)\, dy, \end{aligned} \tag{16.2}$$

where $\hat{f}(\xi)$ is the Fourier transform of f and

$$p(t, x, y) = p(t, x - y) = (4\pi t)^{-n/2}\, e^{-|x-y|^2/4t}. \tag{16.3}$$

A suggestive notation for the solution operator provided by (16.2)–(16.3) is

$$u(t, x) = e^{t\Delta} f(x). \tag{16.4}$$

One property this "exponential" of the operator Δ has in common with the exponential of real numbers is the identity $e^{t\Delta} e^{s\Delta} = e^{(t+s)\Delta}$, which by (16.2)–(16.3) is equivalent to the identity

$$\int p(t, x - y) p(s, y)\, dy = p(t + s, x). \tag{16.5}$$

This identity can be verified directly, by manipulation of Gaussian integrals, as in (9.47)–(9.48), or via the identity $e^{-t|\xi|^2}e^{-s|\xi|^2} = e^{-(t+s)|\xi|^2}$, plus the sort of Fourier analysis behind (16.2).

Some other simple but important properties that can be deduced from (16.3) are

$$p(t,x,y) \geq 0 \tag{16.6}$$

and

$$\int p(t,x,y)\,dy = 1. \tag{16.7}$$

Consequently, for each $x \in \mathbb{R}^n$, $p(t,x,y)\,dy$ defines a probability distribution, which we can interpret as giving the probability that a particle starting at the point x at time 0 will be in a given region in $\mathbb{R}^n$ at time t.

We proceed to construct a probability measure, known as "Wiener measure," on the set of paths $\omega : [0,\infty) \to \mathbb{R}^n$, undergoing a random motion, called Brownian motion, described as follows. Given $t_1 < t_2$ and given that $\omega(t_1) = x_1$, the probability density for the location of $\omega(t_2)$ is

$$p(t, x - x_1) = (4\pi t)^{-n/2} e^{-|x-x_1|^2/4t}, \quad t = t_2 - t_1. \tag{16.8}$$

The motion of a random path for $t_1 \leq t \leq t_2$ is supposed to be independent of its past history. Thus, given $0 < t_1 < t_2 < \cdots < t_k$ and given Borel sets $E_j \subset \mathbb{R}^n$, the probability that a path, starting at $x = 0$ at $t = 0$, lies in E_j at time t_j for each $j \in [1,k]$ is

$$\int_{E_1} \cdots \int_{E_k} p(t_k - t_{k-1}, x_k - x_{k-1}) \cdots p(t_1, x_1)\, dx_k \cdots dx_1. \tag{16.9}$$

It is not obvious that there is a countably additive measure characterized by these properties, and Wiener's result was a great achievement. The construction we give here is a slight modification of one in Appendix A of [**Nel**].

Anticipating that Wiener measure is supported on the set of continuous paths, we will take a path to be characterized by its locations at all positive *rational* t. Thus, we consider the set of "paths"

$$\mathfrak{P} = \prod_{t \in \mathbb{Q}^+} \dot{\mathbb{R}}^n. \tag{16.10}$$

Here, $\dot{\mathbb{R}}^n$ is the one-point compactification of $\mathbb{R}^n$, i.e., $\dot{\mathbb{R}}^n = \mathbb{R}^n \cup \{\infty\}$. Thus $\mathfrak{P}$ is a compact metrizable space. We construct Wiener measure W as a positive Borel measure on $\mathfrak{P}$.

In order to construct this measure, we will construct a certain positive linear functional $E : C(\mathfrak{P}) \to \mathbb{R}$, on the space $C(\mathfrak{P})$ of real-valued continuous functions on $\mathfrak{P}$, satisfying $E(1) = 1$, and a condition motivated by (16.9), which we give in (16.12). We first define E on the subspace $\mathcal{C}^{\#}$ consisting of continuous functions that depend on only finitely many of the factors in (16.10), i.e., functions on $\mathfrak{P}$ of the form

$$\varphi(\omega) = F\big(\omega(t_1), \dots, \omega(t_k)\big), \quad t_1 < \cdots < t_k, \tag{16.11}$$

where F is continuous on $\prod_1^k \dot{\mathbb{R}}^n$ and $t_j \in \mathbb{Q}^+$. Motivated by (16.9), we take

$$\begin{aligned} E(\varphi) = \int \cdots \int p(t_1, x_1) p(t_2 - t_1, x_2 - x_1) \cdots p(t_k - t_{k-1}, x_k - x_{k-1}) \\ \times F(x_1, \dots, x_k)\, dx_k \cdots dx_1. \end{aligned} \tag{16.12}$$

If $\varphi(\omega)$ in (16.11) actually depends only on $\omega(t_\nu)$ for some proper subset $\{t_\nu\}$ of $\{t_1, \dots, t_k\}$, there arises a formula for $E(\varphi)$ with a different appearance from (16.12). The fact that these two expressions are equal follows from the identity (16.5). From this it follows that $E : \mathcal{C}^{\#} \to \mathbb{R}$ is well defined. It is also a positive linear functional, satisfying $E(1) = 1$.

Now, by the Stone-Weierstrass Theorem, $\mathcal{C}^{\#}$ is dense in $C(\mathfrak{P})$. Since $E : \mathcal{C}^{\#} \to \mathbb{R}$ is a positive linear functional and $E(1) = 1$, it follows that E has a unique continuous extension to $C(\mathfrak{P})$, possessing these properties. Theorem 13.5 associates to E the desired probability measure W. Therefore we have

Theorem 16.1. *There is a unique probability measure W on $\mathfrak{P}$ such that* (16.12) *is given by*

$$E(\varphi) = \int_{\mathfrak{P}} \varphi(\omega)\, dW(\omega), \tag{16.13}$$

for each $\varphi(\omega)$ of the form (16.11) *with F continuous on $\prod_1^k \dot{\mathbb{R}}^n$.*

This is the Wiener measure. We note that (16.12) then holds for any bounded Borel function F, and also for any positive Borel function F, on $\prod_1^k \dot{\mathbb{R}}^n$.

REMARK. It is common to define Wiener measure slightly differently, taking $p(t, x)$ to be the integral kernel of $e^{t\Delta/2}$ rather than $e^{t\Delta}$. The path space $\{b\}$ so produced is related to the path space $\{\omega\}$ constructed here by $\omega(t) = b(2t)$.

Some basic examples of calculations of (16.13) include the following. Define functions X_t on $\mathfrak{P}$, taking values in $\mathbb{R}^n$, by

$$X_t(\omega) = \omega(t). \tag{16.14}$$

Then

$$E(|X_t|^2) = \int p(t,x)|x|^2\,dx = 2nt, \tag{16.15}$$

and, if $0 < s < t$,

$$\begin{aligned} E(|X_t - X_s|^2) &= \iint p(s,x_1)p(t-s,x_2-x_1)|x_2-x_1|^2\,dx_1\,dx_2 \\ &= \int p(t-s,y)|y|^2\,dy \\ &= 2n(t-s), \end{aligned} \tag{16.16}$$

a result that works for all $s, t \geq 0$, if $(t-s)$ is replaced by $|t-s|$. Another way to put (16.15)–(16.16) is

$$\|X_t\|_{L^2(\mathfrak{P})} = \sqrt{2nt}, \quad \|X_t - X_s\|_{L^2(\mathfrak{P})} = \sqrt{2n}\,|t-s|^{1/2}. \tag{16.17}$$

Note that the latter result implies $t \mapsto X_t$ is uniformly continuous from $\mathbb{Q}^+$ to $L^2(\mathfrak{P}, W)$ and hence has a unique continuous extension to $\mathbb{R}^+ = [0,\infty)$:

$$\mathfrak{X} : \mathbb{R}^+ \longrightarrow L^2(\mathfrak{P}, W), \tag{16.18}$$

such that $\mathfrak{X}(t) = X_t$, given by (16.14) for $t \in \mathbb{Q}^+$, and then (16.15)–(16.16) are valid for all real $s, t \geq 0$. This is evidence in favor of the assertion made above that W-almost every $\omega \in \mathfrak{P}$ extends continuously from $t \in \mathbb{Q}^+$ to $t \in \mathbb{R}^+$, though it does not prove it. Before we tackle that proof, we make some more observations.

Let us take $t > s > 0$ and calculate

$$\begin{aligned} (X_s, X_t)_{L^2(\mathfrak{P})} &= E(X_s \cdot X_t) \\ &= \int p(s,x_1)p(t-s,x_2-x_1)\,x_1 \cdot x_2\,dx_1\,dx_2 \\ &= \int p(s,x_1)p(t-s,y)\,x_1 \cdot (y+x_1)\,dx_1\,dy. \end{aligned} \tag{16.19}$$

Now $x_1 \cdot (y + x_1) = x_1 \cdot y + |x_1|^2$. The latter contribution is evaluated as in (16.15), and the former contribution is the dot product $A(s) \cdot A(t-s)$, where

$$A(s) = \int p(s,x_1)x_1\,dx_1 = 0. \tag{16.20}$$

So (16.19) is equal to $2ns$ if $t > s > 0$. Hence, by symmetry,

$$(X_s, X_t)_{L^2(\mathfrak{P})} = 2n \min(s,t). \tag{16.21}$$

One can also obtain this by noting $|X_t - X_s|^2 = |X_t|^2 + |X_s|^2 - 2X_s \cdot X_t$ and comparing (16.15) and (16.16). Furthermore, comparing (16.21) and (16.15), we see that

$$t > s \geq 0 \Longrightarrow (X_t - X_s, X_s)_{L^2(\mathfrak{P})} = 0. \tag{16.22}$$

This result is a special case of the following, whose content can be phrased as the statement that if $t > s \geq 0$, then $X_t - X_s$ is independent of X_σ for $\sigma \leq s$, and also that $X_t - X_s$ has the same statistical behavior as X_{t-s}. For more on this independence, see the exercises at the end of this chapter and Chapter 17.

Proposition 16.2. *Assume* $0 < s_1 < \cdots < s_k < s < t$ $(\in \mathbb{Q}^+)$, *and consider functions on* $\mathfrak{P}$ *of the form*

$$\varphi(\omega) = F(\omega(s_1), \dots, \omega(s_k)), \quad \psi(\omega) = G(\omega(t) - \omega(s)). \tag{16.23}$$

Then

$$E(\varphi\psi) = E(\varphi)E(\psi), \tag{16.24}$$

and

$$E(\psi) = E(\tilde{\psi}), \quad \tilde{\psi}(\omega) = G(\omega(t-s)). \tag{16.25}$$

Proof. By (16.12), we have

$$\begin{aligned} E(\psi) &= \int p(s, y_1) p(t-s, y_2 - y_2) G(y_2 - y_1)\, dy_1\, dy_2 \\ &= \int p(s, y_1) p(t-s, z) G(z)\, dy_1\, dz \\ &= \int p(t-s, z) G(z)\, dz, \end{aligned} \tag{16.26}$$

which establishes (16.25). Next, we have

$$\begin{aligned} E(\varphi\psi) = \int p(s_1, x_1) p(s_2 - s_1, x_2 - x_1) \cdots p(s_k - s_{k-1}, x_k - x_{k-1}) \\ p(s - s_k, y_1 - x_k) p(t-s, y_2 - y_1) F(x_1, \dots, x_k) \\ \times G(y_2 - y_1)\, dx_1 \cdots dx_k\, dy_1\, dy_2. \end{aligned} \tag{16.27}$$

If we change variables to $x_1, \dots, x_k, y_1, z = y_2 - y_1$, then comparison with (16.26) shows that $E(\psi)$ factors out of (16.27). Then use of $\int p(s - s_k, y_1 - x_k)\, dy_1 = 1$ shows that the other factor is equal to $E(\varphi)$, so we have (16.24).

Here is the promised result on path continuity.

Proposition 16.3. *The set* $\mathfrak{P}_0$ *of paths from* $\mathbb{Q}^+$ *to* $\mathbb{R}^n$ *that are uniformly continuous on bounded subsets of* $\mathbb{Q}^+$ *(and that hence extend uniquely to continuous paths from* $[0,\infty)$ *to* $\mathbb{R}^n$*) is a Borel subset of* $\mathfrak{P}$ *with Wiener measure* 1.

For a set S, let $\operatorname{osc}_S(\omega)$ denote $\sup_{s,t\in S} |\omega(s)-\omega(t)|$. Set

$$E(a,b,\varepsilon) = \big\{\omega \in \mathfrak{P} : \operatorname{osc}_{[a,b]}(\omega) > 2\varepsilon\big\}; \tag{16.28}$$

here $[a,b]$ denotes $\{s \in \mathbb{Q}^+ : a \le s \le b\}$. The complement is

$$E^c(a,b,\varepsilon) = \bigcap_{t,s\in[a,b]} \{\omega \in \mathfrak{P} : |\omega(s)-\omega(t)| \le 2\varepsilon\}, \tag{16.29}$$

which is closed in $\mathfrak{P}$. Below we will demonstrate the following estimate on the Wiener measure of $E(a,b,\varepsilon)$:

$$W\big(E(a,b,\varepsilon)\big) \le 2\rho\Big(\frac{\varepsilon}{2}, |b-a|\Big), \tag{16.30}$$

where

$$\begin{aligned}\rho(\varepsilon,\delta) &= \sup_{t\le\delta} \int_{|x|>\varepsilon} p(t,x)\,dx \\ &= \sup_{t\le\delta} \int_{|y|>\varepsilon/\sqrt{t}} p(1,y)\,dy,\end{aligned} \tag{16.31}$$

with $p(t,x)$ as in (16.3). Clearly the sup is assumed at $t=\delta$, so

$$\rho(\varepsilon,\delta) = \int_{|y|>\varepsilon/\sqrt{\delta}} p(1,y)\,dy = \psi_n\Big(\frac{\varepsilon}{\sqrt{\delta}}\Big), \tag{16.32}$$

where

$$\psi_n(r) = (4\pi)^{-n/2} \int_{|y|>r} e^{-|y|^2/4}\,dy \le \alpha_n r^n e^{-r^2/4}, \tag{16.33}$$

as $r \to \infty$.

The relevance of the analysis of $E(a,b,\varepsilon)$ is that, if we set

$$F(k,\varepsilon,\delta) = \Big\{\omega \in \mathfrak{P} : \operatorname{osc}_J(\omega) > 4\varepsilon, \text{ for some } J \subset [0,k]\cap\mathbb{Q}^+, \ell(J) \le \frac{\delta}{2}\Big\}, \tag{16.34}$$

where $\ell(J)$ is the length of the interval J, then

$$F(k,\varepsilon,\delta)=\bigcup\Bigl\{E(a,b,2\varepsilon):[a,b]\subset[0,k],|b-a|\le\frac{\delta}{2}\Bigr\} \tag{16.35}$$

is an open set, and, via (16.30), we have

$$W\bigl(F(k,\varepsilon,\delta)\bigr)\le 2k\,\frac{\rho(\varepsilon,\delta)}{\delta}. \tag{16.36}$$

Furthermore, with $F^c(k,\varepsilon,\delta)=\mathfrak{P}\setminus F(k,\varepsilon,\delta)$,

$$\begin{aligned}\mathfrak{P}_0&=\bigl\{\omega:\forall\,k<\infty,\forall\,\varepsilon>0,\exists\,\delta>0\text{ such that }\omega\in F^c(k,\varepsilon,\delta)\bigr\}\\&=\bigcap_k\ \bigcap_{\varepsilon=1/\nu}\ \bigcup_{\delta=1/\mu}F^c(k,\varepsilon,\delta)\end{aligned} \tag{16.37}$$

is a Borel set (in fact, an $\mathcal{F}_{\sigma\delta}$ set, i.e., a countable intersection of $\mathcal{F}_\sigma$ sets), and we can conclude that $W(\mathfrak{P}_0)=1$ from (16.36), given the observation that, for any $\varepsilon>0$,

$$\frac{\rho(\varepsilon,\delta)}{\delta}\longrightarrow 0,\ \text{ as }\ \delta\to 0, \tag{16.38}$$

which follows immediately from (16.32)–(16.33). Thus, to complete the proof of Proposition 16.3, it remains to establish the estimate (16.30). The next lemma goes most of the way towards that goal.

Lemma 16.4. *Given $\varepsilon,\delta>0$, take ν numbers $t_j\in\mathbb{Q}^+$, $0\le t_1<\cdots<t_\nu$, such that $t_\nu-t_1\le\delta$. Let*

$$A=\bigl\{\omega\in\mathfrak{P}:|\omega(t_1)-\omega(t_j)|>\varepsilon,\ \textit{for some } j=1,\dots,\nu\bigr\}. \tag{16.39}$$

Then

$$W(A)\le 2\rho\Bigl(\frac{\varepsilon}{2},\delta\Bigr). \tag{16.40}$$

Proof. Let

$$\begin{aligned}B&=\{\omega:|\omega(t_1)-\omega(t_\nu)|>\varepsilon/2\},\\C_j&=\{\omega:|\omega(t_j)-\omega(t_\nu)|>\varepsilon/2\},\\D_j&=\{\omega:|\omega(t_1)-\omega(t_j)|>\varepsilon\text{ and}\\&\qquad|\omega(t_1)-\omega(t_k)|\le\varepsilon,\text{ for all }k\le j-1\}.\end{aligned} \tag{16.41}$$

Then $A\subset B\cup\bigcup_{j=1}^{\nu}(C_j\cap D_j)$, so

$$W(A)\le W(B)+\sum_{j=1}^{\nu}W\bigl(C_j\cap D_j\bigr). \tag{16.42}$$

Clearly $W(B) \le \rho(\varepsilon/2, \delta)$. Furthermore, we have

$$W(C_j \cap D_j) = W(C_j)W(D_j) \le \rho\Big(\frac{\varepsilon}{2}, \delta\Big) W(D_j), \tag{16.43}$$

the first identity by Proposition 16.2 (i.e., the independence of C_j and D_j) and the subsequent inequality by the easy estimate $W(C_j) \le \rho(\varepsilon/2, \delta)$. Hence

$$\sum_j W(C_j \cap D_j) \le \rho\Big(\frac{\varepsilon}{2}, \delta\Big), \tag{16.44}$$

since the D_j are mutually disjoint. This proves (16.40). Note that this estimate is independent of ν.

We now finish the demonstration of (16.30). Given such t_j as in the statement of Lemma 16.4, if we set

$$E = \big\{\omega : |\omega(t_j) - \omega(t_k)| > 2\varepsilon, \text{ for some } j, k \in [1, \nu]\big\}, \tag{16.45}$$

it follows that

$$W(E) \le 2\rho\Big(\frac{\varepsilon}{2}, \delta\Big), \tag{16.46}$$

since E is a subset of A, given by (16.39). Now, $E(a, b, \varepsilon)$, given by (16.28), is a countable increasing union of sets of the form (16.45), obtained, e.g., by letting $\{t_1, \dots, t_\nu\}$ consist of all $t \in [a, b]$ that are rational with denominator $\le K$ and taking $K \nearrow +\infty$. Thus we have (16.30), and the proof of Proposition 16.3 is complete.

We make the natural identification of paths $\omega \in \mathfrak{P}_0$ with continuous paths $\omega : [0, \infty) \to \mathbb{R}^n$. Note that a function φ on $\mathfrak{P}_0$ of the form (16.11), with $t_j \in \mathbb{R}^+$, not necessarily rational, is a pointwise limit on $\mathfrak{P}_0$ of functions in $\mathcal{C}^\#$, as long as F is continuous on $\prod_1^k \mathbb{R}^n$, and consequently such φ is measurable. Furthermore, (16.12) continues to hold, by the Dominated Convergence Theorem.

An alternative approach to the construction of W would be to replace (16.10) by $\widetilde{\mathfrak{P}} = \prod\{\dot{\mathbb{R}}^n : t \in \mathbb{R}^+\}$. With the product topology, this is compact but not metrizable. The set of continuous paths is a Borel subset of $\widetilde{\mathfrak{P}}$, but not a Baire set, so some extra measure-theoretic considerations arise if one takes this route, which was taken in [**Nel**].

Looking more closely at the estimate (16.36) of the measure of the set $F(k, \varepsilon, \delta)$, defined by (16.34), we note that we can take $\varepsilon = K\sqrt{\delta \log(1/\delta)}$, in which case

$$\rho\Big(\frac{\varepsilon}{2}, \delta\Big) = \psi_n\Big(\frac{K}{2}\sqrt{\log 1/\delta}\Big) \le C_{n,K}\Big(\log \frac{1}{\delta}\Big)^{n/2} \delta^{K^2/16}. \tag{16.47}$$

Then we obtain the following refinement of Proposition 16.3.

Proposition 16.5. *For almost all $\omega \in \mathfrak{P}_0$, we have, for each $T < \infty$,*

$$\limsup_{|s-t|=\delta\to 0} \Bigl(\bigl|\omega(s) - \omega(t)\bigr| - 8\sqrt{\delta \log \frac{1}{\delta}} \Bigr) \le 0, \quad s, t \in [0, T]. \tag{16.48}$$

Consequently, given $T < \infty$,

$$|\omega(s) - \omega(t)| \le C(\omega, T)\sqrt{\delta \log \frac{1}{\delta}}, \quad s, t \in [0, T], \tag{16.49}$$

with $C(\omega, T) < \infty$ for almost all $\omega \in \mathfrak{P}_0$.

In fact, (16.47) gives $W(S_k) = 1$ where S_k is the set of paths satisfying (16.48), with 8 replaced by $8 + 1/k$, since then (16.47) applies with $K > 4$, so (16.38) holds. Then $\bigcap_k S_k$ is precisely the set of paths satisfying (16.48).

The estimate (16.48) is not quite sharp; P. Lévy showed that for almost all $\omega \in \mathfrak{P}$, with $\mu(\delta) = 2\sqrt{\delta \log 1/\delta}$,

$$\limsup_{|s-t|\to 0} \frac{|\omega(s) - \omega(t)|}{\mu(|s - t|)} = 1. \tag{16.50}$$

See [**McK**] for a proof.

Wiener proved that almost all Brownian paths are nowhere differentiable. We refer to [**McK**] for a proof of this. The following result specifies another respect in which Brownian paths are highly irregular.

Proposition 16.6. *Assume $n \ge 2$, and pick $T \in (0, \infty)$. Then, for almost all $\omega \in \mathfrak{P}_0$,*

$$\omega([0, T]) = \{\omega(t) : 0 \le t \le T\} \subset \mathbb{R}^n \ \textit{ has Hausdorff dimension } 2. \tag{16.51}$$

Proof. The fact that $\operatorname{Hdim} \omega([0, T]) \le 2$ for W-a.e. ω follows from the modulus of continuity estimate (16.48), which implies that for each $\delta > 0, \omega$ is Hölder continuous of order $1/(2+\delta)$. This implies by Exercise 9 of Chapter 12 that $\mathcal{H}^r(\omega([0, T])) < \infty$ for $r = 2 + \delta$. (Of course this upper bound is trivial in the case $n = 2$.)

We will obtain the estimate $\operatorname{Hdim} \omega([0, T]) \ge 2$ for a.e. ω as an application of Proposition 12.19. To get this, we start with the following generalization of (16.16): for $0 < s < t$,

$$\begin{aligned} E(\varphi(X_t - X_s)) &= \iint p(s, x_1) p(t - s, x_2 - x_1) \varphi(x_2 - x_1)\, dx_1\, dx_2 \\ &= \int p(t - s, y) \varphi(y)\, dy \\ &= \bigl(4\pi(t - s)\bigr)^{-n/2} \int e^{-|y|^2/4|t-s|} \varphi(y)\, dy. \end{aligned} \tag{16.52}$$

We now assume φ is radial. We switch to spherical polar coordinates. We also allow $t < s$ and obtain

$$E(\varphi(X_t - X_s)) = \frac{A_{n-1}}{(4\pi|t-s|)^{n/2}} \int_0^\infty e^{-r^2/4|t-s|}\, \varphi(r) r^{n-1}\, dr, \tag{16.53}$$

where $A_{n-1} = \text{Area}(S^{n-1})$. We apply this to $\varphi(r) = r^{-a}$ to get

$$\begin{aligned} E(|X_t - X_s|^{-a}) &= C_n |t-s|^{-n/2} \int_0^\infty e^{-r^2/4|t-s|}\, r^{n-1-a}\, dr \\ &= C_{n,a} |t-s|^{-a/2}, \end{aligned} \tag{16.54}$$

where $C_{n,a} < \infty$ provided $a < n$. We deduce that

$$\begin{aligned} &\int\limits_{\mathfrak{P}} \int_0^T \int_0^T \frac{ds\, dt}{|\omega(t) - \omega(s)|^a}\, dW(\omega) \\ &= C_{n,a} \int_0^T \int_0^T \frac{ds\, dt}{|t-s|^{a/2}} = C'_{n,a} < \infty, \quad \text{if } a < 2. \end{aligned} \tag{16.55}$$

Consequently, as long as $a < 2 \le n$,

$$\int_0^T \int_0^T \frac{ds\, dt}{|\omega(t) - \omega(s)|^a} < \infty, \quad \text{for } W\text{-a.e. } \omega. \tag{16.56}$$

We can rewrite this as

$$\int\limits_{\omega([0,T])} \int\limits_{\omega([0,T])} \frac{d\mu^\omega(x)\, d\mu^\omega(y)}{|x-y|^a} < \infty, \quad \text{for } W\text{-a.e. } \omega, \tag{16.57}$$

where μ^ω is the measure on $\omega([0,T])$ given by

$$\mu^\omega(S) = m(\{t \in [0,T] : \omega(t) \in S\}), \tag{16.58}$$

m denoting Lebesgue measure on $\mathbb{R}$. The existence of a nonzero positive Borel measure on $\omega([0,T])$ satisfying (16.57) implies $\operatorname{Hdim} \omega([0,T]) \ge 2$, by Proposition 12.19, so Proposition 16.6 is proven.

So far we have considered Brownian paths starting at the origin in $\mathbb{R}^n$. Via a simple translation of coordinates, we have a similar construction for the set of Brownian paths ω starting at a general point $x \in \mathbb{R}^n$, yielding the positive functional $E_x : C(\mathfrak{P}) \to \mathbb{R}$, and Wiener measure W_x, such that

$$E_x(\varphi) = \int\limits_{\mathfrak{P}} \varphi(\omega)\, dW_x(\omega). \tag{16.59}$$

When $\varphi(\omega)$ is given by (16.11), $E_x(\varphi)$ has the form (16.12), with the function $p(t_1, x_1)$ replaced by $p(t_1, x_1 - x)$. To put it another way, $E_x(\varphi)$ has the form (16.12) with $F(x_1, \dots, x_k)$ replaced by $F(x_1 + x, \dots, x_k + x)$. One often uses such notation as $E_x\big(f(\omega(t))\big)$ instead of $\int_{\mathfrak{P}} f\big(X_t(\omega)\big)\, dW_x(\omega)$ or $E_x\big(f(X_t(\omega))\big)$.

The following simple observation is useful.

Proposition 16.7. *If $\varphi \in C(\mathfrak{P})$, then $E_x(\varphi)$ is continuous in x.*

Proof. Continuity for $\varphi \in \mathcal{C}^{\#}$, the set of functions of the form (16.11), is clear from (16.12), and its extension to $x \neq 0$ discussed above. Since $\mathcal{C}^{\#}$ is dense in $C(\mathfrak{P})$, the result follows easily.

In Chapter 17 we discuss further results on Brownian motion, arising from the study of martingales. For other reading on the topic, we mention the books [**Dur**], [**McK**], and [**Si**] and also Chapter 11 of [**T1**].

The family of random variables X_t on the probability space $(\mathfrak{P}, W)$ is a special case of a *stochastic process*, often called the Wiener process. More general stochastic processes include Lévy processes. They have a characterization similar to (16.12), with $p(t,x)\,dx$ replaced by probability measures ν_t discussed in Exercises 17–18 of Chapter 15. Theorem 16.1 extends to the construction of such Lévy processes. However, the paths are not a.e. continuous in the non-Gaussian case, but rather there are jumps. For material on Lévy processes, see [**Sat**] and references therein.

Exercises

1. With $X_t(\omega) = \omega(t)$ as in (16.14), show that for all $\xi \in \mathbb{R}^n$

$$E(e^{i\xi\cdot X_t}) = e^{-t|\xi|^2}. \tag{16.60}$$

Hence each component of X_t is a Gaussian random variable on $(\mathfrak{P}_0, W)$, of mean 0 and variance $2t$, by (15.41).

2. More generally, if $0 < t_1 < \cdots < t_k$ and $\xi_j \in \mathbb{R}^n$, show that

$$\begin{aligned} &E\big(e^{i\xi_1\cdot X_{t_1} + i\xi_2\cdot(X_{t_2}-X_{t_1}) + \cdots + i\xi_k\cdot(X_{t_k}-X_{t_{k-1}})}\big) \\ &\quad = e^{-t_1|\xi_1|^2 - (t_2-t_1)|\xi_2|^2 - \cdots - (t_k-t_{k-1})|\xi_k|^2}. \end{aligned} \tag{16.61}$$

Deduce that each component of $\zeta_1\cdot X_{t_1} + \cdots + \zeta_k\cdot X_{t_k}$ is a Gaussian random variable on $(\mathfrak{P}_0, W)$, for each $\zeta_1, \dots, \zeta_k \in \mathbb{R}^n$.

3. Show that if $0 < t_1 < \cdots < t_k$, then $X_{t_1}, X_{t_2} - X_{t_1}, \dots, X_{t_k} - X_{t_{k-1}}$ are independent ($\mathbb{R}^n$-valued) random variables on $(\mathfrak{P}_0, W)$.
Hint. Use (16.61) and Exercise 3 of Chapter 15. Alternatively (but less directly), use the orthogonality (16.22), the Gaussian behavior given in

Exercise 2, and the independence result of Exercise 4 in Chapter 15.

4. Compute $E(e^{\lambda |X_t|^2})$. Show this is finite if and only if $\lambda < 1/4t$.

5. Show that

$$E\big([X_t - X_s]^{2k}\big) = E\big(X_{|t-s|}^{2k}\big) = \frac{(2k)!}{k!}\,|t-s|^k.$$

6. Show that $L^p(\mathfrak{P}_0, W)$ is *separable*, for $1 \leq p < \infty$.
Hint. $\mathfrak{P}$ is a compact *metric* space.

7. Given $a > 0$, define a transformation $D_a : \mathfrak{P}_0 \to \mathfrak{P}_0$ by

$$(D_a\omega)(t) = a\omega(t/a^2).$$

Show that D_a preserves the Wiener measure W. This transformation is called Brownian scaling.

8. Let

$$\widetilde{\mathfrak{P}}_0 = \big\{\omega \in \mathfrak{P}_0 : \lim_{s\to\infty} s^{-1}\omega(s) = 0\big\}.$$

Show that $W(\widetilde{\mathfrak{P}}_0) = 1$. Define a transformation $\rho : \widetilde{\mathfrak{P}}_0 \to \mathfrak{P}_0$ by

$$(\rho\omega)(t) = t\omega(1/t),$$

for $t > 0$. Show that ρ preserves the Wiener measure W.

9. Given $a > 0$, define a transformation $R_a : \mathfrak{P}_0 \to \mathfrak{P}_0$ by

$$\begin{aligned}(R_a\omega)(t) &= \omega(t) &&\text{for } 0 \leq t \leq a,\\ &2\omega(a) - \omega(t) &&\text{for } t \geq a.\end{aligned}$$

Show that R_a preserves the Wiener measure W.

Chapter 17

Conditional Expectation and Martingales

There are a number of cases where a set X is endowed with a family of σ-algebras, $\mathfrak{F}_\alpha$, indexed by a partially ordered set A, having the property that

$$\alpha < \beta \Longrightarrow \mathfrak{F}_\alpha \subset \mathfrak{F}_\beta, \tag{17.1}$$

and where there are measures μ_α on $\mathfrak{F}_\alpha$ such that

$$\alpha < \beta \Longrightarrow \mu_\beta\big|_{\mathfrak{F}_\alpha} = \mu_\alpha. \tag{17.2}$$

We will consider some examples here, in which A is either $\mathbb{Z}^+$ or $\mathbb{R}^+$. All the examples we consider have the property that there is a "big" σ-algebra $\mathfrak{F}^\#$ and a measure $\mu^\#$ on $\mathfrak{F}^\#$ such that

$$\forall\, \alpha \in A,\ \mathfrak{F}_\alpha \subset \mathfrak{F}^\# \quad \text{and} \quad \mu^\#\big|_{\mathfrak{F}_\alpha} = \mu_\alpha. \tag{17.3}$$

We assume μ_α (and $\mu^\#$) are probability measures, i.e., μ_α is a positive measure and $\mu_\alpha(X) = 1$.

To take a simple example, let $X = J = [0,1)$. Let $\mathfrak{F}^\# = \mathfrak{B}$, the σ-algebra of Borel sets, with $\mu^\# = m =$ Lebesgue measure. We take $A = \mathbb{Z}^+$, and for $j \in \mathbb{Z}^+$, let $\mathfrak{F}_j$ be the (σ)-algebra generated by the intervals

$$I_{j\nu} = \big[2^{-j}\nu, 2^{-j}(\nu+1)\big), \quad 0 \le \nu \le 2^j - 1, \tag{17.4}$$

with $\mu_j = m\big|_{\mathfrak{F}_j}$. An example with $A = \mathbb{R}^+$, arising from Brownian motion, will be described later in this chapter.

Returning to the general case, we define the *conditional expectation* $E(f|\mathfrak{F}_\alpha)$ of an element $f \in L^1(X, \mathfrak{F}_\beta, \mu_\beta)$, when $\alpha < \beta$. We will have $E(f|\mathfrak{F}_\alpha) \in L^1(X, \mathfrak{F}_\alpha, \mu_\alpha)$. To define it, consider the set function

$$\lambda(S) = \int_S f\, d\mu_\beta, \quad S \in \mathfrak{F}_\alpha. \tag{17.5}$$

This is a measure on $\mathfrak{F}_\alpha$ which is absolutely continuous with respect to μ_α, since, for $\alpha < \beta$,

$$S \in \mathfrak{F}_\alpha,\ \mu_\alpha(S) = 0 \Rightarrow \mu_\beta(S) = 0 \Rightarrow \lambda(S) = 0.$$

Thus, by the Radon-Nikodym Theorem there exists a unique element $f_\alpha \in L^1(X, \mathfrak{F}_\alpha, \mu_\alpha)$ such that $\lambda(S) = \int_S f_\alpha\, d\mu_\alpha$, for all $S \in \mathfrak{F}_\alpha$. We take $E(f|\mathfrak{F}_\alpha) = f_\alpha$. In other words, $E(f|\mathfrak{F}_\alpha)$ is characterized as being $\mathfrak{F}_\alpha$-measurable and satisfying

$$\int_S f\, d\mu_\beta = \int_S E(f|\mathfrak{F}_\alpha)\, d\mu_\alpha, \quad \forall\, S \in \mathfrak{F}_\alpha. \tag{17.6}$$

Note that this implies

$$\int_X gf\, d\mu_\beta = \int_X gE(f|\mathfrak{F}_\alpha)\, d\mu_\alpha, \tag{17.7}$$

for all $g \in L^\infty(X, \mathfrak{F}_\alpha)$.

Note that if $f \in L^2(X, \mathfrak{F}_\beta, \mu_\beta)$, we have a uniquely defined $E(f|\mathfrak{F}_\alpha) \in L^2(X, \mathfrak{F}_\alpha, \mu_\alpha)$, satisfying (17.7) for all $g \in L^2(X, \mathfrak{F}_\alpha)$. In fact, we can regard $L^2(X, \mathfrak{F}_\alpha, \mu_\alpha)$ naturally as a closed linear subspace of $L^2(X, \mathfrak{F}_\beta, \mu_\beta)$, and then $E(f|\mathfrak{F}_\alpha)$ is the image of f under the orthogonal projection of $L^2(X, \mathfrak{F}_\beta, \mu_\beta)$ onto $L^2(X, \mathfrak{F}_\alpha, \mu_\alpha)$. One might compare this observation with the proof of Theorem 4.10.

This construction can also be applied to $f \in L^1(X, \mathfrak{F}^\#, \mu^\#)$, yielding

$$f_\alpha = E(f|\mathfrak{F}_\alpha). \tag{17.8}$$

In this case, it follows easily from the uniqueness of the characterization (17.6) that

$$\alpha < \beta \Longrightarrow f_\alpha = E(f_\beta|\mathfrak{F}_\alpha). \tag{17.9}$$

Generally, a family $f_\alpha \in L^1(X, \mathfrak{F}_\alpha, \mu_\alpha)$ satisfying (17.9) is called a *martingale.* If such a martingale arises via (17.8) for $f \in L^1(X, \mathfrak{F}^\#, \mu^\#)$, we say the martingale (f_α) has a *final element.*

In case $(X, \mathfrak{F}^\#, \mu^\#)$ is $(J, \mathfrak{B}, m)$ as above, with $J = [0, 1)$, we see that, for any $f \in L^1(J, \mathfrak{B}, m)$,

$$f_j = E(f|\mathfrak{F}_j) = \operatorname{Avg}_{I_{j\nu}}(f) \quad \text{on } I_{j\nu}, \tag{17.10}$$

where

$$\operatorname{Avg}_{I_{j\nu}}(f) = 2^{-j} \int_{I_{j\nu}} f \, dx. \tag{17.11}$$

Now this is rather similar to $A_r f$, defined by (11.1), in case $n = 1$, f is supported on J, and $r = 2^{-j}$. The following result is parallel to Theorem 11.1.

Proposition 17.1. *Given $f \in L^1(J, m)$, f_j defined by* (17.10), *we have $f_j \to f$, m-almost everywhere.*

The proof is much like that of Theorem 11.1, once one has a result parallel to the maximal function estimate of Theorem 11.2. We now establish a general result of this nature, known as the Martingale Maximal Inequality.

Proposition 17.2. *Let $(f_\alpha : \alpha \in A)$ be a martingale over $\mathfrak{F}_\alpha$. Assume A is countable and totally ordered. Define the maximal function*

$$f^*(x) = \sup \{f_\alpha(x) : \alpha \in A\}. \tag{17.12}$$

Then, for all $\lambda > 0$,

$$\mu^\#(\{x \in X : f^*(x) > \lambda\}) \le \frac{K}{\lambda}, \quad K = \sup_\alpha \|f_\alpha\|_{L^1}. \tag{17.13}$$

Proof. Note that $\alpha < \beta \Rightarrow \|f_\alpha\|_{L^1} \le \|f_\beta\|_{L^1}$. It suffices to demonstrate (17.13) for an arbitrary finite subset $\{\alpha_j\}$ of A. Thus we can work with $\mathfrak{F}_j = \mathfrak{F}_{\alpha_j}$, $1 \le j \le N$, and $f_j = E(f_k|\mathfrak{F}_j)$ for $j < k$. There is no loss in assuming $f_N \ge 0$, so all $f_j \ge 0$. Now consider

$$S_\lambda = \{x \in X : f^*(x) > \lambda\} = \{x \in X : \text{some } f_j(x) > \lambda\}. \tag{17.14}$$

There is a pairwise disjoint decomposition

$$S_\lambda = \bigcup_{j=1}^N S_{\lambda j}, \quad S_{\lambda j} = \{x : f_j(x) > \lambda \text{ but } f_\ell(x) \le \lambda \text{ for } \ell < j\}. \tag{17.15}$$

Note that $S_{\lambda j} \in \mathfrak{F}_j$. Consequently, we have

$$\begin{aligned} \int\limits_{S_\lambda} f_N \, d\mu_N &= \sum_{j=1}^{N} \int\limits_{S_{\lambda j}} f_N \, d\mu_N = \sum_{j=1}^{N} \int\limits_{S_{\lambda j}} f_j \, d\mu_j \\ &\geq \sum_{j=1}^{N} \lambda \, \mu_j(S_{\lambda j}) = \lambda \, \mu_N(S_\lambda). \end{aligned} \tag{17.16}$$

This yields (17.13) in this special case, and the proposition is hence proved.

REMARK 1. If all $f_\alpha \geq 0$, then $\|f_\alpha\|_{L^1} \equiv K$.

REMARK 2. If $(f_\alpha : \alpha \in A)$ has a final element $f \in L^1(X, \mathfrak{F}^\#, \mu^\#)$, then $K \leq \|f\|_{L^1}$.

The maximal inequality of Proposition 17.2 yields the following (increasing) Martingale Convergence Theorem, which contains Proposition 17.1 as a special case.

Proposition 17.3. *Let $(f_\alpha : \alpha \in A)$ be a martingale over $\mathfrak{F}_\alpha$, with final element $f \in L^1(X, \mathfrak{F}_\omega, \mu)$. Assume A is countable and totally ordered. Assume $\mathfrak{F}_\omega = \sigma\big(\bigcup_{\alpha \in A} \mathfrak{F}_\alpha\big)$. Then*

$$f_\alpha(x) \to f(x) \quad \mu\text{-a.e.}, \qquad \text{as } \alpha \nearrow \omega. \tag{17.17}$$

Proof. We certainly have (17.17) when $f \in \bigcup_{\alpha \in A} L^1(X, \mathfrak{F}_\alpha, \mu) = V$. But V is dense in $L^1(X, \mathfrak{F}_\omega, \mu)$. (See Exercise 6 of Chapter 9.) Thus (17.17) follows for general $f \in L^1(X, \mathfrak{F}_\omega, \mu)$, by the same sort of argument as used in the proof of Theorem 11.1, via the maximal function estimate (17.13).

In detail, given $f \in L^1(X, \mathfrak{F}_\omega, \mu)$, pick $\varepsilon > 0$ and $g \in V$ such that $f = g + h$ with $\|h\|_{L^1} < \varepsilon$. With obvious notation, $f_\alpha = g_\alpha + h_\alpha$ and, for each $\lambda > 0$,

$$\begin{aligned} E_\lambda &= \{x \in X : \limsup_\alpha |f_\alpha(x) - f(x)| > \lambda\} \\ &= \{x \in X : \limsup_\alpha |h_\alpha(x) - h(x)| > \lambda\}. \end{aligned} \tag{17.18}$$

Noting that $|h_\alpha| \leq E(|h| \, |\mathfrak{F}_\alpha)$, we have

$$|h_\alpha| \leq h^*, \quad h^* = \sup_\alpha E(|h| \, |\mathfrak{F}_\alpha), \tag{17.19}$$

and hence
(17.20)

$$\mu(E_\lambda) \le \mu\Bigl(\Bigl\{x \in X : h^*(x) > \frac{\lambda}{2}\Bigr\}\Bigr) + \mu\Bigl(\Bigl\{x \in X : |h(x)| > \frac{\lambda}{2}\Bigr\}\Bigr) \le \frac{4\varepsilon}{\lambda}.$$

Taking $\varepsilon \to 0$ gives $\mu(E_\lambda) = 0$, and (17.17) is proven.

In the exercises we explore situations where $(f_\alpha : \alpha \in A)$ does not have a final element.

In addition to the martingale convergence result of Proposition 17.3, which is related to convergence results of Chapter 11, there are *decreasing* martingale convergence theorems, which as we will see are related to results of Chapter 14. For simplicity we consider martingales $(f_\alpha : \alpha \in A)$ with index set $A = \mathbb{Z}^- = \{k \in \mathbb{Z} : k \le 0\}$. So we have a sequence of σ-algebras of subsets of X:

$$\mathfrak{F}_0 \supset \mathfrak{F}_{-1} \supset \mathfrak{F}_{-2} \supset \cdots . \tag{17.21}$$

We set

$$\mathfrak{F}^b = \bigcap_{k \le 0} \mathfrak{F}_k. \tag{17.22}$$

Let μ be a probability measure on $\mathfrak{F}_0$. Given $f \in L^p(X, \mathfrak{F}_0, \mu)$, we set

$$P_k f = E(f|\mathfrak{F}_k), \quad P^b f = E(f|\mathfrak{F}^b). \tag{17.23}$$

Note that P_k and P^b all have L^p-operator norm 1, for each $p \in [1, \infty]$. Our goal is to produce various results to the effect that $P_k f \to P^b f$ as $k \to -\infty$. The next result is somewhat parallel to the L^2-mean ergodic theorem contained in Proposition 14.2.

Proposition 17.4. *If μ is a probability measure on $\mathfrak{F}_0$, $f \in L^2(X, \mathfrak{F}_0, \mu)$, and $\mathfrak{F}_k, \mathfrak{F}^b, P_k$, and P^b are as in (17.21)–(17.23), then*

$$P_k f \to P^b f \ \text{ in } \ L^2\text{-norm, as } \ k \to -\infty. \tag{17.24}$$

Proof. The key to proving (17.24) is to show that

$$\bigcup_{k \le 0} \operatorname{Ker} P_k \ \text{ is dense in } \ \operatorname{Ker} P^b, \ \text{ in } \ L^2(X, \mu). \tag{17.25}$$

Given this, we set $f = f^b + g$, with $f^b = P^b f$, $g = (I - P^b) f \in \operatorname{Ker} P^b$. Then (17.25) implies $P_k g \to 0$ in $L^2(X, \mu)$, and since $P_k f^b = f^b$, we have (17.24). It remains to establish (17.25). For this, we note that

$$(\operatorname{Ker} P_k)^\perp = L^2(X, \mathfrak{F}_k, \mu), \tag{17.26}$$

so

$$\Big(\bigcup_k \operatorname{Ker} P_k\Big)^{\perp} = \bigcap_k (\operatorname{Ker} P_k)^{\perp} = \bigcap_k L^2(X, \mathfrak{F}_k, \mu) \tag{17.27}$$
$$= L^2(X, \mathfrak{F}^b, \mu) = (\operatorname{Ker} P^b)^{\perp},$$

the third identity by (17.22). This completes the proof.

Given that the operators P_k and P^b all have operator norm 1 on each L^p space, that L^2 is dense in L^p for $p \in [1,2]$, and that $\|f\|_{L^p} \le \|f\|_{L^2}$ for $p \in [1,2]$, the same type of argument used for Proposition 14.3 also establishes the following.

Proposition 17.5. *In the setting of Proposition 17.4, if $p \in [1,2]$,*

$$f \in L^p(X, \mathfrak{F}_0, \mu) \Longrightarrow P_k f \to P^b f \tag{17.28}$$

in L^p-norm, as $k \to -\infty$.

In Proposition 17.7 we will extend this result to $p \in [1,\infty)$.

We next establish a pointwise convergence result, analogous to Theorem 14.6. In analogy with (14.22), we set

$$\mathcal{A}^{\#} f(x) = \sup_{k\le 0} |P_k f(x)| \le \sup_{k \le 0} P_k |f|(x), \tag{17.29}$$

for $f \in L^1(X, \mathfrak{F}_0, \mu)$, and note that Proposition 17.2 gives

$$\mu(\{x \in X : \mathcal{A}^{\#} f(x) > \lambda\}) \le \frac{1}{\lambda} \|f\|_{L^1}. \tag{17.30}$$

Proposition 17.6. *In the setting of Proposition 17.4,*

$$f \in L^1(X, \mathfrak{F}_0, \mu) \Longrightarrow P_k f \to P^b f, \quad \mu\text{-a.e.}, \tag{17.31}$$

as $k \to -\infty$.

Proof. Let us set $V_k = \operatorname{Ker} P_k \cap L^2(X, \mathfrak{F}_0, \mu)$, $V^b = \operatorname{Ker} P^b \cap L^2(X, \mathfrak{F}_0, \mu)$, and recall from the proof of Proposition 17.4 that $\bigcup_k V_k$ is dense in V^b, in L^2-norm. Now let $f \in L^1(X, \mathfrak{F}_0, \mu)$ be given and take $\varepsilon > 0$. Pick $f_1 \in L^2(X, \mathfrak{F}_0, \mu)$ such that $\|f - f_1\|_{L^1} < \varepsilon$. Then write $f_1 = f_1^b + g$, with $f_1^b = P^b f_1$ and $g \in V^b$. Then write $g = g_K + g_\varepsilon$ with $g_K \in V_K$ (for sufficiently large negative K) and $\|g_\varepsilon\|_{L^2} < \varepsilon$ (so also $\|g_\varepsilon\|_{L^1} < \varepsilon$). We hence have

$$f = f_1^b + g_K + f_\varepsilon, \quad f_\varepsilon = (f - f_1) + g_\varepsilon, \quad \|f_\varepsilon\|_{L^1} < 2\varepsilon. \tag{17.32}$$

Now

$$k \le K \Longrightarrow P_k f = f_1^b + g_K + P_k f_\varepsilon,$$

so

$$k, \ell \le K \Longrightarrow P_k f - P_\ell f = P_k f_\varepsilon - P_\ell f_\varepsilon.$$

Hence, for each $\lambda > 0$,

$$\begin{aligned} &\mu\big(\{x \in X : \limsup_{k\to-\infty} P_k f(x) - \liminf_{k\to-\infty} P_k f(x) > \lambda\}\big) \\ &= \mu\big(\{x \in X : \limsup P_k f_\varepsilon(x) - \liminf P_k f_\varepsilon(x) > \lambda\}\big) \\ &\le \mu\Big(\Big\{x \in X : \mathcal{A}^\# f_\varepsilon(x) > \frac{\lambda}{2}\Big\}\Big) \\ &\le \frac{4\varepsilon}{\lambda}, \end{aligned} \tag{17.33}$$

the last inequality by (17.29). Since ε can be taken arbitrarily small, this implies that $P_k f(x)$ converges as $k \to -\infty$, μ-a.e. We already know it converges to $P^b f(x)$ in L^1-norm, so (17.31) follows.

We can use the maximal function estimate (17.29) to extend Proposition 17.5, as follows. There is the obvious estimate $\|\mathcal{A}^\# f\|_{L^\infty} \le \|f\|_{L^\infty}$. Then (17.29) and the Marcinkiewicz Interpolation Theorem (established in Appendix D) yield

$$\|\mathcal{A}^\# f\|_{L^p} \le C_p \|f\|_{L^p}, \quad 1 < p < \infty. \tag{17.34}$$

Using this, we prove the following.

Proposition 17.7. *In the setting of Proposition 17.4, we have, for all $p \in [1,\infty)$,*

$$f \in L^p(X, \mathfrak{F}_0, \mu) \Longrightarrow P_k f \to P^b f, \quad \text{in } L^p\text{-norm}, \tag{17.35}$$

as $k \to -\infty$.

Proof. The proof is identical to that of Proposition 14.7. The convergence (17.35) follows from (17.31) and the Dominated Convergence Theorem, via (17.34), if $p \in (1,\infty)$, and the case $p = 1$ is already done in Proposition 17.5.

We now show how Propositions 17.6–17.7 yield another proof of the strong law of large numbers.

Proposition 17.8. *Let $\{f_k : k \ge 0\}$ be independent, identically distributed random variables on a probability space $(X, \mathfrak{B}, \mu)$. Assume $p \in [1, \infty)$ and $f_k \in L^p(X, \mu)$. Set $a = E(f_k)$. Then, as $k \to \infty$,*

$$S_k = \frac{1}{k+1} \sum_{j=0}^{k} f_j \longrightarrow a, \quad \text{in } L^p\text{-norm and } \mu\text{-a.e.} \tag{17.36}$$

Proof. For $k \ge 0$, let $\mathfrak{F}_{-k}$ be the σ-algebra generated by the collection of sets $\{S_k, S_{k+1}, S_{k+2}, \dots\}$, i.e., by $\{S_k, f_{k+1}, f_{k+2}, \dots\}$. Each S_k is $\mathfrak{F}_{-k}$-measurable. We claim that

$$E(S_0|\mathfrak{F}_{-k}) = S_k. \tag{17.37}$$

Given this, it follows from Propositions 17.6–17.7 that

$$S_k \to E(S_0|\mathfrak{F}^b) \ \text{ in } L^p\text{-norm and } \mu\text{-a.e.}, \tag{17.38}$$

where $\mathfrak{F}^b = \bigcap_k \mathfrak{F}_{-k}$. We next claim that $E(S_0|\mathfrak{F}^b) = E(f_0)$. To see this, take $A \in (0, \infty)$ and set $f_k = f_k^A + g_k^A$, where

$$\begin{aligned} f_k^A(x) = f_k(x), &\quad \text{if} \quad |f_k(x)| \le A, \\ 0 &\quad \text{if} \quad |f_k(x)| > A, \end{aligned} \tag{17.39}$$

and set

$$S_k^A = \frac{1}{k+1} \sum_{j=0}^{k} f_j^A, \quad T_k^A = \frac{1}{k+1} \sum_{j=0}^{k} g_j^A.$$

Note that the random variables f_k^A are independent and identically distributed. Now we can appeal to the weak law of large numbers to say $S_k^A \to E(f_0^A)$, in L^2-norm, and meanwhile $\|T_k^A\|_{L^p} \le \|g_k^A\|_{L^p}$, which is arbitrarily small if A is taken large enough. Passing to the limit, we have $E(S_0|\mathfrak{F}^b) = E(f_0)$, and the proof of Proposition 17.8 is complete, modulo the proof of (17.37).

In order to establish (17.37), we make use of the following symmetry:

$$j, \ell \in \{0, \dots, k\} \Longrightarrow E(f_j|\mathfrak{F}_{-k}) = E(f_\ell|\mathfrak{F}_{-k}), \tag{17.40}$$

which follows from the hypothesis that $\{f_k\}$ are independent and identically distributed. (See the exercises for further comments on this.) Given this, we have

$$\begin{aligned} E(S_0|\mathfrak{F}_{-k}) &= \frac{1}{k+1} \sum_{j=0}^{k} E(f_j|\mathfrak{F}_{-k}) \\ &= E(S_k|\mathfrak{F}_{-k}) = S_k, \end{aligned} \tag{17.41}$$

which gives (17.37).

We now show how the measure space $(\mathfrak{P}_0, W_x)$, constructed in Chapter 16 to describe Brownian motion, gives rise to a family of σ-algebras indexed by $\mathbb{R}^+$, and we examine some martingales that arise fron this family. Given $t \in [0, \infty)$, let $\mathfrak{B}_t$ be the σ-algebra of subsets of $\mathfrak{P}_0$ generated by sets of the form

$$\{\omega \in \mathfrak{P}_0 : \omega(s) \in E\}, \tag{17.42}$$

where $s \in [0, t]$ and E is a Borel subset of $\mathbb{R}^n$. As t increases, $\mathfrak{B}_t$ is an increasing family of σ-algebras, each consisting of sets that are W_x-measurable, for all $x \in \mathbb{R}^n$. Set

$$\mathfrak{B}_\infty = \sigma\Big(\bigcup_{t<\infty} \mathfrak{B}_t\Big).$$

It can be shown that

$$\mathfrak{B}_\infty = \{S \cap \mathfrak{P}_0 : S \in \mathfrak{B}(\mathfrak{P})\} = \{S \in \mathfrak{B}(\mathfrak{P}) : S \subset \mathfrak{P}_0\}. \tag{17.43}$$

Given $f \in L^1(\mathfrak{P}_0, \mathfrak{B}_\infty, W_x)$, we have the associated conditional expectation $E_x\big(f\big|\mathfrak{B}_t\big)$. The following is a statement that Brownian motion possesses the *Markov Property.*

Proposition 17.9. *Given $s, t > 0$, $f \in C(\dot{\mathbb{R}}^n)$,*

$$E_x\big(f(\omega(t+s))\big|\mathfrak{B}_s\big) = E_{\omega(s)}\big(f(\omega(t))\big), \text{ for } W_x\text{-almost all } \omega. \tag{17.44}$$

Proof. The right side of (17.44) is $\mathfrak{B}_s$-measurable, so the identity is equivalent to the statement that

$$\int_S f(\omega(t+s))\, dW_x(\omega) = \int_S \Big(\int f(\widetilde{\omega}(t))\, dW_{\omega(s)}(\widetilde{\omega})\Big)\, dW_x(\omega), \tag{17.45}$$

for all $S \in \mathfrak{B}_s$. It suffices to verify (17.45) for all S of the form

$$S = \{\omega \in \mathfrak{P}_0 : \omega(t_1) \in E_1, \dots, \omega(t_K) \in E_K\},$$

given $t_j \in [0, t]$, E_j Borel sets in $\mathbb{R}^n$. For such S, (17.45) follows directly from the characterization of the Wiener integral given in Chapter 16, i.e., from (16.9)–(16.12) in the case $x = 0$, together with the identity

$$\int f(\widetilde{\omega}(t))\, dW_y(\widetilde{\omega}) = E\big(f(y + \omega(t))\big), \tag{17.46}$$

used to define (16.59).

We can easily extend (17.44) to

$$E_x\big(F(\omega(s+t_1),\dots,\omega(s+t_k))\big|\mathfrak{B}_s\big) = E_{\omega(s)}\big(F(\omega(t_1),\dots,\omega(t_k))\big), \tag{17.47}$$

for W_x-almost all ω, given $t_1,\dots,t_k > 0$, and F continuous on $\prod_1^k \dot{\mathbb{R}}^n$, as in (16.11). Also standard limiting arguments allow us to enlarge the class of functions F for which this works. We then get the following more definitive statement of the Markov Property.

Proposition 17.10. *For $s > 0$, define the map*

$$\sigma_s : \mathfrak{P}_0 \longrightarrow \mathfrak{P}_0, \quad (\sigma_s\omega)(t) = \omega(t+s). \tag{17.48}$$

Then, given φ bounded and $\mathfrak{B}_\infty$-measurable, we have

$$E_x\big(\varphi\circ\sigma_s\big|\mathfrak{B}_s\big) = E_{\omega(s)}(\varphi), \ \textit{for } W_x\textit{-almost all } \omega. \tag{17.49}$$

The Markov Property gives rise to martingales, via the following construction.

Proposition 17.11. *Let $h(t,x)$ be smooth in $t \ge 0, x \in \mathbb{R}^n$, and let it satisfy $|h(t,x)| \le C_\varepsilon e^{\varepsilon|x|^2}$ for all $\varepsilon > 0$ and the backward heat equation*

$$\frac{\partial h}{\partial t} = -\Delta h. \tag{17.50}$$

Then $\mathfrak{h}_t(\omega) = h(t,\omega(t))$ is a martingale over $\mathfrak{B}_t$.

Proof. The hypothesis on $h(t,x)$ implies that, for $t,s > 0$,

$$h(s,x) = \int p(t,y)h(t+s,x-y)\,dy, \tag{17.51}$$

where $p(t,x)$ is given by (16.3). Now

$$\begin{aligned} E_x\big(\mathfrak{h}_{t+s}\big|\mathfrak{B}_s\big) &= E_x\big(h(t+s,\omega(t+s))\big|\mathfrak{B}_s\big) \\ &= E_{\omega(s)}\big(h(t+s,\omega(t))\big), \end{aligned} \tag{17.52}$$

for W_x-almost all ω, by (17.44). This is equal to

$$\int p(t,y-\omega(s))\ h(t+s,y)\,dy, \tag{17.53}$$

by the characterization (16.12) of expectation, adjusted as in (16.59), and by (17.51) this is equal to $h(s,\omega(s)) = \mathfrak{h}_s(\omega)$.

Corollary 17.12. *For one-dimensional Brownian motion, the following are martingales over* $\mathfrak{B}_t$*:*

$$\mathfrak{x}_t(\omega) = \omega(t), \quad \mathfrak{q}_t(\omega) = \omega(t)^2 - 2t, \quad \mathfrak{z}_t(\omega) = e^{a\omega(t)-a^2t}, \tag{17.54}$$

given $a > 0$.

Applying the Martingale Maximal Inequality to $\mathfrak{x}_t(\omega) = \omega(t)$ and to $\mathfrak{z}_t(\omega) = e^{a\omega(t)-a^2t}$, we obtain the following.

Proposition 17.13. *For one-dimensional Brownian motion, given* $t > 0, b > 1, a > 0$, *we have*

$$W_0\Big(\big\{\omega \in \mathfrak{P}_0 : \sup_{0\le s\le t} \omega(s) > b\sqrt{4t/\pi}\big\}\Big) \le \frac{1}{b} \tag{17.55}$$

and

$$W_0\Big(\big\{\omega \in \mathfrak{P}_0 : \sup_{0\le s\le t} \omega(s) - as > \lambda\big\}\Big) \le e^{-a\lambda}. \tag{17.56}$$

Proof. The sets whose measures are estimated in (17.55)–(17.56) are

$$\big\{\omega \in \mathfrak{P}_0 : \sup_{0\le s\le t} \omega(s) > \lambda\big\}, \quad \lambda = b\sqrt{\frac{4t}{\pi}},$$

and

$$\big\{\omega \in \mathfrak{P}_0 : \sup_{0\le s\le t} e^{a\omega(s)-a^2s} > e^{a\lambda}\big\}.$$

Since paths in $\mathfrak{P}_0$ are continuous, one can take the sup over $[0,t]\cap\mathbb{Q}$, which is countable, so Proposition 17.2 applies. Furthermore, we have

$$E_0(|\mathfrak{x}_t|) = \int_{-\infty}^{\infty} |x|\; p(t,x)\; dx = \sqrt{\frac{4t}{\pi}} \tag{17.57}$$

and

$$E_0(\mathfrak{z}_t) = \int_{-\infty}^{\infty} e^{ax-a^2t}\; p(t,x)\; dx = 1, \tag{17.58}$$

yielding the estimates on the right sides of (17.55) and (17.56).

Using more advanced techniques, one can sharpen (17.55) to

$$W_0\Big(\big\{\omega \in \mathfrak{P}_0 : \sup_{0\le s\le t} \omega(s) > \lambda\big\}\Big) = 2\int_{\lambda}^{\infty} p(t,x)\,dx. \tag{17.59}$$

See, e.g., Proposition 3.8 in Chapter 11 of [**T1**].

There is a property more subtle than the Markov Property, called the Strong Markov Property. Material on this can be found in [**Dur**], [**IMc**], [**McK**], and Chapter 11 of [**T1**].

We end this chapter with an application of the estimate (17.56) to the following celebrated Law of the Iterated Logarithm. Our treatment follows [**McK**].

Proposition 17.14. *For one-dimensional Brownian motion, one has*

$$\limsup_{t\to 0} \frac{\omega(t)}{2\sqrt{t\log\log 1/t}} = 1, \quad \textit{for } W_0\textit{-a.e. } \omega. \tag{17.60}$$

Proof. We take this in two steps. First we show

$$\limsup_{t\to 0} \frac{\omega(t)}{2\sqrt{t\log\log 1/t}} \le 1, \quad \text{for } W_0\text{-a.e. } \omega. \tag{17.61}$$

For this, we apply (17.56), with

$$a = a(t) = (1+\delta)t^{-1}h(t), \quad \lambda = \lambda(t) = h(t), \quad h(t) = \sqrt{t\log\log\frac{1}{t}}, \tag{17.62}$$

for small positive t. Here, we pick $\delta \in (0,1)$. Then $a\lambda = (1+\delta)\log\log(1/t)$, so

$$e^{-a\lambda} = \Bigl(\log\frac{1}{t}\Bigr)^{-(1+\delta)}. \tag{17.63}$$

Now pick $\theta \in (0,1)$ and set

$$a_n = a(t_n), \quad \lambda_n = \lambda(t_n), \quad t_n = \theta^n. \tag{17.64}$$

Then

$$e^{-a_n\lambda_n} = C_{\theta,\delta}\, n^{-(1+\delta)}, \tag{17.65}$$

and (17.56) implies

$$W_0\Bigl(\Bigl\{\omega : \sup_{0\le s\le\theta^n} \omega(s) - a_n s > \lambda_n\Bigr\}\Bigr) \le Cn^{-(1+\delta)}. \tag{17.66}$$

Since $\sum_{n\ge 1} n^{-(1+\delta)} < \infty$, the First Borel-Cantelli Lemma (Exercise 5 of Chapter 15) applies to give

$$W_0(A) = 1, \quad A = \Bigl\{\omega \in \mathfrak{P}_0 : \sup_{0\le s\le\theta^n} \omega(s) - a_n s \le \lambda_n, \ \forall \text{ large } n\Bigr\}. \tag{17.67}$$

In particular, for $\omega \in A$ and $\theta^{n+1} < t \le \theta^n$, with n sufficiently large,

$$\begin{aligned} \omega(t) \le \max_{s\le\theta^n} \omega(s) &\le a_n\theta^n + \lambda_n \\ &= (2+\delta)h(\theta^n) \\ &< \frac{2+\delta}{\theta}\, h(t). \end{aligned} \tag{17.68}$$

Taking $\theta \nearrow 1$ and $\delta \searrow 0$ yields (17.61).

Next we show

$$\limsup_{t\to 0} \frac{\omega(t)}{2\sqrt{t \log\log 1/t}} \geq 1, \quad \text{for } W_0\text{-a.e. } \omega. \tag{17.69}$$

For this, take $\theta \in (0,1)$ and consider

$$B_n = \{\omega \in \mathfrak{P}_0 : \omega(\theta^n) - \omega(\theta^{n+1}) \geq 2(1-\theta^{1/2})h(\theta^n)\}, \tag{17.70}$$

with $h(t)$ as in (17.62). By Exercise 3 of Chapter 16, the events $B_n \subset \mathfrak{P}_0$ are independent. By (16.9), we have

$$\begin{aligned} W_0(B_n) &= \int_{2(1-\theta^{1/2})h(\theta^n)}^{\infty} p(\theta^n - \theta^{n+1}, x)\, dx \\ &= \frac{1}{\sqrt{\pi}} \int_{\gamma_n}^{\infty} e^{-y^2}\, dy, \end{aligned} \tag{17.71}$$

where

$$\gamma_n = \frac{2(1-\theta^{1/2})h(\theta^n)}{\sqrt{4(\theta^n - \theta^{n+1})}} = \frac{1-\theta^{1/2}}{(1-\theta)^{1/2}} \sqrt{\log\log \theta^{-n}}. \tag{17.72}$$

Now one has

$$\Big(2a + \frac{1}{a}\Big)^{-1} e^{-a^2} < \int_a^{\infty} e^{-y^2}\, dy < \frac{1}{a}\, e^{-a^2}. \tag{17.73}$$

Hence

$$W_0(B_n) \geq \frac{1}{\sqrt{\pi}} \Big(2\gamma_n + \frac{1}{\gamma_n}\Big)^{-1} e^{-\gamma_n^2} \geq C n^{-1},$$

so

$$\sum_{n\geq 1} W_0(B_n) = +\infty. \tag{17.74}$$

Thus the Second Borel-Cantelli Lemma (Exercise 6 of Chapter 15) gives

$$W_0(B) = 1, \tag{17.75}$$

where

$$B = \{\omega \in \mathfrak{P}_0 : \omega(\theta^n) \geq 2(1-\theta^{1/2})h(\theta^n) \text{ for infinitely many } n\}.$$

The proof of (17.61) implies $\omega(\theta^{n+1}) < 4h(\theta^{n+1})$ for all sufficiently large n, provided $\omega \in A$. Hence, for $\omega \in A \cap B$,

$$\begin{aligned} \omega(\theta^n) &> 2(1-\theta^{1/2})h(\theta^n) - 4h(\theta^{n+1}) \\ &> 2(1-4\theta^{1/2})h(\theta^{n+1}), \end{aligned} \tag{17.76}$$

for infinitely many n. Hence $\limsup_{t\to 0} \omega(t)/h(t) \geq 2(1-4\theta^{1/2})$, which gives (17.69) in the limit $\theta \to 0$. This completes the proof of Proposition 17.14.

Since $\omega(t) \mapsto -\omega(t)$ preserves the measure W_0 and also $\omega(t) \mapsto t\omega(1/t)$ preserves W_0 (cf. Exercise 8, Chapter 16), we have the following.

Corollary 17.15. *For one-dimensional Brownian motion,*

$$\liminf_{t\to 0} \frac{\omega(t)}{2\sqrt{t\log\log 1/t}} = -1, \quad \textit{for } W_0\textit{-a.e. } \omega, \tag{17.77}$$

and

$$\limsup_{t\to\infty} \frac{\omega(t)}{2\sqrt{t\log\log t}} = 1, \quad \textit{for } W_0\textit{-a.e. } \omega. \tag{17.78}$$

Exercises

1. Suppose $(X, \mathfrak{F}, \mu)$ is a probability space and we have a partition of X into a countable family of disjoint sets X_j, with $\mu(X_j) > 0$. Let $\mathcal{P}$ be the σ-algebra generated by $\{X_j\}$. Given $f \in L^1(X, \mathfrak{F}, \mu)$, show that

$$E(f|\mathcal{P})(x) = \frac{1}{\mu(X_j)} \int_{X_j} f\, d\mu, \quad \text{if } x \in X_j.$$

This generalizes (17.10)–(17.11). Show that a sequence of progressively finer such partitions of X gives rise to a martingale.

2. Use Proposition 17.9 to show that, for $s, t \ge 0$, G continuous,

$$E_0\big(G(\omega(t+s) - \omega(s))\big|\mathfrak{B}_s\big) = E_0\big(G(\omega(t))\big).$$

3. Let $\mathfrak{F}_0 \subset \mathfrak{F}_1$ be two σ-algebras of subsets of X, and let μ be a probability measure on $\mathfrak{F}_1$. We say a function $f \in L^1(X, \mathfrak{F}_1, \mu)$ is *independent* of $\mathfrak{F}_0$ provided

$$f \text{ is independent of } g, \quad \forall\, g \in L^1(X, \mathfrak{F}_0, \mu). \tag{17.79}$$

Show that this property is equivalent to

$$E(e^{i\xi f}|\mathfrak{F}_0) = \text{const.} = E(e^{i\xi f}), \quad \forall\, \xi \in \mathbb{R}. \tag{17.80}$$

Hint. For (17.79)⇒(17.80), note that, for $g \in L^1(X, \mathfrak{F}_0, \mu)$,

$$\begin{aligned} E(e^{i\xi f}) \int g\, d\mu &= \int e^{i\xi f} g\, d\mu \qquad \text{(given (17.79))} \\ &= \int E(e^{i\xi f}|\mathfrak{F}_0) g\, d\mu. \end{aligned}$$

For (17.80)⇒(17.79), note that, for such g and $\xi, \eta \in \mathbb{R}$,

$$\begin{aligned} E(e^{i\xi f + i\eta g}) &= \int E(e^{i\xi f} | \mathfrak{F}_0) e^{i\eta g}\, d\mu \\ &= E(e^{i\xi f}) E(e^{i\eta g}) \qquad \text{(given (17.80))}. \end{aligned}$$

4. Deduce for Brownian paths that $\omega(t+s) - \omega(s)$ is independent of $\mathfrak{B}_s$, for $s, t \geq 0$.

5. Let $X = [0,1)$ and let $\mathfrak{F}_j$ be the family of σ-algebras of subsets of X given by (15.4), and let $\mu_j = m\big|_{\mathfrak{F}_j}$, where m denotes Lebesgue measure. Let μ_K be the measure on X, supported on the Cantor middle third set K, constructed in Exercise 11 of Chapter 6. Show that there exist $f_j \in L^1(X, \mathfrak{F}_j, \mu_j)$ such that

$$S \in \mathfrak{F}_j \Longrightarrow \mu_K(S) = \int_S f_j \, d\mu_j.$$

Show that the family (f_j) is a martingale, without a final element in $L^1([0,1), m)$.

In Exercises 6–9, we have a set X and σ-algebras $\mathcal{F}_k \subset \mathcal{F}_{k+1}$, $k \in \mathbb{Z}^+$. We take

$$\mathcal{A} = \bigcup_k \mathcal{F}_k, \quad \mathcal{F}^\# = \sigma(\mathcal{A}).$$

We assume given probability measures μ_k on $\mathcal{F}_k$, satisfying $\mu_{k+1}\big|_{\mathcal{F}_k} = \mu_k$. We aim to produce a probability measure $\mu^\#$ on $\mathcal{F}^\#$ such that

$$\mu^\#\Big|_{\mathcal{F}_k} = \mu_k. \tag{17.81}$$

6. Define μ^b on $\mathcal{S}$ by

$$\mu^b(S) = \mu_k(s), \quad S \in \mathcal{F}_k.$$

Show that μ^b is a *premeasure* on $\mathcal{A}$, as defined by (5.15)–(5.16).

7. Use Proposition 5.3, Theorem 5.4, and Proposition 5.5 to obtain a measure $\mu^\#$ on $\mathcal{F}^\#$ such that (17.81) holds.

In Exercises 8–9, we retain the setting of Exercises 6–7. Let λ be a probability measure on $\mathcal{F}^{\#}$, and assume each $\mu_k << \lambda\big|_{\mathcal{F}_k}$; say

$$\mu_k = f_k\lambda, \quad f_k \in L^1(X, \mathcal{F}_k, \lambda).$$

Also use the Lebesgue-Radon-Nikodym decomposition to write

$$\mu^{\#} = f\lambda + \nu, \quad f \in L^1(X, \mathcal{F}^{\#}, \lambda), \quad \nu \perp \lambda.$$

8. Show that $f\lambda\big|_{\mathcal{F}_k} << \lambda$ and $\nu\big|_{\mathcal{F}_k} << \lambda$, so

$$f\lambda\Big|_{\mathcal{F}_k} = g_k\lambda, \quad \nu\Big|_{\mathcal{F}_k} = h_k\lambda, \quad g_k, h_k \in L^1(X, \mathcal{F}_k, \lambda).$$

Note that $f_k = g_k + h_k$. Show that (g_k) is a martingale with final element $f \in L^1(X, \mathcal{F}^{\#}, \lambda)$.

9. A refinement of the Martingale Convergence Theorem, Proposition 17.3, states that

$$f_k \longrightarrow f, \quad \lambda\text{-a.e.} \tag{17.82}$$

Deduce this from the two results

$$g_k \longrightarrow f, \quad \lambda\text{-a.e.}, \tag{17.83}$$

$$h_k \longrightarrow 0, \quad \lambda\text{-a.e.} \tag{17.84}$$

Show that (17.83) follows from Proposition 17.3. Try to prove (17.84).

REMARK 1. See [**Doo**], pp. 631–632, for further discussion of this result.

REMARK 2. Applying the result of Exercise 9 to the martingale (f_j) in Exercise 5, one has $f_j \to 0$, m-a.e. Compare this with Proposition 11.10.

REMARK 3. For an example of a family of singular martingales, where Exercises 6–9 apply and $f = 0$, see [**RST**].

10. Let $(X, \mathfrak{F}, \mu)$ be a probability space and $\varphi : X \to X$ a measure-preserving transformation. Let $\mathfrak{I} \subset \mathfrak{F}$ be the σ-algebra of invariant sets, defined as in (14.35), and let $P : L^p(X, \mu) \to L^p(X, \mu)$ be the projection arising in the Mean Ergodic Theorem, via (14.7). Show that

$$f \in L^1(X, \mathfrak{F}, \mu) \Longrightarrow Pf = E(f|\mathfrak{I}).$$

11. Assume $(X, \mathfrak{F}, \mu)$ and $(Z, \mathfrak{Z}, \nu)$ are probability spaces and $F : X \to Z$ is a measure-preserving map, i.e.,

$$S \in \mathfrak{Z} \Rightarrow F^{-1}(S) \in \mathfrak{F} \ \text{ and } \ \mu(F^{-1}(S)) = \nu(S).$$

(Do not assume F is bijective.) Let $\mathfrak{I} \subset \mathfrak{Z}$ be a σ-algebra, and let $\mathfrak{A} \subset \mathfrak{F}$ be the σ-algebra

$$\mathfrak{A} = \{F^{-1}(S) : S \in \mathfrak{I}\}.$$

Show that, for all $g \in L^1(Z, \nu)$,

$$E(g \circ F|\mathfrak{A}) = E(g|\mathfrak{I}) \circ F.$$

12. Establish the following complement to Proposition 15.6. Let $\{f_j : j \in \mathbb{N}\}$ be a stationary process, consisting of $f_j \in L^p(X, \mu)$, with $p \in [1, \infty)$. Then

$$\frac{1}{k}\sum_{j=1}^{k} f_j \to E(f_1|\mathfrak{A}), \quad \mu\text{-a.e. and in } L^p\text{-norm},$$

where $\mathfrak{A} = \{F^{-1}(S) : S \in \mathfrak{I}\}$, $\mathfrak{I}$ consists of elements of $\mathfrak{B}(\mathbb{R}^\infty)$ invariant under the shift θ in (15.55), and the map $F : X \to \mathbb{R}^\infty$ is given by $F(x) = (f_1(x), f_2(x), f_3(x), \dots)$.

13. Let ν be a probability measure on $\mathbb{R}$, and let ω be the associated product measure on $\mathbb{R}^\infty$. Define

$$\xi_j : \mathbb{R}^\infty \to \mathbb{R}, \quad \xi_j(x) = x_j, \quad x = (x_0, x_1, x_2, \dots).$$

Let $S_k = (\xi_0 + \cdots + \xi_k)/(k+1)$ and let $\mathcal{G}_{-k}$ be the σ-algebra generated by $\{S_k, S_{k+1}, S_{k+2}, \dots\}$. Show that

$$j, \ell \le k \Longrightarrow E(\xi_j|\mathcal{G}_{-k}) = E(\xi_\ell|\mathcal{G}_{-k}).$$

14. If $\{f_k : k \ge 0\}$ are independent, identically distributed random variables on $(X, \mathfrak{B}, \mu)$, define $F : X \to \mathbb{R}^\infty$ by $F(x) = (f_0(x), f_1(x), f_2(x), \dots)$. Combine the results of Exercises 11 and 13 to establish the identity (17.40).

Appendix A

Metric Spaces, Topological Spaces, and Compactness

A metric space is a set X, together with a distance function $d : X \times X \to [0, \infty)$, having the properties that

$$\begin{aligned} d(x,y) = 0 &\Longleftrightarrow x = y, \\ d(x,y) &= d(y,x), \\ d(x,y) &\le d(x,z) + d(y,z). \end{aligned} \tag{A.1}$$

The third of these properties is called the triangle inequality. An example of a metric space is the set of rational numbers $\mathbb{Q}$, with $d(x, y) = |x - y|$. Another example is $X = \mathbb{R}^n$, with

$$d(x,y) = \sqrt{(x_1 - y_1)^2 + \cdots + (x_n - y_n)^2}.$$

If (x_ν) is a sequence in X, indexed by $\nu = 1, 2, 3, \dots$, i.e., by $\nu \in \mathbb{Z}^+$, one says $x_\nu \to y$ if $d(x_\nu, y) \to 0$, as $\nu \to \infty$. One says (x_ν) is a Cauchy sequence if $d(x_\nu, x_\mu) \to 0$ as $\mu, \nu \to \infty$. One says X is a *complete* metric space if every Cauchy sequence converges to a limit in X. Some metric spaces are not complete; for example, $\mathbb{Q}$ is not complete. You can take a sequence (x_ν) of rational numbers such that $x_\nu \to \sqrt{2}$, which is not rational. Then (x_ν) is Cauchy in $\mathbb{Q}$, but it has no limit in $\mathbb{Q}$.

If a metric space X is not complete, one can construct its completion $\widehat{X}$ as follows. Let an element ξ of $\widehat{X}$ consist of an *equivalence class* of Cauchy sequences in X, where we say $(x_\nu) \sim (y_\nu)$ provided $d(x_\nu, y_\nu) \to 0$. We write the equivalence class containing (x_ν) as $[x_\nu]$. If $\xi = [x_\nu]$ and $\eta = [y_\nu]$, we can set $d(\xi, \eta) = \lim_{\nu\to\infty} d(x_\nu, y_\nu)$ and verify that this is well defined and that it makes $\widehat{X}$ a complete metric space.

If the completion of $\mathbb{Q}$ is constructed by this process, we get $\mathbb{R}$, the set of real numbers. This construction provides a good way to develop the basic theory of the real numbers.

There are a number of useful concepts related to the notion of closeness. We define some of them here. First, if p is a point in a metric space X and $r \in (0, \infty)$, the set

$$B_r(p) = \{x \in X : d(x, p) < r\} \tag{A.2}$$

is called the open ball about p of radius r. Generally, a *neighborhood* of $p \in X$ is a set containing such a ball, for some $r > 0$.

A set $U \subset X$ is called *open* if it contains a neighborhood of each of its points. The complement of an open set is said to be *closed.* The following result characterizes closed sets.

Proposition A.1. *A subset $K \subset X$ of a metric space X is closed if and only if*

$$x_j \in K,\ x_j \to p \in X \Longrightarrow p \in K. \tag{A.3}$$

Proof. Assume K is closed, $x_j \in K$, $x_j \to p$. If $p \notin K$, then $p \in X \setminus K$, which is open, so some $B_\varepsilon(p) \subset X \setminus K$, and $d(x_j, p) \geq \varepsilon$ for all j. This contradiction implies $p \in K$.

Conversely, assume (A.3) holds, and let $q \in U = X \setminus K$. If $B_{1/n}(q)$ is not contained in U for any n, then there exists $x_n \in K \cap B_{1/n}(q)$; hence $x_n \to q$, contradicting (A.3). This completes the proof.

The following is straightforward.

Proposition A.2. *If U_α is a family of open sets in X, then $\bigcup_\alpha U_\alpha$ is open. If K_α is a family of closed subsets of X, then $\bigcap_\alpha K_\alpha$ is closed.*

Given $S \subset X$, we denote by $\overline{S}$ (the *closure* of S) the smallest closed subset of X containing S, i.e., the intersection of all the closed sets $K_\alpha \subset X$ containing S. The following result is also straightforward.

Proposition A.3. *Given $S \subset X$, $p \in \overline{S}$ if and only if there exist $x_j \in S$ such that $x_j \to p$.*

Given $S \subset X$, $p \in X$, we say p is an *accumulation point* of S if and only if, for each $\varepsilon > 0$, there exists $q \in S \cap B_\varepsilon(p)$, $q \neq p$. It follows that p is an accumulation point of S if and only if each $B_\varepsilon(p)$, $\varepsilon > 0$, contains infinitely many points of S. One straightforward observation is that all points of $\overline{S} \setminus S$ are accumulation points of S.

The *interior* of a set $S \subset X$ is the largest open set contained in S, i.e., the union of all the open sets contained in S. Note that the complement of the interior of S is equal to the closure of $X \setminus S$.

We now turn to the notion of compactness. We say a metric space X is *compact* provided the following property holds:

(A) Each sequence (x_k) in X has a convergent subsequence.

We will establish various properties of compact metric spaces and provide various equivalent characterizations. For example, it is easily seen that (A) is equivalent to the following:

(B) Each infinite subset $S \subset X$ has an accumulation point.

The following property is known as total boundedness:

Proposition A.4. *If X is a compact metric space, then*

$$\text{(C)} \qquad \begin{array}{l} \textit{given } \varepsilon > 0, \ \exists \ \textit{a finite set } \{x_1, \dots, x_N\} \\ \textit{such that } B_\varepsilon(x_1), \dots, B_\varepsilon(x_N) \ \textit{covers } X. \end{array}$$

Proof. Take $\varepsilon > 0$ and pick $x_1 \in X$. If $B_\varepsilon(x_1) = X$, we are done. If not, pick $x_2 \in X \setminus B_\varepsilon(x_1)$. If $B_\varepsilon(x_1) \cup B_\varepsilon(x_2) = X$, we are done. If not, pick $x_3 \in X \setminus [B_\varepsilon(x_1) \cup B_\varepsilon(x_2)]$. Continue, taking $x_{k+1} \in X \setminus [B_\varepsilon(x_1) \cup \cdots \cup B_\varepsilon(x_k)]$, if $B_\varepsilon(x_1) \cup \cdots \cup B_\varepsilon(x_k) \neq X$. Note that, for $1 \le i, j \le k$,

$$i \neq j \Longrightarrow d(x_i, x_j) \ge \varepsilon.$$

If one never covers X this way, consider $S = \{x_j : j \in \mathbb{N}\}$. This is an infinite set with no accumulation point, so property (B) is contradicted.

Corollary A.5. *If X is a compact metric space, it has a countable dense subset.*

Proof. Given $\varepsilon = 2^{-n}$, let S_n be a finite set of points x_j such that $\{B_\varepsilon(x_j)\}$ covers X. Then $\mathcal{C} = \bigcup_n S_n$ is a countable dense subset of X.

Here is another useful property of compact metric spaces, which will eventually be generalized even further, in (E) below.

Proposition A.6. *Let X be a compact metric space. Assume $K_1 \supset K_2 \supset K_3 \supset \cdots$ form a decreasing sequence of closed subsets of X. If each $K_n \neq \emptyset$, then $\bigcap_n K_n \neq \emptyset$.*

Proof. Pick $x_n \in K_n$. If (A) holds, (x_n) has a convergent subsequence, $x_{n_k} \to y$. Since $\{x_{n_k} : k \geq \ell\} \subset K_{n_\ell}$, which is closed, we have $y \in \bigcap_n K_n$.

Corollary A.7. *Let X be a compact metric space. Assume $U_1 \subset U_2 \subset U_3 \subset \cdots$ form an increasing sequence of open subsets of X. If $\bigcup_n U_n = X$, then $U_N = X$ for some N.*

Proof. Consider $K_n = X \setminus U_n$.

The following is an important extension of Corollary A.7.

Proposition A.8. *If X is a compact metric space, then it has the property:*

(D) *every open cover* $\{U_\alpha : \alpha \in \mathcal{A}\}$ *of* X *has a finite subcover.*

Proof. Each U_α is a union of open balls, so it suffices to show that (A) implies the following:

(D′) Every cover $\{B_\alpha : \alpha \in \mathcal{A}\}$ of X by open balls has a finite subcover.

Let $\mathcal{C} = \{z_j : j \in \mathbb{N}\} \subset X$ be a countable dense subset of X, as in Corollary A.5. Each B_α is a union of balls $B_{r_j}(z_j)$, with $z_j \in \mathcal{C} \cap B_\alpha$, r_j rational. Thus it suffices to show that

(D″) every countable cover $\{B_j : j \in \mathbb{N}\}$ of X by open balls has a finite subcover.

For this, we set

$$U_n = B_1 \cup \cdots \cup B_n$$

and apply Corollary A.7.

The following is a convenient alternative to property (D):

(E) If $K_\alpha \subset X$ are closed and $\bigcap_\alpha K_\alpha = \emptyset$, then some finite intersection is empty.

Considering $U_\alpha = X \setminus K_\alpha$, we see that

$$(D) \Longleftrightarrow (E).$$

The following result completes Proposition A.8.

Theorem A.9. *For a metric space X,*

$$(A) \Longleftrightarrow (D).$$

Proof. By Proposition A.8, $(A) \Rightarrow (D)$. To prove the converse, it will suffice to show that $(E) \Rightarrow (B)$. So let $S \subset X$ and assume S has no accumulation point. We claim

$$\text{such } S \text{ must be closed.}$$

Indeed, if $z \in \overline{S}$ and $z \notin S$, then z would have to be an accumulation point. Say $S = \{x_\alpha : \alpha \in \mathcal{A}\}$. Set $K_\alpha = S \setminus \{x_\alpha\}$. Then each K_α has no accumulation point hence $K_\alpha \subset X$ is closed. Also $\bigcap_\alpha K_\alpha = \emptyset$. Hence there exists a finite set $\mathcal{F} \subset \mathcal{A}$ such that $\bigcap_{\alpha\in\mathcal{F}} K_\alpha = \emptyset$ if (E) holds. Hence $S = \bigcup_{\alpha\in\mathcal{F}}\{x_\alpha\}$ is finite, so indeed $(E) \Rightarrow (B)$.

Remark. So far we have that for any metric space X,

$$(A) \Longleftrightarrow (B) \Longleftrightarrow (D) \Longleftrightarrow (E) \Longrightarrow (C).$$

We claim that (C) implies the other conditions if X is *complete*. Of course, compactness implies completeness, but (C) may hold for incomplete X, e.g., $X = (0,1) \subset \mathbb{R}$.

Proposition A.10. *If X is a complete metric space with property* (C), *then X is compact.*

Proof. It suffices to show that $(C) \Rightarrow (B)$ if X is a complete metric space. So let $S \subset X$ be an infinite set. Cover X by balls $B_{1/2}(x_1), \dots, B_{1/2}(x_N)$. One of these balls contains infinitely many points of S, and so does its closure, say $X_1 = \overline{B_{1/2}(y_1)}$. Now cover X by finitely many balls of radius $1/4$; their intersection with X_1 provides a cover of X_1. One such set contains infinitely many points of S, and so does its closure $X_2 = \overline{B_{1/4}(y_2)} \cap X_1$. Continue in this fashion, obtaining

$$X_1 \supset X_2 \supset X_3 \supset \cdots \supset X_k \supset X_{k+1} \supset \cdots, \quad X_j \subset \overline{B_{2^{-j}}(y_j)},$$

each containing infinitely many points of S. One sees that (y_j) forms a Cauchy sequence. If X is complete, it has a limit, $y_j \to z$, and z is seen to be an accumulation point of S.

If X_j, $1 \le j \le m$, is a finite collection of metric spaces, with metrics d_j, we can define a Cartesian product metric space

$$X = \prod_{j=1}^{m} X_j, \quad d(x,y) = d_1(x_1,y_1) + \cdots + d_m(x_m,y_m). \tag{A.4}$$

Another choice of metric is $\delta(x,y) = \sqrt{d_1(x_1,y_1)^2 + \cdots + d_m(x_m,y_m)^2}$. The metrics d and δ are *equivalent*; i.e., there exist constants $C_0, C_1 \in (0,\infty)$ such that

$$C_0\delta(x,y) \le d(x,y) \le C_1\delta(x,y), \quad \forall\, x,y \in X. \tag{A.5}$$

A key example is $\mathbb{R}^m$, the Cartesian product of m copies of the real line $\mathbb{R}$.

We describe some important classes of compact spaces.

Proposition A.11. *If X_j are compact metric spaces, $1 \le j \le m$, so is $X = \prod_{j=1}^{m} X_j$.*

Proof. If (x_ν) is an infinite sequence of points in X, say $x_\nu = (x_{1\nu}, \dots, x_{m\nu})$, pick a convergent subsequence of $(x_{1\nu})$ in X_1 and consider the corresponding subsequence of (x_ν), which we relabel (x_ν). Using this, pick a convergent subsequence of $(x_{2\nu})$ in X_2. Continue. Having a subsequence such that $x_{j\nu} \to y_j$ in X_j for each $j = 1, \dots, m$, we then have a convergent subsequence in X.

The following result is useful for calculus on $\mathbb{R}^n$.

Proposition A.12. *If K is a closed bounded subset of $\mathbb{R}^n$, then K is compact.*

Proof. The discussion above reduces the problem to showing that any closed interval $I = [a,b]$ in $\mathbb{R}$ is compact. This compactness is a corollary of Proposition A.10. For pedagogical purposes, we redo the argument here, since in this concrete case it can be streamlined.

Suppose S is a subset of I with infinitely many elements. Divide I into two equal subintervals, $I_1 = [a,b_1]$, $I_2 = [b_1,b]$, $b_1 = (a+b)/2$. Then either I_1 or I_2 must contain infinitely many elements of S. Say I_j does. Let x_1 be any element of S lying in I_j. Now divide I_j into two equal pieces, $I_j = I_{j1} \cup I_{j2}$. One of these intervals (say I_{jk}) contains infinitely many points of S. Pick $x_2 \in I_{jk}$ to be one such point (different from x_1). Then subdivide I_{jk} into two equal subintervals, and continue. We get an infinite sequence of distinct points $x_\nu \in S$, and $|x_\nu - x_{\nu+k}| \le 2^{-\nu}(b-a)$, for $k \ge 1$. Since $\mathbb{R}$ is complete, (x_ν) converges, say to $y \in I$. Any neighborhood of y contains infinitely many points in S, so we are done.

If X and Y are metric spaces, a function $f : X \to Y$ is said to be continuous provided $x_\nu \to x$ in X implies $f(x_\nu) \to f(x)$ in Y. An equivalent condition, which the reader can verify, is

$$U \text{ open in } Y \Longrightarrow f^{-1}(U) \text{ open in } X.$$

Proposition A.13. *If X and Y are metric spaces, $f : X \to Y$ continuous, and $K \subset X$ compact, then $f(K)$ is a compact subset of Y.*

Proof. If (y_ν) is an infinite sequence of points in $f(K)$, pick $x_\nu \in K$ such that $f(x_\nu) = y_\nu$. If K is compact, we have a subsequence $x_{\nu_j} \to p$ in X, and then $y_{\nu_j} \to f(p)$ in Y.

If $F : X \to \mathbb{R}$ is continuous, we say $f \in C(X)$. A useful corollary of Proposition A.13 is

Proposition A.14. *If X is a compact metric space and $f \in C(X)$, then f assumes a maximum and a minimum value on X.*

Proof. We know from Proposition A.13 that $f(X)$ is a compact subset of $\mathbb{R}$. Hence $f(X)$ is bounded, say $f(X) \subset I = [a, b]$. Repeatedly subdividing I into equal halves, as in the proof of Proposition A.12, at each stage throwing out intervals that do not intersect $f(X)$ and keeping only the leftmost and rightmost interval amongst those remaining, we obtain points $\alpha \in f(X)$ and $\beta \in f(X)$ such that $f(X) \subset [\alpha, \beta]$. Then $\alpha = f(x_0)$ for some $x_0 \in X$ is the minimum and $\beta = f(x_1)$ for some $x_1 \in X$ is the maximum.

At this point, the reader might take a look at the proof of the Mean Value Theorem, given in Chapter 1, which applies this result.

A function $f \in C(X)$ is said to be *uniformly continuous* provided that, for any $\varepsilon > 0$, there exists $\delta > 0$ such that

$$x, y \in X, \ d(x, y) \le \delta \Longrightarrow |f(x) - f(y)| \le \varepsilon. \tag{A.6}$$

An equivalent condition is that f have a *modulus of continuity*, i.e., a monotonic function $\omega : [0, 1) \to [0, \infty)$ such that $\delta \searrow 0 \Rightarrow \omega(\delta) \searrow 0$ and such that

$$x, y \in X, \ d(x, y) \le \delta \le 1 \Longrightarrow |f(x) - f(y)| \le \omega(\delta). \tag{A.7}$$

Not all continuous functions are uniformly continuous. For example, if $X = (0, 1) \subset \mathbb{R}$, then $f(x) = \sin 1/x$ is continuous, but not uniformly continuous, on X. The following result is useful, for example, in the development of the Riemann integral in Chapter 1.

Proposition A.15. *If X is a compact metric space and $f \in C(X)$, then f is uniformly continuous.*

Proof. If not, there exist $x_\nu, y_\nu \in X$ and $\varepsilon > 0$ such that $d(x_\nu, y_\nu) \le 2^{-\nu}$ but

$$|f(x_\nu) - f(y_\nu)| \ge \varepsilon. \tag{A.8}$$

Taking a convergent subsequence $x_{\nu_j} \to p$, we also have $y_{\nu_j} \to p$. Now continuity of f at p implies $f(x_{\nu_j}) \to f(p)$ and $f(y_{\nu_j}) \to f(p)$, contradicting (A.8).

If X and Y are metric spaces, the space $C(X, Y)$ of continuous maps $f : X \to Y$ has a natural metric structure, under some additional hypotheses. We use

$$D(f, g) = \sup_{x \in X} d\big(f(x), g(x)\big). \tag{A.9}$$

This sup exists provided $f(X)$ and $g(X)$ are *bounded* subsets of Y, where to say $B \subset Y$ is bounded is to say $d : B \times B \to [0, \infty)$ has bounded image. In particular, this supremum exists if X is compact. The following result is frequently useful.

Proposition A.16. *If X is a compact metric space and Y is a complete metric space, then $C(X, Y)$, with the metric (A.9), is complete.*

Proof. That $D(f, g)$ satisfies the conditions to define a metric on $C(X, Y)$ is straightforward. We check completeness. Suppose (f_ν) is a Cauchy sequence in $C(X, Y)$, so, as $\nu \to \infty$,

$$\sup_{k \ge 0} \sup_{x \in X} d\big(f_{\nu+k}(x), f_\nu(x)\big) \le \varepsilon_\nu \to 0.$$

Then in particular $(f_\nu(x))$ is a Cauchy sequence in Y for each $x \in X$, so it converges, say to $g(x) \in Y$. It remains to show that $g \in C(X, Y)$ and that $f_\nu \to g$ in the metric (A.9).

In fact, taking $k \to \infty$ in the estimate above, we have

$$\sup_{x \in X} d\big(g(x), f_\nu(x)\big) \le \varepsilon_\nu \to 0,$$

i.e., $f_\nu \to g$ uniformly. It remains only to show that g is continuous. For this, let $x_j \to x$ in X and fix $\varepsilon > 0$. Pick N so that $\varepsilon_N < \varepsilon$. Since f_N is continuous, there exists J such that $j \ge J \Rightarrow d(f_N(x_j), f_N(x)) < \varepsilon$. Hence

$$\begin{aligned} j \ge J \Rightarrow d\big(g(x_j), g(x)\big) \le\ & d\big(g(x_j), f_N(x_j)\big) \\ & + d\big(f_N(x_j), f_N(x)\big) + d\big(f_N(x), g(x)\big) < 3\varepsilon. \end{aligned}$$

This completes the proof.

We next give a couple of slightly more sophisticated results on compactness. The following extension of Proposition A.11 is a special case of Tychonov's Theorem.

Proposition A.17. *If $\{X_j : j \in \mathbb{Z}^+\}$ are compact metric spaces, so is $X = \prod_{j=1}^\infty X_j$.*

Here, we can make X a metric space by setting

$$d(x, y) = \sum_{j=1}^\infty \frac{2^{-j}\, d_j(x, y)}{1 + d_j(x, y)}. \tag{A.10}$$

It is easy to verify that, if $x_\nu \in X$, then $x_\nu \to y$ in X, as $\nu \to \infty$, if and only if, for each j, $p_j(x_\nu) \to p_j(y)$ in X_j, where $p_j : X \to X_j$ is the projection onto the jth factor.

Proof. Following the argument in Proposition A.11, if (x_ν) is an infinite sequence of points in X, we obtain a nested family of subsequences

$$(x_\nu) \supset ({x^1}_\nu) \supset ({x^2}_\nu) \supset \cdots \supset ({x^j}_\nu) \supset \cdots \tag{A.11}$$

such that $p_\ell({x^j}_\nu)$ converges in X_ℓ, for $1 \le \ell \le j$. The next step is a *diagonal construction*. We set

$$\xi_\nu = {x^\nu}_\nu \in X. \tag{A.12}$$

Then, for each j, after throwing away a finite number $N(j)$ of elements, one obtains from (ξ_ν) a subsequence of the sequence $({x^j}_\nu)$ in (A.11), so $p_\ell(\xi_\nu)$ converges in X_ℓ for all ℓ. Hence (ξ_ν) is a convergent subsequence of (x_ν).

The next result is a special case of Ascoli's Theorem.

Proposition A.18. *Let X and Y be compact metric spaces, and fix a modulus of continuity $\omega(\delta)$. Then*

$$\mathcal{C}_\omega = \big\{ f \in C(X, Y) : d\big(f(x_1), f(x_2)\big) \le \omega\big(d(x_1, x_2)\big),\ \forall\, x_1, x_2 \in X \big\} \tag{A.13}$$

is a compact subset of $C(X, Y)$.

Proof. Let (f_ν) be a sequence in $\mathcal{C}_\omega$. Let Σ be a countable dense subset of X, as in Corollary A.5. For each $x \in \Sigma$, $(f_\nu(x))$ is a sequence in Y, which hence has a convergent subsequence. Using a diagonal construction similar to that in the proof of Proposition A.17, we obtain a subsequence (φ_ν) of (f_ν) with the property that $\varphi_\nu(x)$ converges in Y, for *each* $x \in \Sigma$, say

$$x \in \Sigma \Longrightarrow \varphi_\nu(x) \to \psi(x), \tag{A.14}$$

where $\psi : \Sigma \to Y$.

So far, we have not used (A.13), but this hypothesis readily yields

$$d\big(\psi(x), \psi(y)\big) \le \omega\big(d(x,y)\big), \tag{A.15}$$

for all $x, y \in \Sigma$. Using the denseness of $\Sigma \subset X$, we can extend ψ uniquely to a continuous map of $X \to Y$, which we continue to denote by ψ. Also, (A.15) holds for all $x, y \in X$, i.e., $\psi \in \mathcal{C}_\omega$.

It remains to show that $\varphi_\nu \to \psi$ uniformly on X. Pick $\varepsilon > 0$. Then pick $\delta > 0$ such that $\omega(\delta) < \varepsilon/3$. Since X is compact, we can cover X by finitely many balls $B_\delta(x_j)$, $1 \le j \le N$, $x_j \in \Sigma$. Pick M so large that $\varphi_\nu(x_j)$ is within $\varepsilon/3$ of its limit for all $\nu \ge M$ (when $1 \le j \le N$). Now, for any $x \in X$, picking $\ell \in \{1, \dots, N\}$ such that $d(x, x_\ell) \le \delta$, we have, for $k \ge 0$, $\nu \ge M$,

$$\begin{aligned} d\big(\varphi_{\nu+k}(x), \varphi_\nu(x)\big) &\le d\big(\varphi_{\nu+k}(x), \varphi_{\nu+k}(x_\ell)\big) + d\big(\varphi_{\nu+k}(x_\ell), \varphi_\nu(x_\ell)\big) \\ &\quad + d\big(\varphi_\nu(x_\ell), \varphi_\nu(x)\big) \\ &\le \varepsilon/3 + \varepsilon/3 + \varepsilon/3, \end{aligned} \tag{A.16}$$

proving the proposition.

We next define the notion of a *connected* space. A metric space X is said to be connected provided that it cannot be written as the union of two disjoint open subsets. The following is a basic class of examples.

Proposition A.19. *An interval I in $\mathbb{R}$ is connected.*

Proof. Suppose $A \subset I$ is nonempty, with nonempty complement $B \subset I$, and both sets are open. Take $a \in A$, $b \in B$; we can assume $a < b$. Let $\xi = \sup\{x \in [a, b] : x \in A\}$. This exists as a consequence of the basic fact that $\mathbb{R}$ is complete.

Now we obtain a contradiction, as follows. Since A is closed, $\xi \in A$. But then, since A is open, there must be a neighborhood $(\xi - \varepsilon, \xi + \varepsilon)$ contained in A; this is not possible.

We say X is path-connected if, given any $p, q \in X$, there is a continuous map $\gamma : [0, 1] \to X$ such that $\gamma(0) = p$ and $\gamma(1) = q$. It is an easy consequence of Proposition A.19 that X is connected whenever it is path-connected.

The next result, known as the Intermediate Value Theorem, is frequently useful.

Proposition A.20. *Let X be a connected metric space and $f : X \to \mathbb{R}$ continuous. Assume $p, q \in X$ and $f(p) = a < f(q) = b$. Then, given any $c \in (a, b)$, there exists $z \in X$ such that $f(z) = c$.*

Proof. Under the hypotheses, $A = \{x \in X : f(x) < c\}$ is open and contains p, while $B = \{x \in X : f(x) > c\}$ is open and contains q. If X is connected, then $A \cup B$ cannot be all of X; so any point in its complement has the desired property.

We turn now to the notion of a topological space. This is a set X, together with a family $\mathcal{O}$ of subsets, called *open*, satisfying the following conditions:

$$
\begin{aligned}
&X, \emptyset \text{ open}, \\
&U_j \text{ open}, 1 \le j \le N \Rightarrow \bigcap_{j=1}^{N} U_j \text{ open}, \\
&U_\alpha \text{ open}, \alpha \in A \Rightarrow \bigcup_{\alpha \in A} U_\alpha \text{ open},
\end{aligned}
\tag{A.17}
$$

where A is *any* index set. It is obvious that the collection of open subsets of a metric space, defined above, satisfies these conditions. As before, a set $S \subset X$ is closed provided $X \setminus S$ is open. Also, we say a subset $N \subset X$ containing p is a *neighborhood* of p provided N contains an open set U which in turn contains p.

If X is a topological space and S is a subset, S gets a topology as follows. For each U open in X, $U \cap S$ is declared to be open in S. This is called the induced topology.

A topological space X is said to be *Hausdorff* provided that any distinct $p, q \in X$ have disjoint neighborhoods. Clearly any metric space is Hausdorff. Most important topological spaces are Hausdorff.

A Hausdorff topological space is said to be compact provided the following condition holds. If $\{U_\alpha : \alpha \in A\}$ is any family of open subsets of X, covering X, i.e., $X = \bigcup_{\alpha \in A} U_\alpha$, then there is a finite subcover, i.e., a finite subset $\{U_{\alpha_1}, \dots, U_{\alpha_N} : \alpha_j \in A\}$ such that $X = U_{\alpha_1} \cup \cdots \cup U_{\alpha_N}$. An equivalent formulation is the following, known as the finite intersection property. Let $\{S_\alpha : \alpha \in A\}$ be any collection of closed subsets of X. If each finite collection of these closed sets has nonempty intersection, then the complete intersection $\bigcap_{\alpha \in A} S_\alpha$ is nonempty. In the first part of this chapter, we have shown that any compact metric space satisfies this condition.

Any closed subset of a compact space is compact. Furthermore, any compact subset of a Hausdorff space is necessarily closed.

Most of the propositions stated above for compact metric spaces have extensions to compact Hausdorff spaces. We mention one nontrivial result, which is the general form of Tychonov's Theorem. One can find a proof in many topology texts, such as [**Dug**] or [**Mun**].

Theorem A.21. *If S is any nonempty set (possibly uncountable) and, for any $\alpha \in S$, X_α is a compact Hausdorff space, then so is $X = \prod_{\alpha \in S} X_\alpha$.*

A Hausdorff space X is said to be *locally compact* provided every $p \in X$ has a neighborhood N which is compact (with the induced topology).

A Hausdorff space is said to be *paracompact* provided every open cover $\{U_\alpha : \alpha \in A\}$ has a locally finite refinement, i.e., an open cover $\{V_\beta : \beta \in B\}$ such that each V_β is contained in some U_α and each $p \in X$ has a neighborhood N_p such that $N_p \cap V_\beta$ is nonempty for only finitely many $\beta \in B$. A typical example of a paracompact space is a locally compact Hausdorff space X which is also σ-compact, i.e., $X = \bigcup_{n=1}^{\infty} X_n$ with X_n compact. Paracompactness is a natural condition under which to construct partitions of unity.

A map $F : X \to Y$ between two topological spaces is said to be *continuous* provided $F^{-1}(U)$ is open in X whenever U is open in Y. If $F : X \to Y$ is one-to-one and onto and both F and F^{-1} are continuous, F is said to be a *homeomorphism*. For a bijective map $F : X \to Y$, the continuity of F^{-1} is equivalent to the statement that $F(V)$ is open in Y whenever V is open in X; another equivalent statement is that $F(S)$ is closed in Y whenever S is closed in X.

If X and Y are Hausdorff, $F : X \to Y$ continuous, then $F(K)$ is compact in Y whenever K is compact in X. In view of the discussion above, there arises the following useful sufficient condition for a continuous map $F : X \to Y$ to be a homeomorphism. Namely, if X is compact, Y Hausdorff, and F one-to-one and onto, then F is a homeomorphism.

We turn to a discussion of the Stone-Weierstrass Theorem, which gives a sufficient condition for a family of functions on a compact Hausdorff space X to be dense in the space $C(X)$ of continuous functions on X, with the sup norm. This result is used several times in the text. It is an extension of the Weierstrass Approximation Theorem, which we state first.

Theorem A.22. *If $I = [a, b]$ is an interval in $\mathbb{R}$, the space $\mathcal{P}$ of polynomials in one variable is dense in $C(I)$.*

There are many proofs of this. One close to Weierstrass' original (and my favorite) goes as follows. Given $f \in C(I)$, extend it to be continuous and compactly supported on $\mathbb{R}$, convolve this with a highly peaked Gaussian, and approximate the result by power series.

Such an argument involves a little use of complex variable theory. We sketch another proof of Theorem A.22, which avoids complex variable theory. Our starting point will be the result that the power series for $(1-x)^a$ converges uniformly on $[-1,1]$, for any $a > 0$. This is a simple consequence of Taylor's formula with remainder. We will use it with $a = 1/2$.

From the identity $x^{1/2} = (1-(1-x))^{1/2}$, we have $x^{1/2} \in \overline{\mathcal{P}}([0,2])$. More to the point, from the identity

$$|x| = \big(1-(1-x^2)\big)^{1/2}, \tag{A.18}$$

we have $|x| \in \overline{\mathcal{P}}([-\sqrt{2},\sqrt{2}])$. Using $|x| = b^{-1}|bx|$, for any $b > 0$, we see that $|x| \in \overline{\mathcal{P}}(I)$ for any interval $I = [-c,c]$ and also for any closed subinterval, hence for any compact interval I. By translation, we have

$$|x-a| \in \overline{\mathcal{P}}(I) \tag{A.19}$$

for any compact interval I. Using the identities

$$\max(x,y) = \frac{1}{2}(x+y) + \frac{1}{2}|x-y|, \quad \min(x,y) = \frac{1}{2}(x+y) - \frac{1}{2}|x-y|, \tag{A.20}$$

we see that for any $a \in \mathbb{R}$ and any compact I,

$$\max(x,a),\ \min(x,a) \in \overline{\mathcal{P}}(I). \tag{A.21}$$

We next note that $\overline{\mathcal{P}}(I)$ is an algebra of functions, i.e.,

$$f,g \in \overline{\mathcal{P}}(I),\ c \in \mathbb{R} \Longrightarrow f+g, fg, cf \in \overline{\mathcal{P}}(I). \tag{A.22}$$

Using this, one sees that, given $f \in \overline{\mathcal{P}}(I)$, with range in a compact interval J, one has $h \circ f \in \overline{\mathcal{P}}(I)$ for all $h \in \overline{\mathcal{P}}(J)$. Hence $f \in \overline{\mathcal{P}}(I) \Rightarrow |f| \in \overline{\mathcal{P}}(I)$, and, via (A.20), we deduce that

$$f,g \in \overline{\mathcal{P}}(I) \Longrightarrow \max(f,g),\ \min(f,g) \in \overline{\mathcal{P}}(I). \tag{A.23}$$

Suppose now that $I' = [a',b']$ is a subinterval of $I = [a,b]$. With the notation $x_+ = \max(x,0)$, we have

$$f_{II'}(x) = \min\big((x-a')_+, (b'-x)_+\big) \in \overline{\mathcal{P}}(I). \tag{A.24}$$

This is a piecewise linear function, equal to zero off $I \setminus I'$, with slope 1 from a' to the midpoint m' of I' and slope -1 from m' to b'.

Now if I is divided into N equal subintervals, any continuous function on I that is linear on each such subinterval can be written as a linear combination of such "tent functions," so it belongs to $\overline{\mathcal{P}}(I)$. Finally, any $f \in C(I)$ can be uniformly approximated by such piecewise linear functions, so we have $f \in \overline{\mathcal{P}}(I)$, proving the theorem.

We now state the Stone-Weierstrass Theorem.

Theorem A.23. *Let X be a compact Hausdorff space, $\mathcal{A}$ a subalgebra of $C_{\mathbb{R}}(X)$, the algebra of real-valued continuous functions on X. Suppose $1 \in \mathcal{A}$ and that $\mathcal{A}$ separates points of X, i.e., for distinct $p, q \in X$, there exists $h_{pq} \in \mathcal{A}$ with $h_{pq}(p) \neq h_{pq}(q)$. Then the closure $\overline{\mathcal{A}}$ is equal to $C_{\mathbb{R}}(X)$.*

We give a proof of Theorem A.23, making use of the argument above, which implies that, if $f \in \overline{\mathcal{A}}$ and $\varphi : \mathbb{R} \to \mathbb{R}$ is continuous, then $\varphi \circ f \in \overline{\mathcal{A}}$. Consequently, if $f_j \in \overline{\mathcal{A}}$, then $\sup(f_1, f_2) = (1/2)|f_1 - f_2| + (1/2)(f_1 + f_2) \in \overline{\mathcal{A}}$. Similarly $\inf(f_1, f_2) \in \overline{\mathcal{A}}$.

The hypothesis of separating points implies that, for distinct $p, q \in X$, there exists $f_{pq} \in \overline{\mathcal{A}}$, equal to 1 at p, 0 at q. Applying appropriate φ, we can arrange also that $0 \le f_{pq}(x) \le 1$ on X and that f_{pq} is 1 *near* p and 0 *near* q. Taking infima, we can obtain $f_{pU} \in \overline{\mathcal{A}}$, equal to 1 on a neighborhood of p and equal to 0 off a given neighborhood U of p. Applying sups to these, we obtain for each compact $K \subset X$ and open $U \supset K$ a function $g_{KU} \in \overline{\mathcal{A}}$ such that g_{KU} is 1 on K, 0 off U, and $0 \le g_{KU}(x) \le 1$ on X. Once we have gotten this far, it is easy to approximate any continuous $u \ge 0$ on X by a sup of (positive constants times) such g_{KU}'s and from there to prove the theorem.

Theorem A.23 has a complex analogue. In that case, we add the assumption that $f \in \mathcal{A} \Rightarrow \overline{f} \in \mathcal{A}$ and conclude that $\overline{\mathcal{A}} = C(X)$. This is easily reduced to the real case.

The next result is known as the Contraction Mapping Principle, and it has many uses in analysis. In particular, we will use it in the proof of the Inverse Function Theorem, in Appendix B.

Theorem A.24. *Let X be a complete metric space, and let $T : X \to X$ satisfy*

$$\operatorname{dist}(Tx, Ty) \le r \operatorname{dist}(x, y), \tag{A.25}$$

for some $r < 1$. (We say T is a contraction.) Then T has a unique fixed point x. For any $y_0 \in X$, $T^k y_0 \to x$ as $k \to \infty$.

Proof. Pick $y_0 \in X$ and let $y_k = T^k y_0$. Then $\operatorname{dist}(y_{k+1}, y_k) \le r^k \operatorname{dist}(y_1, y_0)$, so

$$\begin{aligned} \operatorname{dist}(y_{k+m}, y_k) &\le \operatorname{dist}(y_{k+m}, y_{k+m-1}) + \cdots + \operatorname{dist}(y_{k+1}, y_k) \\ &\le \left(r^k + \cdots + r^{k+m-1}\right) \operatorname{dist}(y_1, y_0) \\ &\le r^k (1 - r)^{-1} \operatorname{dist}(y_1, y_0). \end{aligned} \tag{A.26}$$

It follows that (y_k) is a Cauchy sequence, so it converges; $y_k \to x$. Since $Ty_k = y_{k+1}$ and T is continuous, it follows that $Tx = x$, i.e., x is a fixed point. Uniqueness of the fixed point is clear from the estimate $\text{dist}(Tx, Tx') \leq r \text{ dist}(x, x')$, which implies $\text{dist}(x, x') = 0$ if x and x' are fixed points. This proves Theorem A.24.

Appendix B

Derivatives, Diffeomorphisms, and Manifolds

To start this chapter off, we define the derivative and discuss some of its basic properties. Let $\mathcal{O}$ be an open subset of $\mathbb{R}^n$ and $F : \mathcal{O} \to \mathbb{R}^m$ a continuous function. We say F is differentiable at a point $x \in \mathcal{O}$, with derivative L, if $L : \mathbb{R}^n \to \mathbb{R}^m$ is a linear transformation such that, for $y \in \mathbb{R}^n$, small,

$$F(x+y) = F(x) + Ly + R(x,y) \tag{B.1}$$

with

$$\frac{\|R(x,y)\|}{\|y\|} \to 0 \quad \text{as} \quad y \to 0. \tag{B.2}$$

We denote the derivative at x by $DF(x) = L$. With respect to the standard bases of $\mathbb{R}^n$ and $\mathbb{R}^m$, $DF(x)$ is simply the matrix of partial derivatives,

$$DF(x) = \left(\frac{\partial F_j}{\partial x_k}\right), \tag{B.3}$$

so that, if $v = (v_1, \dots, v_n)^t$ (regarded as a column vector), then

$$DF(x)v = \Bigl(\sum_k \frac{\partial F_1}{\partial x_k} v_k, \dots, \sum_k \frac{\partial F_m}{\partial x_k} v_k\Bigr)^t. \tag{B.4}$$

It will be shown below that F is differentiable whenever all the partial derivatives exist and are *continuous* on $\mathcal{O}$. In such a case we say F is a C^1

function on $\mathcal{O}$. More generally, F is said to be C^k if all its partial derivatives of order $\leq k$ exist and are continuous. If F is C^k for all k, we say F is C^∞.

In (B.2), we can use the *Euclidean* norm on $\mathbb{R}^n$ and $\mathbb{R}^m$. This norm is defined by

$$\|x\| = \left(x_1^2 + \cdots + x_n^2\right)^{1/2} \tag{B.5}$$

for $x = (x_1, \ldots, x_n) \in \mathbb{R}^n$. Any other norm would do equally well.

We now derive the *chain rule* for the derivative. Let $F : \mathcal{O} \to \mathbb{R}^m$ be differentiable at $x \in \mathcal{O}$, as above, let U be a neighborhood of $z = F(x)$ in $\mathbb{R}^m$, and let $G : U \to \mathbb{R}^k$ be differentiable at z. Consider $H = G \circ F$. We have

$$\begin{aligned} H(x+y) &= G(F(x+y)) \\ &= G\big(F(x) + DF(x)y + R(x,y)\big) \\ &= G(z) + DG(z)\big(DF(x)y + R(x,y)\big) + R_1(x,y) \\ &= G(z) + DG(z)DF(x)y + R_2(x,y) \end{aligned} \tag{B.6}$$

with

$$\frac{\|R_2(x,y)\|}{\|y\|} \to 0 \ \text{ as } \ y \to 0.$$

Thus $G \circ F$ is differentiable at x, and

$$D(G \circ F)(x) = DG(F(x)) \cdot DF(x). \tag{B.7}$$

Another useful remark is that, by the Fundamental Theorem of Calculus, applied to $\varphi(t) = F(x + ty)$,

$$F(x+y) = F(x) + \int_0^1 DF(x+ty)y\,dt, \tag{B.8}$$

provided F is C^1. A closely related application of the Fundamental Theorem of Calculus is that, if we assume $F : \mathcal{O} \to \mathbb{R}^m$ is differentiable in each variable separately and that each $\partial F/\partial x_j$ is continuous on $\mathcal{O}$, then
(B.9)

$$F(x+y) = F(x) + \sum_{j=1}^n \big[F(x+z_j) - F(x+z_{j-1})\big] = F(x) + \sum_{j=1}^n A_j(x,y)y_j,$$

$$A_j(x,y) = \int_0^1 \frac{\partial F}{\partial x_j}\big(x + z_{j-1} + ty_je_j\big)\,dt,$$

where $z_0 = 0$, $z_j = (y_1, \ldots, y_j, 0, \ldots, 0)$, and $\{e_j\}$ is the standard basis of $\mathbb{R}^n$. Now (B.9) implies F is differentiable on $\mathcal{O}$, as we stated below (B.4). Thus we have established the following.

Proposition B.1. *If $\mathcal{O}$ is an open subset of $\mathbb{R}^n$ and $F : \mathcal{O} \to \mathbb{R}^m$ is of class C^1, then F is differentiable at each point $x \in \mathcal{O}$.*

As is shown in many calculus texts, one can use the Mean Value Theorem instead of the Fundamental Theorem of Calculus and obtain a slightly sharper result.

For the study of higher order derivatives of a function, the following result is fundamental. We will not take the space to provide a proof, which can be found in many advanced calculus texts.

Proposition B.2. *Assume $F : \mathcal{O} \to \mathbb{R}^m$ is of class C^2, with $\mathcal{O}$ open in $\mathbb{R}^n$. Then, for each $x \in \mathcal{O}$, $1 \le j, k \le n$,*

$$\frac{\partial}{\partial x_j}\frac{\partial F}{\partial x_k}(x) = \frac{\partial}{\partial x_k}\frac{\partial F}{\partial x_j}(x). \tag{B.10}$$

If U and V be open subsets of $\mathbb{R}^n$ and $F : U \to V$ is a C^1 map, we say F is a diffeomorphism of U onto V provided F maps U one-to-one and onto V and its inverse $G = F^{-1}$ is a C^1 map. If F is a diffeomorphism, it follows from the chain rule that $DF(x)$ is invertible for each $x \in U$. We now present a partial converse of this, the Inverse Function Theorem, which is a fundamental result in multi-variable calculus.

Theorem B.3. *Let F be a C^k map from an open neighborhood Ω of $p_0 \in \mathbb{R}^n$ to $\mathbb{R}^n$, with $q_0 = F(p_0)$. Assume $k \ge 1$. Suppose the derivative $DF(p_0)$ is invertible. Then there is a neighborhood U of p_0 and a neighborhood V of q_0 such that $F : U \to V$ is one-to-one and onto and $F^{-1} : V \to U$ is a C^k map. (So $F : U \to V$ is a diffeomorphism.)*

First we show that F is one-to-one on a neighborhood of p_0, under these hypotheses. In fact, we establish the following result, of interest in its own right.

Proposition B.4. *Assume $\Omega \subset \mathbb{R}^n$ is open and convex, and let $f : \Omega \to \mathbb{R}^n$ be C^1. Assume that the symmetric part of $Df(u)$ is positive-definite, for each $u \in \Omega$. Then f is one-to-one on Ω.*

Proof. Take distinct points $u_1,\ u_2 \in \Omega$, and set $u_2 - u_1 = w$. Consider $\varphi : [0,1] \to \mathbb{R}$, given by

$$\varphi(t) = w \cdot f(u_1 + tw).$$

Then $\varphi'(t) = w \cdot Df(u_1 + tw)w > 0$ for $t \in [0,1]$, so $\varphi(0) \neq \varphi(1)$. But $\varphi(0) = w \cdot f(u_1)$ and $\varphi(1) = w \cdot f(u_2)$, so $f(u_1) \neq f(u_2)$.

To continue the proof of Theorem B.3, let us set

$$f(u) = A\big(F(p_0 + u) - q_0\big), \quad A = DF(p_0)^{-1}. \tag{B.11}$$

Then $f(0) = 0$ and $Df(0) = I$, the identity matrix. We show that f maps a neighborhood of 0 one-to-one and onto some neighborhood of 0. Proposition B.4 applies, so we know f is one-to-one on some neighborhood $\mathcal{O}$ of 0. We next show that the image of $\mathcal{O}$ under f contains a neighborhood of 0.

Note that

$$f(u) = u + R(u), \quad R(0) = 0, \ DR(0) = 0. \tag{B.12}$$

For v small, we want to solve

$$f(u) = v. \tag{B.13}$$

This is equivalent to $u + R(u) = v$, so let

$$T_v(u) = v - R(u). \tag{B.14}$$

Thus solving (B.13) is equivalent to solving

$$T_v(u) = u. \tag{B.15}$$

We look for a *fixed point* $u = K(v) = f^{-1}(v)$. Also, we want to prove that $DK(0) = I$, i.e., that $K(v) = v + r(v)$ with $r(v) = o(\|v\|)$, i.e., $r(v)/\|v\| \to 0$ as $v \to 0$. If we succeed in doing this, it follows easily that, for general x close to q_0, $G(x) = F^{-1}(x)$ is defined, and

$$DG(x) = \Big(DF\big(G(x)\big)\Big)^{-1}. \tag{B.16}$$

Then a simple inductive argument shows that G is C^k if F is C^k.

A tool we will use to solve (B.15) is the Contraction Mapping Principle, Theorem A.24, which states that if X is a complete metric space and if $T : X \to X$ satisfies

$$\operatorname{dist}(Tx, Ty) \le r \operatorname{dist}(x, y), \tag{B.17}$$

for some $r < 1$ (we say T is a contraction), then T has a unique fixed point x.

In order to implement this, we consider

$$T_v : X_v \longrightarrow X_v \tag{B.18}$$

with

$$X_v = \{u \in \Omega : \|u - v\| \le A_v\} \tag{B.19}$$

where we set

$$A_v = \sup_{\|w\| \le 2\|v\|} \|R(w)\|. \tag{B.20}$$

We claim that (B.18) holds if $\|v\|$ is sufficiently small. To prove this, note that $T_v(u) - v = -R(u)$, so we need to show that, provided $\|v\|$ is small, $u \in X_v$ implies $\|R(u)\| \le A_v$. But indeed, if $u \in X_v$, then $\|u\| \le \|v\| + A_v$, which is $\le 2\|v\|$ if $\|v\|$ is small, so then

$$\|R(u)\| \le \sup_{\|w\| \le 2\|v\|} \|R(w)\| = A_v.$$

This establishes (B.18).

Note that $T_v(u_1) - T_v(u_2) = R(u_2) - R(u_1)$ and R is a C^k map, satisfying $DR(0) = 0$. It follows that, if $\|v\|$ is small enough, the map (B.18) is a contraction map. Hence there exists a unique fixed point $u = K(v) \in X_v$. Also, since $u \in X_v$,

$$\|K(v) - v\| \le A_v = o(\|v\|), \tag{B.21}$$

so the Inverse Function Theorem is proved.

Thus if DF is invertible on the domain of F, F is a local diffeomorphism. Stronger hypotheses are needed to guarantee that F is a global diffeomorphism onto its range. Proposition B.4 provides one tool for doing this. Here is a slight strengthening.

Corollary B.5. *Assume $\Omega \subset \mathbb{R}^n$ is open and convex and that $F : \Omega \to \mathbb{R}^n$ is C^1. Assume there exist $n \times n$ matrices A and B such that the symmetric part of $A\,DF(u)\,B$ is positive-definite for each $u \in \Omega$. Then F maps Ω diffeomorphically onto its image, an open set in $\mathbb{R}^n$.*

Proof. Exercise.

We turn to a discussion of surfaces in Euclidean space. A smooth m-dimensional surface $M \subset \mathbb{R}^n$ is characterized by the following property. Given $p \in M$, there is a neighborhood U of p in M and a smooth map $\varphi : \mathcal{O} \to U$, from an open set $\mathcal{O} \subset \mathbb{R}^m$ bijectively to U, with injective derivative at each point. Such a map φ is called a *coordinate chart* on M.

We call $U \subset M$ a coordinate patch. If all such maps φ are smooth of class C^k, we say M is a surface of class C^k.

There is an abstraction of the notion of a surface, namely the notion of a "manifold," which we will discuss a little later in this appendix.

If $\varphi : \mathcal{O} \to U$ is a C^k coordinate chart, such as described above, and $\varphi(x_0) = p$, we set

$$T_pM = \operatorname{Range} D\varphi(x_0), \tag{B.22}$$

a linear subspace of $\mathbb{R}^n$ of dimension m, and we denote by N_pM its orthogonal complement. It is useful to consider the following map. Pick a linear isomorphism $A : \mathbb{R}^{n-m} \to N_pM$, and define

$$\Phi : \mathcal{O} \times \mathbb{R}^{n-m} \longrightarrow \mathbb{R}^n, \quad \Phi(x, z) = \varphi(x) + Az. \tag{B.23}$$

Thus Φ is a C^k map defined on an open subset of $\mathbb{R}^n$. Note that

$$D\Phi(x_0, 0)\begin{pmatrix} v \\ w \end{pmatrix} = D\varphi(x_0)v + Aw, \tag{B.24}$$

so $D\Phi(x_0, 0) : \mathbb{R}^n \to \mathbb{R}^n$ is surjective, hence bijective, so the Inverse Function Theorem applies; Φ maps some neighborhood of $(x_0, 0)$ diffeomorphically onto a neighborhood of $p \in \mathbb{R}^n$.

Suppose there is another C^k coordinate chart, $\psi : \Omega \to U$. Since φ and ψ are by hypothesis one-to-one and onto, it follows that $F = \psi^{-1} \circ \varphi : \mathcal{O} \to \Omega$ is a well-defined map, which is one-to-one and onto. See Figure B.1. In fact, we can say more.

Proposition B.6. *Under the hypotheses above, F is a C^k diffeomorphism.*

Proof. It suffices to show that F and F^{-1} are C^k on a neighborhood of x_0 and y_0, respectively, where $\varphi(x_0) = \psi(y_0) = p$. Let us define a map Ψ in a fashion similar to (B.23). To be precise, we set $\widetilde{T}_pM = \operatorname{Range} D\psi(y_0)$ and let $\widetilde{N}_pM$ be its orthogonal complement. (Shortly we will show that $\widetilde{T}_pM = T_pM$, but we are not quite ready for that.) Then pick a linear isomorphism $B : \mathbb{R}^{n-m} \to \widetilde{N}_pM$ and set $\Psi(y, z) = \psi(y) + Bz$, for $(y, z) \in \Omega \times \mathbb{R}^{n-m}$. Again, Ψ is a C^k diffeomorphism from a neighborhood of $(y_0, 0)$ onto a neighborhood of p.

It follows that $\Psi^{-1} \circ \Phi$ is a C^k diffeomeophism from a neighborhood of $(x_0, 0)$ onto a neighborhood of $(y_0, 0)$. Now note that, for x close to x_0 and y close to y_0,

$$\Psi^{-1} \circ \Phi(x, 0) = \big(F(x), 0\big), \quad \Phi^{-1} \circ \Psi(y, 0) = \big(F^{-1}(y), 0\big). \tag{B.25}$$

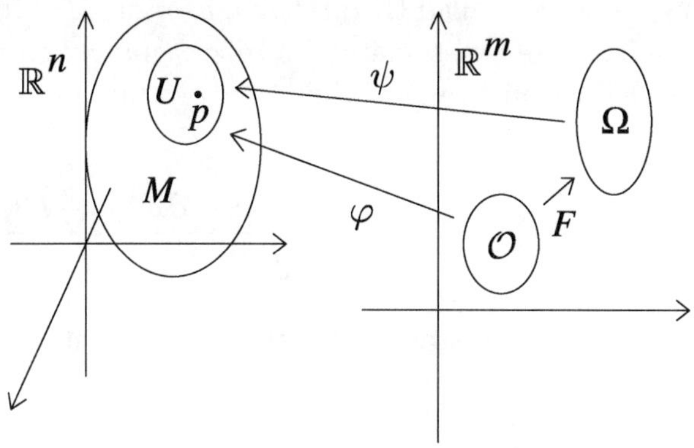

Figure B.1

These identities imply that F and F^{-1} have the desired regularity.

Thus, when there are two such coordinate charts, $\varphi : \mathcal{O} \to U$, $\psi : \Omega \to U$, we have a C^k diffeomorphism $F : \mathcal{O} \to \Omega$ such that

(B.26) $$\varphi = \psi \circ F.$$

By the chain rule,

(B.27) $$D\varphi(x) = D\psi(y)\, DF(x), \quad y = F(x).$$

In particular this implies that $\operatorname{Range} D\varphi(x_0) = \operatorname{Range} D\psi(y_0)$, so T_pM in (B.22) is independent of the choice of coordinate chart. It is called the *tangent space* to M at p.

The following result is a useful complement to Proposition B.6. It identifies another way to produce C^k surfaces in $\mathbb{R}^n$.

Proposition B.7. *Assume $n \geq m \geq 1$ and let $F : \mathbb{R}^n \to \mathbb{R}^m$ be a C^k map, $k \geq 1$. Fix $p \in \mathbb{R}^m$ and consider*

(B.28) $$S = \{x \in \mathbb{R}^n : F(x) = p\}.$$

Assume that for each $x \in S$, $DF(x) : \mathbb{R}^n \to \mathbb{R}^m$ is surjective. Then S is a C^k submanifold of $\mathbb{R}^n$. Furthermore, for each $x \in S$,

(B.29) $$T_xS = \operatorname{Ker} DF(x).$$

This result follows from the Inverse Function Theorem in a fashion similar to Proposition B.6. We leave the details to the reader.

We next define an object called the *metric tensor* on M. Given a coordinate chart $\varphi : \mathcal{O} \to U$, there is associated an $m \times m$ matrix $G(x) = \big(g_{jk}(x)\big)$ of functions on $\mathcal{O}$, defined in terms of the inner product of vectors tangent to M :

$$g_{jk}(x) = D\varphi(x)e_j \cdot D\varphi(x)e_k = \frac{\partial \varphi}{\partial x_j} \cdot \frac{\partial \varphi}{\partial x_k} = \sum_{\ell=1}^{n} \frac{\partial \varphi_\ell}{\partial x_j} \frac{\partial \varphi_\ell}{\partial x_k}, \tag{B.30}$$

where $\{e_j : 1 \le j \le m\}$ is the standard orthonormal basis of $\mathbb{R}^m$. Equivalently,

$$G(x) = D\varphi(x)^t \, D\varphi(x). \tag{B.31}$$

We call (g_{jk}) the metric tensor of M, on U, with respect to the coordinate chart $\varphi : \mathcal{O} \to U$. Note that this matrix is positive-definite. From a coordinate-independent point of view, the metric tensor on M specifies inner products of vectors tangent to M, using the inner product of $\mathbb{R}^n$.

If we take another coordinate chart $\psi : \Omega \to U$, we want to compare (g_{jk}) with $H = (h_{jk})$, given by

$$h_{jk}(y) = D\psi(y)e_j \cdot D\psi(y)e_k, \quad \text{i.e.,} \quad H(y) = D\psi(y)^t \, D\psi(y). \tag{B.32}$$

As seen above, we have a diffeomorphism $F : \mathcal{O} \to \Omega$ such that (B.26)–(B.27) hold. Consequently,

$$G(x) = DF(x)^t \, H(y) \, DF(x) \tag{B.33}$$

or, equivalently,

$$g_{jk}(x) = \sum_{i,\ell} \frac{\partial F_i}{\partial x_j} \frac{\partial F_\ell}{\partial x_k} \, h_{i\ell}(y). \tag{B.34}$$

Having discussed surfaces in $\mathbb{R}^n$, we turn to the more general concept of a *manifold*, which crops up starting in Chapter 7. A manifold is a Hausdorff topological space with an "atlas," i.e., a covering by open sets U_j together with homeomorphisms $\varphi_j : U_j \to V_j$, V_j open in $\mathbb{R}^n$. The number n is called the dimension of M. We say that M is a smooth manifold of class C^k provided the atlas has the following property. If $U_{jk} = U_j \cap U_k \neq \emptyset$, then the map

$$\psi_{jk} : \varphi_j(U_{jk}) \to \varphi_k(U_{jk})$$

given by $\varphi_k \circ \varphi_j^{-1}$ is a smooth diffeomorphism of class C^k from the open set $\varphi_j(U_{jk})$ to the open set $\varphi_k(U_{jk})$ in $\mathbb{R}^n$. By this, we mean that ψ_{jk} is C^k, with a C^k inverse. The pairs (U_j, φ_j) are called local coordinate charts.

A continuous map from M to another smooth manifold N is said to be smooth of class C^k if it is smooth of class C^k in local coordinates. Two different atlases on M, giving a priori two structures of M as a smooth manifold, are said to be equivalent if the identity map on M is smooth (of class C^k) from each one of these two manifolds to the other. Really a smooth manifold is considered to be defined by equivalence classes of such atlases, under this equivalence relation.

It follows from Proposition B.6 that a C^k surface in $\mathbb{R}^n$ is a smooth manifold of class C^k. For our purposes, that is a rich enough class of examples. One can find more material on manifolds in [**Spi**]. We mention that the notion of a metric tensor generalizes readily from surfaces in $\mathbb{R}^n$ to smooth manifolds; compare (7.18)–(7.20). As shown in (7.21)–(7.23), this leads to the notion of an integral on a manifold with a metric tensor (i.e., a Riemannian manifold).

We next discuss partitions of unity. Let M be a manifold, and supose M is paracompact. That is to say, every open cover $\{U_\alpha : \alpha \in A\}$ of M has a locally finite refinement, a locally finite open cover $\{V_\beta : \beta \in B\}$ such that each V_β is contained in some U_α. (The most prominent case is that M is compact.) Under this hypothesis, using a locally finite covering of M by coordinate neighborhoods, we can construct $\psi_j \in C_0^k(M)$ such that for any compact $K \subset M$, only finitely many ψ_j are nonzero on K (we say the sequence ψ_j is locally finite) and such that for any $p \in M$, some $\psi_j(p) \neq 0$. Then

$$\varphi_j(x) = \psi_j(x)^2 \Big/ \sum_k \psi_k(x)^2$$

is a locally finite sequence of functions in $C_0^k(M)$, satisfying $\sum_j \varphi_j(x) = 1$. Such a sequence is called a partition of unity. The construction is a variant of the construction of continuous partitions of unity, outlined in Exercise 9 of Chapter 13. It has many uses. For example, it is used in the proof of the Whitney Extension Theorem in Appendix C, in the proof of the general Stokes formula, in Proposition G.2, and in the proof of the Gauss-Green formula on Lipschitz domains, in Appendix I.

Appendix C

The Whitney Extension Theorem

The following result, known as the Whitney Extension Theorem, is used in the proof of Theorem 11.9.

Theorem C.1. *Let $K \subset \mathbb{R}^n$ be a compact set in $\mathbb{R}^n$. Let $f : K \to \mathbb{R}$ and $f^\# : K \to \mathbb{R}^n$ be continuous. For $x \in K$, $\delta > 0$, set*

$$R(x,y) = f(x) - f(y) - f^\#(y) \cdot (x-y) \tag{C.1}$$

and

$$\eta(\delta) = \sup \{|R(x,y)|/|x-y| : x, y \in K,\ 0 < |x-y| < \delta\}. \tag{C.2}$$

If $\eta(\delta) \to 0$ as $\delta \to 0$, then there exists $g \in C^1(\mathbb{R}^n)$ such that

$$g(x) = f(x), \quad \nabla g(x) = f^\#(x), \quad \forall\, x \in K. \tag{C.3}$$

Furthermore, we can arrange

$$\sup\ |g| \le C(n) \sup_K\ |f|, \quad \sup\ |\nabla g| \le C(n) \sup_K\ |f^\#|. \tag{C.4}$$

The function g will be produced using a carefully constructed partition of unity, described as follows. Suppose K is contained in the interior of a cube $\mathcal{Q}$, of edge ℓ.

Lemma C.2. *There is a partition of unity* $\{\Phi_j : j \ge 0\}$ *on* $\mathbb{R}^n \setminus K$, *such that*

$$\Phi_0 = 1 \text{ for } x \notin \mathcal{Q}, \quad \Phi_0 = 0 \text{ near } K,$$

and $\Phi_j \in C_0^\infty(\mathbb{R}^n \setminus K)$ *for* $j \ge 1$, *with the following properties, for some* $M, C < \infty$:
(1) Each $x \in \mathbb{R}^n$ *is in the support of at most* M *of the* Φ_j.
(2) For any $\delta \in (0, R]$, *if* $x \in$ *supp* Φ_j, $j \ge 1$, *and* $d(x, K) = \delta$, *then*

$$\text{diam supp } \Phi_j \le \frac{1}{2}\delta \tag{C.5}$$

and

$$|\nabla \Phi_j(x)| \le C\delta^{-1}. \tag{C.6}$$

Proof. We will write $\mathcal{Q}\setminus K = U$ as a countable union of closed cubes, whose interiors are disjoint. We start the process with a collection $\mathcal{C}$ consisting of one cube, $\mathcal{Q}$. We successively alter our collection of cubes as follows.

If Q is a cube in our collection, let $\widehat{Q}$ denote the concentric cube of three times the linear size (hence 3^n times the volume). If $\widehat{Q}$ has nonempty intersection with K, chop Q into 2^n smaller cubes, each of half the linear size. Throw away any of these cubes that are contained in K. Continue this process on each cube with nonempty intersection with U, ad infinitum. We obtain

$$\mathcal{Q} \setminus K = \bigcup_{j \ge 1} Q_j, \quad \overset{\circ}{Q}_j \cap \overset{\circ}{Q}_k = \emptyset, \text{ for } j \ne k, \quad \widehat{Q}_j \subset \mathbb{R}^n \setminus K, \ \forall\, j \ge 1.$$

If $\widetilde{Q}_j$ denotes the cube concentric with Q, with 1.1 times its linear size, then $\widetilde{Q}_j$ cannot intersect more than 6^n other $\widetilde{Q}_k$, $k \ne j$.

Having such a collection of cubes, we proceed as follows. Let Q be a unit cube, and pick $\psi \in C^\infty(\widetilde{Q})$, supported in the interior, such that $\psi = 1$ on Q. Via translation and scaling of coordinates, we obtain $\psi_j \in C^\infty(\widetilde{Q}_j)$, supported in the interior, such that $\psi_j = 1$ on Q_j, for $j \ge 1$. Clearly $|\nabla \psi_j| \le C/(\text{diam } Q_j)$. Take $\psi_0 \in C^\infty(\mathbb{R}^n)$ such that $\psi_0 = 1$ on $\mathbb{R}^n \setminus \mathcal{Q}$ and $\psi_0 = 0$ on a neighborhood of K. Now set

$$\Psi(x) = \sum_{j \ge 0} \psi_j(x), \quad \Phi_j(x) = \psi_j(x)/\Psi(x).$$

The properties stated for Φ_j follow from this construction.

We are now prepared to construct g. For each j, let y_j be a point in K of minimal distance from Q_j. Then we set

$$g(x) = \sum_j \Big[f(y_j) + f^\#(y_j)(x - y_j)\Big]\Phi_j(x), \quad x \in \mathbb{R}^n \setminus K. \tag{C.7}$$

Clearly $g \in C^\infty(\mathbb{R}^n \setminus K)$. We claim that, as $x \to z \in K$,

$$g(x) \to f(x) \text{ and } \nabla g(x) \to f^\#(x). \tag{C.8}$$

To verify the first part, suppose $|x - z| = \delta$. Then

$$x \in \text{ supp } \Phi_j \Rightarrow |x - y_j| \le C\delta \Rightarrow |f(y_j) - f(z)| \le c\delta. \tag{C.9}$$

Thus, since there are at most M nonzero terms in the sum (C.7), when evaluated at x, we have

$$g(x) = f(z) + \sum \big[f(y_j) - f(z) + f^\#(y_j)(x - y_j)\big]\Phi_j(x) = f(z) + O(\delta). \tag{C.10}$$

To demonstrate the second part of (C.8), write

$$\nabla g(x) = \sum f^\#(y_j)\Phi_j(x) + \sum \big[f(y_j) + f^\#(y_j)(x - y_j)\big]\nabla\Phi_j(x). \tag{C.11}$$

We can write the first sum on the right side of (C.11) as

$$f^\#(z) + \sum \big[f^\#(y_j) - f^\#(z)\big]\Phi_j(x) \to f^\#(z) \tag{C.12}$$

as $\delta \to 0$, by the continuity of $f^\#$ on K. We next examine the last sum in (C.11). Since $\sum \nabla\Phi_j = \nabla 1 = 0$, we can write this sum as

$$\sum \big[f(y_j) - f(z) - f^\#(y_j)(y_j - z)\big]\nabla\Phi_j(x) + \sum f^\#(y_j)(x - z)\nabla\Phi_j(x). \tag{C.13}$$

By (C.6) plus the estimate assumed on (C.1), the first sum is seen to tend to 0 as $\delta \to 0$. Again using $\sum \nabla\Phi_j = 0$, we write the second sum in (C.13) as

$$(x - z)\sum \big[f^\#(y_j) - f^\#(z)\big]\nabla\Phi_j(x), \tag{C.14}$$

and again (C.6) plus continuity of $f^\#$ on K implies that this tends to 0 as $\delta \to 0$.

We have now defined g on $U = \mathbb{R}^n \setminus K$. Setting $g = f$ on K, we see that g is continuous. It is clear that g is C^∞ on U, and we have shown that $\nabla g(x) \to f^\#(z)$ as $x \to z \in K$, $x \in U$. In fact, if we set $g^\#(x) = \nabla g(x)$ for $x \in U$, $f^\#(x)$ for $x \in K$, we have that $g^\#$ is continuous on $\mathbb{R}^n$.

To finish the proof of Theorem C.1, it remains to show that

$$g(x) = f(z) + f^{\#}(z)(x-z) + o(|x-z|), \quad x \in \mathbb{R}^n, \ z \in \partial K. \tag{C.15}$$

If $x \in K$, this is part of the hypothesis. On the other hand, suppose $x \in \mathbb{R}^n \setminus K$, $|x-z| = \delta << 1$. Comparing (C.7), we need

$$f(z) + f^{\#}(z)(x-z) = \sum \big[f(y_j) + f^{\#}(y_j)(x-y_j)\big]\Phi_j(x) + o(\delta), \tag{C.16}$$

where $|y_j - z| \le C\delta$. Note that

$$\begin{aligned} f(z) &= f(y_j) + f^{\#}(y_j)(z-y_j) + o(\delta), \\ f^{\#}(z)(x-z) &= f^{\#}(y_j)(x-z) + o(\delta). \end{aligned} \tag{C.17}$$

Multiplying by $\Phi_j(x)$ and summing over j, we get (C.16), and the proof of Theorem C.1 is complete.

Having the partition of unity Φ_j, we apply it to a result a bit simpler than Theorem C.1, which is nevertheless useful in Chapter 12.

Proposition C.3. *If $w : K \to \mathbb{R}$ is Lipschitz, then there exists a Lipschitz map $v : \mathbb{R}^n \to \mathbb{R}$ such that $v = w$ on K.*

Proof. In place of (C.7), take

$$v(x) = \sum_j w(y_j)\Phi_j(x), \quad x \in \mathbb{R}^n \setminus K. \tag{C.18}$$

Suppose $x \in \mathbb{R}^n \setminus K, \ z \in K, \ |x-z| = \delta$. Then, as in (C.9),

$$x \in \ \text{supp}\ \Phi_j \Rightarrow |x - y_j| \le C\delta \Rightarrow |w(y_j) - w(z)| \le c\delta, \tag{C.19}$$

so

$$v(x) = w(z) + \sum_j \big[w(y_j) - w(z)\big]\Phi_j(x) = w(z) + O(\delta). \tag{C.20}$$

We next estimate $|\nabla v(x)|$, for $x \in \mathbb{R}^n \setminus K$. By an argument similar to that used to write (C.13), we have

$$\nabla v(x) = \sum \big[w(y_j) - w(z)\big]\nabla\Phi_j(x), \tag{C.21}$$

for any $z \in K$. Say z is one of the y_j with $x \in$ supp Φ_j. Note that, for any $x \in U = \mathbb{R}^n \setminus K$, all the $\widetilde{Q}_j$ from Lemma C.2 containing x (at most M in number) have comparable diameters, in turn comparable to their distance to K. In view of the estimate (C.6), we have a bound

$$|\nabla v(x)| \leq \sum |w(y_j) - w(z)| \cdot |\nabla \Phi_j(x)| \leq C. \tag{C.22}$$

If we define $v(x)$ by (C.18) for $x \notin K$ and set $v = w$ on K, then (C.20) implies that v is continuous on $\mathbb{R}^n$ and Lipschitz on K. It remains to estimate $|v(x_1) - v(x_2)|$ when $x_1, x_2 \in U = \mathbb{R}^n \setminus K$. There are two possibilities; either there exists $z \in K$ on the line segment from x_1 to x_2 or there does not exist such a z. In the first case, (C.20) yields

$$|v(x_1) - v(x_2)| \leq |v(x_1) - v(z)| + |v(z) - v(x_2)| \leq C|x_1 - x_2|.$$

In the second case, we have

$$|v(x_1) - v(x_2)| \leq \Bigl(\sup_U |\nabla v|\Bigr)|x_1 - x_2|.$$

Thus Proposition C.3 is proved.

REMARK. H. Whitney established a more general result than Theorem C.1, characterizing when $f : K \to \mathbb{R}$ has a C^m extension to $\mathbb{R}^n$. A similar extension of Proposition C.3, telling when f has a $C^{m-1,1}$ extension to $\mathbb{R}^n$ (i.e., $m - 1$ order derivatives are Lipschitz), is given in [**St**], Chapter 6, and a further extension, not requiring K to be compact, or even closed, is given in [**Fef**].

Appendix D

The Marcinkiewicz Interpolation Theorem

Let $(X, \mathfrak{F}, \mu)$ be a measure space. We say a transformation T acting on functions on X is sublinear provided

$$|Tu + Tv| \le |Tu| + |Tv|, \quad |T(au)| = |a| \cdot |Tu|, \tag{D.1}$$

where a is a constant. The Marcinkiewicz Interpolation Theorem provides a sufficient condition that T is bounded on $L^p(X, \mu)$, i.e.,

$$\|Tu\|_{L^p} \le C\|u\|_{L^p}, \quad \forall u \in L^p(X, \mu), \tag{D.2}$$

given weaker bounds on L^r and L^q, with $r < p < q$. In detail, we say T is weak type (p, p) provided that, for any $\lambda > 0$,

$$\mu(\{x \in X : |Tu(x)| > \lambda\}) \le C\lambda^{-p}\|u\|_{L^p}^p. \tag{D.3}$$

In view of the simple estimate (Tchebychev's inequality)

$$\mu(\{x \in X : |u(x)| > \lambda\}) \le \lambda^{-1}\|u\|_{L^1},$$

it readily follows that (D.2)$\Rightarrow$(D.3). Here is part of the Marcinkiewicz Interpolation Theorem.

Proposition D.1. *If $1 \le r < p < q < \infty$ and if T is a sublinear operator that is both of weak type (r, r) and of weak type (q, q), then T is bounded on L^p, i.e., the estimate* (D.2) *holds.*

Proof. Given $\lambda \in (0, \infty)$, write $u = u_1 + u_2$, with $u_1(x) = u(x)$ for $|u(x)| > \lambda$ and $u_2(x) = u(x)$ for $|u(x)| \le \lambda$. With the notation

$$\nu_f(\lambda) = \mu(\{x \in X : |f(x)| \ge \lambda\}), \tag{D.4}$$

we have

$$\begin{aligned} \nu_{Tu}(2\lambda) &\le \nu_{Tu_1}(\lambda) + \nu_{Tu_2}(\lambda) \\ &\le C_1 \lambda^{-r} \|u_1\|_{L^r}^r + C_2 \lambda^{-q} \|u_2\|_{L^q}^q. \end{aligned} \tag{D.5}$$

Also,

$$\int_X |f|^p \, d\mu = p \int_0^\infty \nu_f(\lambda) \lambda^{p-1} \, d\lambda. \tag{D.6}$$

Hence

$$\begin{aligned} \int_X |Tu|^p \, d\mu &= p \int_0^\infty \nu_{Tu}(\lambda) \lambda^{p-1} \, d\lambda \\ &\le C_1 p \int_0^\infty \lambda^{p-1-r} \Big(\int_{|u|>\lambda} |u|^r \, d\mu \Big) d\lambda \\ &\quad + C_2 p \int_0^\infty \lambda^{p-1-q} \Big(\int_{|u|\le\lambda} |u|^q \, d\mu \Big) d\lambda. \end{aligned} \tag{D.7}$$

Now

$$\int_0^\infty \lambda^{p-1-r} \Big(\int_{|u|>\lambda} |u|^r \, d\mu \Big) d\lambda = \frac{1}{p-r} \int |u|^p \, d\mu, \tag{D.8}$$

and similarly

$$\int_0^\infty \lambda^{p-1-q} \Big(\int_{|u|\le\lambda} |u|^q \, d\mu \Big) d\lambda = \frac{1}{q-p} \int |u|^p \, d\mu. \tag{D.9}$$

Combining these gives the desired estimate on $\|Tu\|_{L^p}^p$.

The following complement to Proposition D.1 is applied in the proofs of Proposition 14.7 and Proposition 17.7.

Proposition D.2. *If $1 \le r < p < \infty$ and T is a sublinear operator that is of weak type (r, r) and satisfies*

$$\|Tu\|_{L^\infty} \le A \|u\|_{L^\infty}, \tag{D.10}$$

then T is bounded on L^p.

Proof. It suffices to show that T is of weak type (q,q) for each $q \in (r,\infty)$, since then we can take $q \in (p,\infty)$ and apply Proposition D.1. This time, given $\lambda \in (0,\infty)$, write $u = u_1 + u_2$ with $u_1(x) = u(x)$ for $|u(x)| > \lambda/A$ and $u_2(x) = u(x)$ for $|u(x)| \le \lambda/A$, where $A \in (0,\infty)$ is as in (D.10). It follows that $\mu(\{x \in X : |Tu_2(x)| > \lambda\}) = 0$. It remains to estimate $Tu_1(x)$. We have

$$\begin{aligned} \mu(\{x \in X : |Tu_1(x)| > \lambda\}) &\le C\lambda^{-r}\|u_1\|_{L^r}^r \\ &= C\lambda^{-r} \int\limits_{|u|>\lambda/A} |u|^r \, d\mu. \end{aligned} \tag{D.11}$$

Now taking this last integral as the inner product of $|u|^r$ and 1 and using Hölder's inequality, we have

$$\begin{aligned} \int\limits_{|u|>\lambda/A} |u|^r \, d\mu &\le \Bigl(\int |u|^q \, d\mu\Bigr)^{r/q} \mu\Bigl(\Bigl\{x \in X : |u(x)| > \frac{\lambda}{A}\Bigr\}\Bigr)^{(q-r)/q} \\ &\le \|u\|_{L^q}^r \bigl(\lambda^{-q}\|u\|_{L^q}^q\bigr)^{(q-r)/q} \\ &= \lambda^{-(q-r)}\|u\|_{L^q}^q, \end{aligned} \tag{D.12}$$

the second estimate by Tchebychev's inequality. Plugging the estimate (D.12) into (D.11) yields the desired estimate (D.3) (with p replaced by q) and completes the proof.

There are more general versions of Marcinkiewicz interpolation, involving operators of weak type (p,q). We refer to [**Fol**] for a treatment.

Sard's Theorem

Let $F : \Omega \to \mathbb{R}^n$ be a C^1 map, with Ω open in $\mathbb{R}^n$. If $p \in \Omega$ and $DF(p) : \mathbb{R}^n \to \mathbb{R}^n$ is not surjective, then p is said to be a *critical point* and $F(p)$ a *critical value*. The set C of critical points can be a large subset of Ω, even all of it, but the set of critical values $F(C)$ must be small in $\mathbb{R}^n$. This is part of Sard's Theorem.

Theorem E.1. *If $F : \Omega \to \mathbb{R}^n$ is a C^1 map, then the set of critical values of F has measure 0 in $\mathbb{R}^n$.*

Proof. If $K \subset \Omega$ is compact, cover $K \cap C$ with m-dimensional cubes Q_j, with disjoint interiors, of side δ_j. Pick $p_j \in C \cap Q_j$, so $L_j = DF(p_j)$ has rank $\leq n-1$. Then, for $x \in Q_j$,

$$F(p_j + x) = F(p_j) + L_j x + R_j(x), \quad \|R_j(x)\| \leq \rho_j = \eta_j \delta_j,$$

where $\eta_j \to 0$ as $\delta_j \to 0$. Now $L_j(Q_j)$ is certainly contained in an $(n-1)$-dimensional cube of side $C_0\delta_j$, where C_0 is an upper bound for $\sqrt{m}\|DF\|$ on K. Since all points of $F(Q_j)$ are a distance $\leq \rho_j$ from (a translate of) $L_j(Q_j)$, this implies

$$\text{meas } F(Q_j) \leq 2\rho_j(C_0\delta_j + 2\rho_j)^{n-1} \leq C_1 \eta_j \delta_j^n,$$

provided δ_j is sufficiently small that $\rho_j \leq \delta_j$. Now $\sum_j \delta_j^n$ is the volume of the cover of $K \cap C$. For fixed K this can be assumed to be bounded. Hence

$$\text{meas } F(C \cap K) \leq C_K \, \eta,$$

where $\eta = \max \{\eta_j\}$. Picking a cover by small cubes, we make η arbitrarily small, so meas $F(C \cap K) = 0$. Letting $K_j \nearrow \Omega$, we complete the proof.

Sard's Theorem also treats the more difficult case when Ω is open in $\mathbb{R}^m, m > n$. Then a more elaborate argument is needed, and one requires more differentiability, namely that F is class C^k, with $k = m - n + 1$. A proof can be found in [**Stb**].

The main application of Sard's Theorem in this text is to the proof of the change of variable theorem we present in the next appendix. Here, we give another application of Sard's Theorem, to the existence of lots of Morse functions. This application gives the typical flavor of how one uses Sard's Theorem. We begin with a special case:

Proposition E.2. *Let $\Omega \subset \mathbb{R}^n$ be open, $f \in C^\infty(\Omega)$. For $a \in \mathbb{R}^n$, set $f_a(x) = f(x) - a \cdot x$. Then, for almost every $a \in \mathbb{R}^n$, f_a is a Morse function, i.e., it has only nondegenerate critical points.*

Proof. Consider $F(x) = \nabla f(x); F : \Omega \to \mathbb{R}^n$. A point $x \in \Omega$ is a critical point of f_a if and only if $F(x) = a$, and this critical point is degenerate only if, in addition, a is a critical value of F. Hence the desired conclusion holds for all $a \in \mathbb{R}^n$ that are not critical values of F.

Now for the result on manifolds:

Proposition E.3. *Let M be an n-dimensional manifold, embedded in $\mathbb{R}^K$. Let $f \in C^\infty(M)$, and, for $a \in \mathbb{R}^K$, let $f_a(x) = f(x) - a \cdot x$, for $x \in M \subset \mathbb{R}^K$. Then, for almost all $a \in \mathbb{R}^K$, f_a is a Morse function.*

Proof. Each $p \in M$ has a neighborhood Ω_p such that some n of the coordinates x_ν on $\mathbb{R}^K$ produce coordinates on Ω_p. Let us say $x_1, \dots, x_n$ do it. Let $(a_{n+1}, \dots, a_K)$ be fixed but arbitrary. Then, by Proposition E.2, for almost every $(a_1, \dots, a_n) \in \mathbb{R}^n$, f_a has only nondegenerate critical points on Ω_p. By Fubini's Theorem, we deduce that, for almost every $a \in \mathbb{R}^K$, f_a has only nondegenerate critical points on Ω_p. (The set of bad $a \in \mathbb{R}^K$ is readily seen to be a countable union of closed sets, hence measurable.) Covering M by a countable family of such sets Ω_p, we finish the proof.

Appendix F

A Change of Variable Theorem for Many-to-one Maps

Here we present a change of variable theorem for a C^1 map defined on an open set in $\mathbb{R}^n$ that is not assumed to be a diffeomorphism onto its range. Theorem F.1 below is hence a generalization of Theorem 7.2. There are further generalizations, involving Lipschitz maps and maps between spaces of different dimensions, that can be found in [**EG**] and [**Fed**].

Theorem F.1. *Let $\Omega \subset \mathbb{R}^n$ be open and let $F : \Omega \to \mathbb{R}^n$ be a C^1 map. For $x \in \mathbb{R}^n$ set $n(x) = \operatorname{card} F^{-1}(x)$. Then n is measurable and for any measurable $u \geq 0$ on $\mathbb{R}^n$,*

$$\int_{\Omega} u(F(x))\, |\det DF(x)|\, dx = \int_{\mathbb{R}^n} u(x)\, n(x)\, dx. \tag{F.1}$$

This is an extension of the standard change of variable formula, in which one assumes that F is a diffeomorphism of Ω onto its image. (Then $n(x)$ is the characteristic function of $F(\Omega)$.) We will make use of this standard result in the proof of the theorem.

We make a sequence of reductions. Let

$$K = \{x \in \Omega : \det DF(x) = 0\}, \quad \widetilde{\Omega} = \Omega \setminus K,$$
$$\widetilde{F} = F\big|_{\widetilde{\Omega}}, \quad \widetilde{n}(x) = \operatorname{card} \widetilde{F}^{-1}(x).$$

Sard's Theorem implies $F(K)$ has measure zero. Hence $n(x) = \widetilde{n}(x)$ a.e. on $\mathbb{R}^n$, so if we can show that $\widetilde{n}$ is measurable and (F.1) holds with Ω, F, n replaced by $\widetilde{\Omega}, \widetilde{F}, \widetilde{n}$, we will have the desired result. Thus we will henceforth assume that $\det DF(x) \neq 0$ for all $x \in \Omega$.

Next, for $k \in \mathbb{Z}^+$, let

$$D_k = \{x \in \Omega : |x| \le k,\ \mathrm{dist}(x, \partial\Omega) \ge 1/k\}.$$

Each D_k is a compact subset of Ω. Furthermore, $D_k \subset D_{k+1}$ and $\Omega = \bigcup D_k$. Let $F_k = F\big|_{D_k}$ and $n_k(x) = \mathrm{card}\, F_k^{-1}(x)$. Then, for each $x \in \mathbb{R}^n$, $n_k(x) \nearrow n(x)$ as $k \to \infty$. Suppose we can show that, for each k, n_k is measurable and

$$\int_{D_k} u(F(x))\,|\det DF(x)|\,dx = \int_{\mathbb{R}^n} u(x)\,n_k(x)\,dx. \tag{F.2}$$

Now the Monotone Convergence Theorem implies that as $k \to \infty$ the left side of (F.2) tends to the left side of (F.1) and the right side of (F.2) tends to the right side of (F.1). Hence (F.1) follows from (F.2), so it remains to prove (F.2) (and the measurability of n_k).

For notational simplicity, drop the index k. We have a compact set $D \subset \mathbb{R}^n$, F is C^1 on a neighborhood Ω of D, and $\det DF(x)$ is nowhere vanishing. We set $n_D(x) = \mathrm{card}\, D \cap F^{-1}(x)$ and desire to prove that n_D is measurable and

$$\int_{D} u(F(x))\,|\det DF(x)|\,dx = \int_{\mathbb{R}^n} u(x)\,n_D(x)\,dx. \tag{F.3}$$

By the Inverse Function Theorem, each $x \in \Omega$ has a neighborhood $\mathcal{O}_x$ such that F is a diffeomorphism of $\mathcal{O}_x$ onto its image. Since D is compact, we can cover D with a finite number of open sets $\mathcal{O}_j$ on each of which F is a diffeomorphism. Then there exists $\delta > 0$ such that any subset of D of diameter $\le \delta$ is contained in one of these sets $\mathcal{O}_j$. Tile $\mathbb{R}^n$ with closed cubes of edge $\delta/\sqrt{n}$ (so they intersect only along their faces). Let $\{Q_k : 1 \le k \le N\}$ denote those cubes that intersect D, and let $E_k = D \cap Q_k$. Then E_k is compact and the standard change of variables theorem gives

$$\int_{E_k} u(F(x))\,|\det DF(x)|\,dx = \int_{F(E_k)} u(x)\,dx. \tag{F.4}$$

If we sum the left side of (F.4) over k, we get the left side of (F.3). Since the faces of each cube, and also their images under F, all have measure zero, we also have

$$n_D(x) = \text{card}\,\{k : x \in F(E_k)\} = \sum_{k=1}^{N} \chi_{F(E_k)}(x) \quad \text{a.e. on } \mathbb{R}^n.$$

Since $F(E_k)$ is compact, this proves that n_D is measurable and also implies that if we sum the right side of (F.4) over k, we get the right side of (F.3). The theorem is hence proven.

Appendix G

Integration of Differential Forms

The calculus of differential forms provides a convenient setting for integration on manifolds, as we will explain in this appendix, due to the efficient way it keeps track of changes of variables.

A k-form β on an open set $\mathcal{O} \subset \mathbb{R}^n$ has the form

$$\beta = \sum_j b_j(x)\, dx_{j_1} \wedge \cdots \wedge dx_{j_k}. \tag{G.1}$$

Here $j = (j_1, \dots, j_k)$ is a k-multi-index. We write $\beta \in \Lambda^k(\mathcal{O})$. The wedge product used in (G.1) has the anti-commutative property

$$dx_\ell \wedge dx_m = -dx_m \wedge dx_\ell, \tag{G.2}$$

so that if σ is a permutation of $\{1, \dots, k\}$, we have

$$dx_{j_1} \wedge \cdots \wedge dx_{j_k} = (\operatorname{sgn} \sigma)\, dx_{j_{\sigma(1)}} \wedge \cdots \wedge dx_{j_{\sigma(k)}}. \tag{G.3}$$

In particular, an n-form α on $\Omega \subset \mathbb{R}^n$ can be written

$$\alpha = A(x)\, dx_1 \wedge \cdots \wedge dx_n. \tag{G.4}$$

If $A \in L^1(\mathcal{O}, dx)$, we write

$$\int_{\mathcal{O}} \alpha = \int_{\mathcal{O}} A(x)\, dx, \tag{G.5}$$

the right side being the usual Lebesgue integral, developed in Chapter 7.

Suppose now $\Omega \subset \mathbb{R}^n$ is open and there is a C^1 diffeomorphism $F : \Omega \to \mathcal{O}$. We define the *pull-back* $F^*\beta$ of the k-form β in (G.1) as

$$F^*\beta = \sum_j b_j(F(x))\,(F^*dx_{j_1}) \wedge \cdots \wedge (F^*dx_{j_k}), \tag{G.6}$$

where

$$F^*dx_j = \sum_\ell \frac{\partial F_j}{\partial x_\ell}\,dx_\ell, \tag{G.7}$$

the algebraic computation in (G.6) being performed using the rule (G.3).

If $B = (b_{\ell m})$ is an $n \times n$ matrix, then, by (G.3) and the formula for the determinant given in (7.77) (and (7.83)),

$$\begin{aligned}\Big(\sum_m b_{1m}\,dx_m\Big) \wedge \Big(\sum_m b_{2m}\,dx_m\Big) \wedge \cdots \wedge \Big(\sum_m b_{nm}\,dx_m\Big) \\ = \Big(\sum_\sigma (\operatorname{sgn}\sigma)\, b_{1\sigma(1)} b_{2\sigma(2)} \cdots b_{n\sigma(n)}\Big)\,dx_1 \wedge \cdots \wedge dx_n \\ = (\det B)\,dx_1 \wedge \cdots \wedge dx_n.\end{aligned} \tag{G.8}$$

Hence, if $F : \Omega \to \mathcal{O}$ is a C^1 map and α is an n-form on $\mathcal{O}$, as in (G.4), then

$$F^*\alpha = \det DF(x)\, A(F(x))\,dx_1 \wedge \cdots \wedge dx_n. \tag{G.9}$$

This formula is especially significant in light of the change of variable formula

$$\int_{\mathcal{O}} A(x)\,dx = \int_\Omega A(F(x))\,|\det DF(x)|\,dx, \tag{G.10}$$

when $F : \Omega \to \mathcal{O}$ is a C^1 diffeomorphism, given in Theorem 7.2. The only difference between the right side of (G.10) and $\int_\Omega F^*\alpha$ is the absolute value sign around $\det DF(x)$. We say a C^1 map $F : \Omega \to \mathcal{O}$ is *orientation preserving* when $\det DF(x) > 0$ for all $x \in \Omega$. In such a case, Theorem 7.2 yields

Proposition G.1. *If $F : \Omega \to \mathcal{O}$ is a C^1 orientation-preserving diffeomorphism and α an integrable n-form on $\mathcal{O}$, then*

$$\int_{\mathcal{O}} \alpha = \int_\Omega F^*\alpha. \tag{G.11}$$

In Appendix H we will present another proof of the change of variable formula, making direct use of basic results on differential forms developed in this appendix.

In addition to the pull-back, there are some other operations on differential forms. The wedge product of dx_ℓ's extends to a wedge product on forms as follows. If $\beta \in \Lambda^k(\mathcal{O})$ has the form (G.1) and if

$$\alpha = \sum_i a_i(x)\, dx_{i_1} \wedge \cdots \wedge dx_{i_\ell} \in \Lambda^\ell(\mathcal{O}), \tag{G.12}$$

define

$$\alpha \wedge \beta = \sum_{i,j} a_i(x) b_j(x)\, dx_{i_1} \wedge \cdots \wedge dx_{i_\ell} \wedge dx_{j_1} \wedge \cdots \wedge dx_{j_k} \tag{G.13}$$

in $\Lambda^{k+\ell}(\mathcal{O})$. We retain the equivalences (G.3). It follows that

$$\alpha \wedge \beta = (-1)^{k\ell} \beta \wedge \alpha. \tag{G.14}$$

It is also readily verified that

$$F^*(\alpha \wedge \beta) = (F^*\alpha) \wedge (F^*\beta). \tag{G.15}$$

Another important operator on forms is the *exterior derivative*:

$$d : \Lambda^k(\mathcal{O}) \longrightarrow \Lambda^{k+1}(\mathcal{O}), \tag{G.16}$$

defined as follows. If $\beta \in \Lambda^k(\mathcal{O})$ is given by (G.1), then

$$d\beta = \sum_{j,\ell} \frac{\partial b_j}{\partial x_\ell}\, dx_\ell \wedge dx_{j_1} \wedge \cdots \wedge dx_{j_k}. \tag{G.17}$$

The antisymmetry $dx_m \wedge dx_\ell = -dx_\ell \wedge dx_m$, together with the identity $\partial^2 b_j / \partial x_\ell \partial x_m = \partial^2 b_j / \partial x_m \partial x_\ell$, implies

$$d(d\beta) = 0, \tag{G.18}$$

for any smooth differential form β. We also have a product rule:

$$d(\alpha \wedge \beta) = (d\alpha) \wedge \beta + (-1)^j \alpha \wedge (d\beta), \quad \alpha \in \Lambda^j(\mathcal{O}),\ \beta \in \Lambda^k(\mathcal{O}). \tag{G.19}$$

The exterior derivative has the following important property under pull-backs:

$$F^*(d\beta) = dF^*\beta, \tag{G.20}$$

if $\beta \in \Lambda^k(\mathcal{O})$ and $F : \Omega \to \mathcal{O}$ is a smooth map. To see this, extending (G.19) to a formula for $d(\alpha \wedge \beta_1 \wedge \cdots \wedge \beta_\ell)$ and using this to apply d to $F^*\beta$, we have

(G.21)

$$\begin{aligned} dF^*\beta = &\sum_{j,\ell} \frac{\partial}{\partial x_\ell}(b_j \circ F(x))\, dx_\ell \wedge (F^* dx_{j_1}) \wedge \cdots \wedge (F^* dx_{j_k}) \\ &+ \sum_{j,\nu} (\pm) b_j(F(x)) (F^* dx_{j_1}) \wedge \cdots \wedge d(F^* dx_{j_\nu}) \wedge \cdots \wedge (F^* dx_{j_k}). \end{aligned}$$

Now the definition (G.6)–(G.7) of pull-back gives directly that

$$F^* dx_i = \sum_\ell \frac{\partial F_i}{\partial x_\ell}\, dx_\ell = dF_i, \tag{G.22}$$

and hence $d(F^* dx_i) = ddF_i = 0$, so only the first sum in (G.21) contributes to $dF^*\beta$. Meanwhile,

$$F^* d\beta = \sum_{j,m} \frac{\partial b_j}{\partial x_m}(F(x))\ (F^* dx_m) \wedge (F^* dx_{j_1}) \wedge \cdots \wedge (F^* dx_{j_k}), \tag{G.23}$$

so (G.20) follows from the identity

$$\sum_\ell \frac{\partial}{\partial x_\ell}(b_j \circ F(x))\, dx_\ell = \sum_m \frac{\partial b_j}{\partial x_m}(F(x))\, F^* dx_m,$$

which in turn follows from the chain rule.

Here is another important consequence of the chain rule. Suppose $F : \Omega \to \mathcal{O}$ and $\psi : \mathcal{O} \to U$ are smooth maps between open subsets of $\mathbb{R}^n$. We claim that for any form α of any degree,

$$\psi \circ F = \varphi \Longrightarrow \varphi^*\alpha = F^*\psi^*\alpha. \tag{G.24}$$

It suffices to check (G.24) for $\alpha = dx_j$. Then (G.7) gives the basic identity $\psi^*\, dx_j = \sum (\partial\psi_j/\partial x_\ell)\, dx_\ell$. Consequently,

$$F^*\psi^*\, dx_j = \sum_{\ell,m} \frac{\partial F_\ell}{\partial x_m}\frac{\partial \psi_j}{\partial x_\ell}\, dx_m, \quad \varphi^*\, dx_j = \sum_m \frac{\partial \varphi_j}{\partial x_m}\, dx_m; \tag{G.25}$$

but the identity of these forms follows from the chain rule:

$$D\varphi = (D\psi)(DF) \Longrightarrow \frac{\partial \varphi_j}{\partial x_m} = \sum_\ell \frac{\partial \psi_j}{\partial x_\ell}\frac{\partial F_\ell}{\partial x_m}. \tag{G.26}$$

One can define a k-form on an n-dimensional manifold M as follows. Say M is covered by open sets $\mathcal{O}_j$ and there are coordinate charts $F_j : \Omega_j \to \mathcal{O}_j$, with $\Omega_j \subset \mathbb{R}^n$ open. A collection of forms $\beta_j \in \Lambda^k(\Omega_j)$ is said to define a k-form on M provided the following compatibility condition holds. If $\mathcal{O}_i \cap \mathcal{O}_j \neq \emptyset$ and we consider $\Omega_{ij} = F_i^{-1}(\mathcal{O}_i \cap \mathcal{O}_j)$ and diffeomorphisms

$$\varphi_{ij} = F_j^{-1} \circ F_i : \Omega_{ij} \longrightarrow \Omega_{ji}, \tag{G.27}$$

we require

$$\varphi_{ij}^* \beta_j = \beta_i. \tag{G.28}$$

The fact that this is a consistent definition is a consequence of (G.24). For example, if $G : M \to R^m$ is a smooth map and γ is a k-form on $\mathbb{R}^m$, then there is a well-defined k-form $\beta = G^*\gamma$ on M, represented in such coordinate charts by $\beta_j = (G \circ F_j)^*\gamma$. Similarly, if β is a k-form on M as defined above and $G : U \to M$ is smooth, with $U \subset \mathbb{R}^m$ open, then $G^*\beta$ is a well-defined k-form on U.

We give an intrinsic definition of $\int_M \alpha$ when α is an n-form on M, provided M is *oriented*, i.e., there is a coordinate cover as above such that $\det D\varphi_{jk} > 0$. The object called an "orientation" on M can be identified as an equivalence class of nowhere vanishing n-forms on M, two such forms being equivalent if one is a multiple of another by a positive function in $C^\infty(\Omega)$. A member of this equivalence class, say ω, defines the orientation. The standard orientation on $\mathbb{R}^n$ is determined by $dx_1 \wedge \cdots \wedge dx_n$. The equivalence class of positive multiples $a(x)\omega$ is said to consist of "positive" forms. A smooth map $\psi : S \to M$ between oriented n-dimensional manifolds preserves orientation provided $\psi^*\sigma$ is positive on S whenever $\sigma \in \Lambda^n(M)$ is positive. We mention that there exist surfaces that cannot be oriented, such as the famous "Möbius strip."

We define the integral of an n-form over an oriented n-dimensional manifold as follows. First, if α is an n-form supported on an open set $\mathcal{O} \subset \mathbb{R}^n$, given by (G.4), then we define $\int_{\mathcal{O}} \alpha$ by (G.5).

More generally, if M is an n-dimensional manifold with an orientation, say the image of an open set $\mathcal{O} \subset \mathbb{R}^n$ by $\varphi : \mathcal{O} \to M$, carrying the natural orientation of $\mathcal{O}$, we can set

$$\int_M \alpha = \int_{\mathcal{O}} \varphi^*\alpha \tag{G.29}$$

for an n-form α on M. If it takes several coordinate patches to cover M, define $\int_M \alpha$ by writing α as a sum of forms, each supported on one patch.

We need to show that this definition of $\int_M \alpha$ is independent of the choice of coordinate system on M (as long as the orientation of M is respected). Thus, suppose $\varphi : \mathcal{O} \to U \subset M$ and $\psi : \Omega \to U \subset M$ are both coordinate patches, so that $F = \psi^{-1} \circ \varphi : \mathcal{O} \to \Omega$ is an orientation-preserving diffeomorphism. We need to check that, if α is an n-form on M, supported on U, then

$$\int\limits_{\mathcal{O}} \varphi^* \alpha = \int\limits_{\Omega} \psi^* \alpha. \tag{G.30}$$

To establish this, we use (G.24). This implies that the left side of (G.30) is equal to

$$\int\limits_{\mathcal{O}} F^*(\psi^* \alpha), \tag{G.31}$$

which is equal to the right side of (G.30), by (G.11) (with slightly altered notation). Thus the integral of an n-form over an oriented n-dimensional manifold is well defined.

We turn now to the Gauss-Green-Stokes formula for differential forms, commonly called simply the Stokes formula. This involves integrating a k-form over a k-dimensional manifold with boundary. We first define that concept. Let S be a smooth k-dimensional manifold, and let M be an open subset of S, such that its closure $\overline{M}$ (in $\mathbb{R}^N$) is contained in S. Its boundary is $\partial M = \overline{M} \setminus M$. We say $\overline{M}$ is a smooth surface with boundary if also ∂M is a smooth $(k-1)$-dimensional surface. In such a case, any $p \in \partial M$ has a neighborhood $U \subset S$ with a coordinate chart $\varphi : \mathcal{O} \to U$, where $\mathcal{O}$ is an open neighborhood of 0 in $\mathbb{R}^k$, such that $\varphi(0) = p$ and φ maps $\{x \in \mathcal{O} : x_1 = 0\}$ onto $U \cap \partial M$.

If S is oriented, then $\overline{M}$ is oriented, and ∂M inherits an orientation, uniquely determined by the following requirement: if

$$\overline{M} = \mathbb{R}^k_- = \{x \in \mathbb{R}^k : x_1 \le 0\}, \tag{G.32}$$

then $\partial M = \{(x_2, \dots, x_k)\}$ has the orientation determined by $dx_2 \wedge \cdots \wedge dx_k$.

We can now state the Stokes formula.

Proposition G.2. *Given a compactly supported $(k-1)$-form β of class C^1 on an oriented k-dimensional surface $\overline{M}$ (of class C^2) with boundary ∂M, with its natural orientation,*

$$\int\limits_{M} d\beta = \int\limits_{\partial M} \beta. \tag{G.33}$$

Proof. Using a partition of unity and invariance of the integral and the exterior derivative under coordinate transformations, it suffices to prove this when $\overline{M}$ has the form (G.32). In that case, we will be able to deduce (G.33) from the Fundamental Theorem of Calculus. Indeed, if

$$\beta = b_j(x)\,dx_1 \wedge \cdots \wedge \widehat{dx_j} \wedge \cdots \wedge dx_k, \tag{G.34}$$

with $b_j(x)$ of bounded support, we have

$$d\beta = (-1)^{j-1}\frac{\partial b_j}{\partial x_j}\,dx_1 \wedge \cdots \wedge dx_k. \tag{G.35}$$

If $j > 1$, we have

$$\int\limits_M d\beta = (-1)^{j-1}\int\Big\{\int_{-\infty}^{\infty}\frac{\partial b_j}{\partial x_j}\,dx_j\Big\}\,dx' = 0, \tag{G.36}$$

and also $\kappa^*\beta = 0$, where $\kappa : \partial M \to \overline{M}$ is the inclusion. On the other hand, for $j = 1$, we have

$$\begin{aligned}\int\limits_M d\beta &= \int\Big\{\int_{-\infty}^{0}\frac{\partial b_1}{\partial x_1}\,dx_1\Big\}\,dx_2\cdots dx_k\\ &= \int b_1(0,x')\,dx'\\ &= \int\limits_{\partial M}\beta.\end{aligned} \tag{G.37}$$

This proves Stokes' formula (G.33).

The reason we required $\overline{M}$ to be a surface of class C^2 in Proposition G.2 is the following. Due to the formulas (G.6)–(G.7) for a pull-back, if β is of class C^j and F is of class C^ℓ, then $F^*\beta$ is generally of class C^μ, with $\mu = \min(j, \ell - 1)$. Thus, if $j = \ell = 1$, $F^*\beta$ might be only of class C^0, so there is not a well-defined notion of a differential form of class C^1 on a C^1 surface, though such a notion is well defined on a C^2 surface. This problem can be overcome, and one can extend Proposition G.2 to the case where $\overline{M}$ is a C^1 surface and β is a $(k-1)$-form with the property that both β and $d\beta$ are continuous. One can go further and formulate (G.33) for a $(k-1)$-form β with the property that

$$\beta,\ d\beta \in L^\infty(M), \quad \iota^*\beta \in L^\infty(\partial M), \tag{G.38}$$

where $\iota : \partial M \to \overline{M}$ is the natural inclusion, a class of forms that can be shown to be invariant under bi-Lipschitz maps. (It can be shown that the first two conditions in (G.38) imply $\iota^*\beta \in H^{1,1}(\partial M)'$.) We will not go into the details. However, in Appendix I we will present an elementary treatment of (G.33), stated in a more classical language, when M is an open domain in $\mathbb{R}^k$ whose boundary is locally the graph of a Lipschitz function. A far reaching extension, due to H. Federer, can be found in [**Fed**]; see also [**EG**].

The calculus of differential forms has many applications to differential equations, differential geometry, and topology. More on this can be found in [**Spi**] and also in [**T1**] (particularly Chapters 1, 5, and 10). To end this appendix, we make use of the calculus of differential forms to provide simple proofs of some important topological results of Brouwer. The first two results concern *retractions.* If Y is a subset of X, by definition a retraction of X onto Y is a map $\varphi : X \to Y$ such that $\varphi(x) = x$ for all $x \in Y$.

Proposition G.3. *There is no smooth retraction $\varphi : B \to S^{n-1}$ of the closed unit ball B in $\mathbb{R}^n$ onto its boundary S^{n-1}.*

In fact, it is just as easy to prove the following more general result. The approach we use is adapted from [**Kan**].

Proposition G.4. *If $\overline{M}$ is a compact oriented n-dimensional manifold with nonempty boundary ∂M, there is no smooth retraction $\varphi : \overline{M} \to \partial M$.*

Proof. You can pick $\omega \in \Lambda^{n-1}(\partial M)$ to be an $(n-1)$-form on ∂M such that $\int_{\partial M} \omega > 0$. Now apply Stokes' Theorem to $\beta = \varphi^*\omega$. If φ is a retraction, then $\varphi \circ j(x) = x$, where $j : \partial M \hookrightarrow \overline{M}$ is the natural inclusion. Hence $j^*\varphi^*\omega = \omega$, so we have

$$\int\limits_{\partial M} \omega = \int\limits_{M} d\varphi^*\omega. \tag{G.39}$$

But $d\varphi^*\omega = \varphi^* d\omega = 0$, so the integral (G.39) is zero. This is a contradiction, so there can be no retraction.

A simple consequence of this is the famous Brouwer Fixed-Point Theorem.

Theorem G.5. *If $F : B \to B$ is a continuous map on the closed unit ball in $\mathbb{R}^n$, then F has a fixed point.*

Proof. First, an approximation argument shows that if there is a continuous such F without a fixed point, then there is a smooth one, so assume $F : B \to B$ is smooth. We are claiming that $F(x) = x$ for some $x \in B$. If not, then for each $x \in B$ define $\varphi(x)$ to be the endpoint of the ray from $F(x)$ to x, continued until it hits $\partial B = S^{n-1}$. It is clear that φ would be a smooth retraction, contradicting Proposition G.3.

REMARK. Typical proofs of the Brouwer Fixed-Point Theorem use concepts of algebraic topology; cf. [**Spa**]. In fact, the proof of Proposition G.4 contains a germ of de Rham cohomology. See [**T1**], Chapter 1, §19 for more on this.

An integral calculus proof of the Brouwer Fixed-Point Theorem that does not involve differential forms is given in [**DS**], Vol. 1, pp. 467–470. One might compare it with the proof given above.

Change of Variables Revisited

As indicated in Exercise 10 of Chapter 1, the change of variable formula for a one-variable integral,

$$\int_{\varphi(a)}^{\varphi(b)} f(x)\,dx = \int_a^b f(\varphi(x))\varphi'(x)\,dx, \tag{H.1}$$

given f continuous and φ of class C^1, satisfying $\varphi' > 0$, can be established via the Fundamental Theorem of Calculus and the chain rule. By comparison, let us recall the change of variable formula for multiple integrals, as given in Theorem 7.2.

Theorem H.1. *Let $\mathcal{O}$, Ω be open sets on $\mathbb{R}^n$ and let $\varphi : \mathcal{O} \to \Omega$ be a C^1 diffeomorphism. Given $f \in \mathcal{M}^+(\Omega)$ or $f \in \mathcal{L}^1(\Omega, dx)$, we have*

$$\int_{\mathcal{O}} f\big(\varphi(x)\big)\big|\det D\varphi(x)\big|\,dx = \int_{\Omega} f(x)\,dx. \tag{H.2}$$

The proof of Theorem 7.2 given in Chapter 7 was very different from the argument indicated above, and it is our goal here to provide a proof of Theorem H.1 more closely parallel to that of (H.1). Our proof is based on one of P. Lax [**La**], who found a fresh approach to the proof of the multi-dimensional change of variable formula. More precisely, [**La**] established the following result, from which Theorem H.1 can be deduced. Our proof, adapted from [**T2**], will differ from that of [**La**] in that we make use of basic results on differential forms, from Appendix G.

Theorem H.2. *Let $\varphi : \mathbb{R}^n \to \mathbb{R}^n$ be a C^1 map. Assume $\varphi(x) = x$ for $|x| \ge R$. Let f be a continuous function on $\mathbb{R}^n$ with compact support. Then*

$$\int f\big(\varphi(x)\big) \det D\varphi(x)\, dx = \int f(x)\, dx. \tag{H.3}$$

Proof. Via standard approximation arguments, it suffices to prove this when φ is C^2 and $f \in C_0^1(\mathbb{R}^n)$, which we will assume from here on.

To begin, pick $A > 0$ such that $f(x - Ae_1)$ is supported in $\{x : |x| > R\}$, where $e_1 = (1, 0, \dots, 0)$. Also take A large enough that the image of $\{x : |x| \le R\}$ under φ does not intersect the support of $f(x - Ae_1)$. We can set

$$F(x) = f(x) - f(x - Ae_1) = \frac{\partial \psi}{\partial x_1}(x), \tag{H.4}$$

where

$$\psi(x) = \int_0^A f(x - se_1)\, ds, \quad \psi \in C_0^1(\mathbb{R}^n). \tag{H.5}$$

Then we have the following identities involving n-forms:

$$\begin{aligned} \alpha = F(x)\, dx_1 \wedge \cdots \wedge dx_n &= \frac{\partial \psi}{\partial x_1}\, dx_1 \wedge \cdots \wedge dx_n \\ &= d\psi \wedge dx_2 \wedge \cdots \wedge dx_n \\ &= d(\psi\, dx_2 \wedge \cdots \wedge dx_n), \end{aligned} \tag{H.6}$$

i.e., $\alpha = d\beta$, with $\beta = \psi\, dx_2 \wedge \cdots \wedge dx_n$ a compactly supported $(n-1)$-form of class C^1. Now the pull-back of α under φ is given by

$$\varphi^*\alpha = F\big(\varphi(x)\big) \det D\varphi(x)\, dx_1 \wedge \cdots \wedge dx_n. \tag{H.7}$$

Furthermore, the right side of (H.7) is equal to

$$f\big(\varphi(x)\big) \det D\varphi(x)\, dx_1 \wedge \cdots \wedge dx_n - f(x - Ae_1)\, dx_1 \wedge \cdots \wedge dx_n. \tag{H.8}$$

Hence we have

$$\begin{aligned} \int f\big(\varphi(x)\big) \det D\varphi(x)\, dx_1 \cdots dx_n &- \int f(x)\, dx_1 \cdots dx_n \\ &= \int \varphi^*\alpha = \int \varphi^* d\beta = \int d(\varphi^*\beta), \end{aligned} \tag{H.9}$$

where we use the identity

$$\varphi^* d\beta = d(\varphi^*\beta), \tag{H.10}$$

established in Apendix G. On the other hand, a very special case of Stokes' Theorem applies to

$$\varphi^*\beta = \gamma = \sum_j \gamma_j(x)\, dx_1 \wedge \cdots \wedge \widehat{dx_j} \wedge \cdots \wedge dx_n, \tag{H.11}$$

with $\gamma_j \in C_0^1(\mathbb{R}^n)$. Namely

$$d\gamma = \sum_j (-1)^{j-1} \frac{\partial \gamma_j}{\partial x_j}\, dx_1 \wedge \cdots \wedge dx_n, \tag{H.12}$$

and hence, by the Fundamental Theorem of Calculus,

$$\int d\gamma = 0. \tag{H.13}$$

This gives the desired identity (H.3), from (H.9).

We make some remarks on Theorem H.2. Note that φ is not assumed to be one-to-one or onto. In fact, as noted in [**La**], the identity (H.3) implies that such φ must be onto, and this has important topological implications. In more detail, it is readily seen that the range of φ must be closed, so if φ is not onto, it must omit some nonempty open set U. One can take a function f supported in U and with a positive integral and obtain a contradiction in (H.3), since the left side would have to vanish. One then easily observes that, if there is a smooth retraction of the ball B onto its boundary, such a φ would exist that is not onto. This contradiction then gives another proof of the Brouwer no-retraction result, Proposition G.3.

We recall that, if one puts absolute values around $\det D\varphi(x)$ in (H.3), the appropriate formula is

$$\int f(\varphi(x))\, \big|\det D\varphi(x)\big|\, dx = \int f(x)\, n(x)\, dx, \tag{H.14}$$

where $n(x) = \#\{y : \varphi(y) = x\}$. This result was proven in Appendix F.

We want to extend Theorem H.2 to treat more general maps φ and functions f. The fact that Theorem H.2 holds for maps that do not have to be diffeomorphisms simplifies the task of making the extension to more singular maps. We begin with the following:

Proposition H.3. *Let $\varphi : \mathbb{R}^n \to \mathbb{R}^n$ be continuous. Assume $\varphi(x) = x$ for $|x| \ge R$. Furthermore, assume*

$$D\varphi \in L^n_{\text{loc}}(\mathbb{R}^n), \tag{H.15}$$

where $D\varphi$ is the weak derivative of φ. Then (H.3) *holds for all continuous f with compact support.*

Proof. Using a mollifier, we can produce C^∞ maps $\varphi_\nu : \mathbb{R}^n \to \mathbb{R}^n$, such that $\varphi_\nu(x) = x$ for $|x| \geq R+1$, such that $\varphi_\nu \to \varphi$ uniformly, and such that

$$D\varphi_\nu \to D\varphi \quad \text{in} \quad L^n(K)\text{-norm}, \tag{H.16}$$

for all compact $K \subset \mathbb{R}^n$. By Theorem H.2 we have

$$\int f\big(\varphi_\nu(x)\big) \det D\varphi_\nu(x)\, dx = \int f(x)\, dx \tag{H.17}$$

for each ν. Under our hypotheses, $f \circ \varphi_\nu \to f \circ \varphi$ uniformly on $\mathbb{R}^n$, and we can pick a compact $K \subset \mathbb{R}^n$ containing the support of $f \circ \varphi_\nu$ for all ν. By (H.16) we have

$$\det D\varphi_\nu \longrightarrow \det D\varphi \quad \text{in} \quad L^1(K), \tag{H.18}$$

and hence

$$f \circ \varphi_\nu \, \det D\varphi_\nu \longrightarrow f \circ \varphi \, \det D\varphi \quad \text{in} \quad L^1(\mathbb{R}^n). \tag{H.19}$$

Thus we can pass to the limit in (H.17) to obtain (H.3) in this setting.

We mention that, if the hypothesis (H.15) is strengthened to $D\varphi \in L^p_{\text{loc}}(\mathbb{R}^n)$, for some $p > n$, then, by Proposition 11.6, φ is differentiable a.e. and its pointwise derivative coincides a.e. with the weak derivative. In light of this, one can compare Proposition H.3 (and Proposition H.4 below) with Propositions 7.4 and 7.6.

Now we establish a result valid for more singular f.

Proposition H.4. *Retain the hypotheses of Proposition H.3. Furthermore, assume*

$$\det D\varphi(x) \geq 0 \quad \textit{for a.e.} \quad x. \tag{H.20}$$

Then (H.3) *holds for all positive measurable f.*

Proof. We have (H.3) for compactly supported continuous f. An application of the Monotone Convergence Theorem then gives (H.3) for all characteristic functions $f = \chi_K$, $K \subset \mathbb{R}^n$ compact. A second application of monotone convergence gives (H.3) for all characteristic functions $f = \chi_U$, $U \subset \mathbb{R}^n$ open and bounded. Now if $S \subset \mathbb{R}^n$ is a bounded measurable set, we can write $K_\nu \nearrow S_0$, $U_\nu \searrow S_1$, $S_0 \subset S \subset S_1$, with $\text{meas}(S_0) = \text{meas}(S_1) = A$, say, and monotone convergence gives

$$\begin{aligned} &\int \chi_{S_0}\big(\varphi(x)\big) \det D\varphi(x)\, dx = A, \\ &\int \chi_{S_1}\big(\varphi(x)\big) \det D\varphi(x)\, dx = A, \end{aligned} \tag{H.21}$$

so (H.3) holds for $f = \chi_S$. Hence (H.3) holds for positive simple functions f. By one more application of monotone convergence it hence holds for all positive measurable functions (with the standard convention that $+\infty \cdot 0 = 0$).

We now discuss how Theorem H.1 can be deduced from Theorem H.2 (or more precisely from the extension to allow arbitrary $f \in \mathcal{M}^+(\Omega)$, as in Proposition H.4). It suffices to obtain (H.2) for $f \in \mathcal{M}^+(\Omega)$ supported in some compact $K \subset \Omega$, by the Monotone Convergence Theorem. Also, to prove Theorem H.1, it suffices to assume $\det D\varphi > 0$. We will use the following lemma.

Lemma H.5. *In the setting of Theorem H.1 (and with* $\det D\varphi > 0$*), given* $p \in \Omega$*, there is a neighborhood* U *of* p *and a* C^1 *map* $\Phi : \mathbb{R}^n \to \mathbb{R}^n$ *such that* $\Phi = \varphi$ *on* U*,* $\Phi(x) = x$ *for* $|x|$ *large, and such that*

$$\operatorname{supp} f \subset U \Longrightarrow f(\varphi(x)) = f(\Phi(x)). \tag{H.22}$$

Granted the lemma, we can proceed as follows. Given $f \in \mathcal{M}^+(\Omega)$, $\operatorname{supp} f \subset K$, cover K with a finite number of neighborhoods U_j as in Lemma H.5, and write $f = \sum_j f_j$, $\operatorname{supp} f_j \subset U_j$. Also, let Φ_j have the obvious significance. By Theorem H.2 we have

$$\int f_j(\Phi_j(x)) \det D\Phi_j(x)\, dx = \int f_j(x)\, dx. \tag{H.23}$$

But we also have

$$\int f_j(\Phi_j(x)) \det D\Phi_j(x)\, dx = \int f_j(\varphi(x)) \det D\varphi(x)\, dx. \tag{H.24}$$

Now summing over j gives (H.2).

The proof of Lemma H.5 is essentially an exercise in differential calculus. We indicate how to do this. Say $\det D\varphi(p) = A \in Gl_+(n, \mathbb{R})$, i.e., $A \in Gl(n, \mathbb{R})$ and $\det A > 0$. Let us assume for simplicity that $p = 0$ and $\varphi(p) = 0$. We can write

$$\Psi(x) = \beta(x)\varphi(x) + (1 - \beta(x))Ax,$$

where $\beta \in C_0^\infty(\mathbb{R}^n)$ has support in a small neighborhood of p and $\beta \equiv 1$ on a smaller neighborhood U, and use Corollary B.5 to show that $\Psi : \mathbb{R}^n \to \mathbb{R}^n$ is a diffeomorphism, for conveniently chosen β. Now we can exploit the fact that $Gl_+(n, \mathbb{R})$ is connected (cf. (7.95)) to alter Ψ to Φ, as follows.

Pick a smooth path $\Gamma : [0,1] \to Gl_+(n,\mathbb{R})$, with $\Gamma(t) = A$ for $t \in [0,1/4]$ and $\Gamma(t) = I$ for $t \in [3/4,1]$. Pick $B > 0$, sufficiently large that, for each $t \in [0,1]$, the image of $\{x : |x| < B\}$ under $\Gamma(t)$ contains $\operatorname{supp}\beta$, and set

$$\begin{aligned} \Phi(x) = \Psi(x) &\quad \text{for } |x| \le B, \\ \Gamma(t)x &\quad \text{for } |x| = B + t,\ 0 \le t \le 1, \\ x &\quad \text{for } |x| \ge B + 1. \end{aligned} \tag{H.25}$$

This works.

The Gauss-Green Formula on Lipschitz Domains

Let Ω be a bounded open subset of $\mathbb{R}^n$. We say Ω is Lipschitz if, for each $p \in \partial\Omega$, there is a neighborhood U of p in $\mathbb{R}^n$, a rotation of coordinate axes, and a Lipschitz function $u : \mathcal{O} \to \mathbb{R}$, defined on an open set $\mathcal{O} \subset \mathbb{R}^{n-1}$, such that

$$\Omega \cap U = \{x \in \mathbb{R}^n : x_n \le u(x'), x' \in \mathcal{O}\} \cap U, \tag{I.1}$$

where $x = (x', x_n)$, $x' = (x_1, \dots, x_{n-1})$.

Our goal is to prove the following version of the Gauss-Green formula.

Theorem I.1. *If Ω is a bounded Lipschitz domain in $\mathbb{R}^n$ and $F = (f_1, \dots, f_n)$ is a C^1 vector field on $\overline{\Omega}$, then*

$$\int_{\Omega} (\operatorname{div} F)\, dx = \int_{\partial\Omega} N \cdot F\, dS. \tag{I.2}$$

Here $dS = d\mathcal{H}^{n-1}$ is $(n-1)$-dimensional Hausdorff measure, studied in detail in Chapter 12, and N is the outgoing unit normal to $\partial\Omega$, defined by (12.91)–(12.92), which is well defined for $\mathcal{H}^{n-1}$-a.e. $x \in \partial\Omega$, as discussed in Exercises 7–8 of Chapter 12. In fact, for a.e. $x' \in \mathcal{O}$, the function u is differentiable at x', by Rademacher's Theorem, and, at $x = (x', u(x')) \in \partial\Omega$, we have

$$N = (1 + |\nabla u|^2)^{-1/2} (-\nabla u, 1). \tag{I.3}$$

Theorem I.1 is a consequence of the following result.

Proposition I.2. *In the setting of Theorem I.1, if $f \in C^1(\overline{\Omega})$ and e is an element of $\mathbb{R}^n$,*

$$\int_{\Omega} e \cdot \nabla f(x)\, dx = \int_{\partial\Omega} (e \cdot N) f\, dS. \tag{I.4}$$

In fact, taking e_j to be the standard orthonormal basis of $\mathbb{R}^n$, replacing e by e_j and f by f_j in (I.4) and summing, one obtains (I.2).

To prove (I.4), after applying a partition of unity, we may as well suppose f is supported in such a neighborhood U as appears in (I.1). Then we have

$$\begin{aligned} \int_{\Omega} \frac{\partial f}{\partial x_n}\, dx &= \int_{\mathcal{O}} \Big(\int_{x_n \le u(x')} \partial_n f(x', x_n)\, dx_n \Big)\, dx' \\ &= \int_{\mathcal{O}} f(x', u(x'))\, dx' \\ &= \int_{\partial\Omega} (e_n \cdot N) f\, dS. \end{aligned} \tag{I.5}$$

The first identity in (I.5) follows from Fubini's Theorem, the second identity from the Fundamental Theorem of Calculus, and the third identity from the identification $dS = \sqrt{1 + |\nabla u|^2}\, dx'$, established in Proposition 12.10, and the formula (I.3). This establishes (I.4) for $e = e_n$, when f is supported in a region for which (I.1) holds.

Now if Ω is Lipschitz, the region $\Omega \cap U$ has a representation of the form (I.1) in new coordinates obtained by any rotation sufficiently close to the identity. Hence the identity (I.4) holds when $e = e_n$ is replaced by any sufficiently close element of $\mathbb{R}^n$. In particular, it works for $e = e_n + ae_j$, for $1 \le j \le n-1$ and for $|a|$ sufficiently small. Thus, we have

$$\int_{\Omega} (e_n + ae_j) \cdot \nabla f(x)\, dx = \int_{\partial\Omega} (e_n + ae_j) \cdot N\, f\, dS. \tag{I.6}$$

If we subtract (I.5) from this and divide the result by a, we obtain (I.4) for $e = e_j$, for all j, and hence (I.4) holds in general. This completes the proof.

Bibliography

[AG] M. Adams and V. Guillemin, *Measure Theory and Probability*, Wadsworth, Monterey, California, 1986.

[BS] R. Bartle and D. Sherbert, *Introduction to Real Analysis*, J. Wiley, New York, 1992.

[Bil] P. Billingsley, *Probability and Measure*, Wiley, New York, 1979.

[Car] L. Carleson, *Selected Problems on Exceptional Sets*, Van Nostrand, Princeton, 1968.

[Che] J. Cheeger, *Differentiability of Lipschitz functions on metric spaces*, Geom. Funct. Anal. **9** (1999), 428–517.

[Cho] G. Choquet, *Theory of Capacities*, Ann. Inst. Fourier, Grenoble **5** (1953-54), 131–292.

[Doo] J. Doob, *Stochastic Processes*, Wiley, New York, 1953.

[Dud] R. Dudley, *Real Analysis and Probability*, Cambridge Univ. Press, Cambridge, 2002.

[Dug] J. Dugundji, *Topology*, Allyn and Bacon, New York, 1966.

[DS] N. Dunford and J. Schwartz, *Linear Operators*, Wiley, New York, 1958.

[Dur] R. Durrett, *Brownian Motion and Martingales in Analysis*, Wadsworth, Belmont, California, 1984.

[Ed] G. Edgar, *Measure, Topology, and Fractal Geometry*, Springer, New York, 1990.

[EG] L. Evans and R. Gariepy, *Measure Theory and Fine Properties of Functions*, CRC Press, Boca Raton, 1992.

[Fal] K. Falconer, *The Geometry of Fractal Sets*, Cambridge Univ. Press, 1985.

[Fed] H. Federer, *Geometric Measure Theory*, Springer-Verlag, Berlin, 1969.

[Fef] C. Fefferman, *A sharp form of Whitney's extension theorem*, Annals of Math. **161** (2005), 509–577.

[Fol] G. Folland, *Real Analysis: Modern Techniques and Applications*, Wiley-Interscience, New York, 1984.

[Gar] A. Garsia, *Topics in Almost Everywhere Convergence*, Markham, Chicago, 1970.

[Hal] P. Halmos, *Measure Theory*, van Nostrand, New York, 1950.

[HaS] R. Hardt and L. Simon, *Seminar on Geometric Measure Theory*, Birkhauser, Basel, 1986.

[Haw] T. Hawkins, *Lebesgue's Theory of Integration*, Univ. of Wisconsin Press, 1970.

[Hei] J. Heinonen, *Lectures on Analysis on Metric Spaces*, Springer-Verlag, New York, 2001.

[HS] E. Hewitt and K. Stromberg, *Real and Abstract Analysis*, Springer-Verlag, New York, 1965.

[Hut] J. Hutchinson, *Fractals and self-similarity*, Indiana Univ. Math. J. **30** (1981), 713–747.

[IMc] K. Ito and H. McKean, *Diffusion Processes and Their Sample Paths*, Springer-Verlag, New York, 1974.

[Jac] K. Jacobs, *Measure and Integral*, Academic Press, New York, 1978.

[Kan] Y. Kannai, *An elementary proof of the no-retraction theorem*, Amer. Math. Monthly **88** (1981), 264–268.

[KS] I. Karatzas and S. Shreve, *Brownian Motion and Stochastic Calculus*, Springer-Verlag, New York, 1988.

[Kr] U. Krengel, *Ergodic Theorems*, Walter de Gruyter, Berlin, 1985.

[La] P. Lax, *Change of variables in multiple integrals*, Amer. Math. Monthly **106** (1999), 497–501.

[La2] P. Lax, *Functional Analysis*, Wiley, New York, 2002.

[Leb] H. Lebesgue, *Leçons sur l'Intégration (reprint of 1928 orig.)*, Chelsea, New York, 1973.

[LL] E. Lieb and M. Loss, *Analysis*, Amer. Math. Soc., Providence, R. I., 1997.

[Lo] L. Loomis, *Abstract Harmonic Analysis*, van Nostrand, New York, 1953.

[LS] L. Loomis and S. Sternberg, *Advanced Calculus*, Addison-Wesley, New York, 1968.

[McK] H. McKean, *Stochastic Integrals*, Academic Press, New York, 1969.

[Mdb] B. Mandelbrot, *The Fractal Geometry of Nature*, W. H. Freeman, San Francisco, 1977.

[Mey] P. Meyer, *Probability and Potentials*, Blaisdell, Boston, 1966.

[Mor] F. Morgan, *Geometric Measure Theory: A Beginner's Guide*, Academic Press, Boston, 1988.

[Mun] J. Munkres, *Topology, a First Course*, Prentice-Hall, Englewood Cliffs, NJ, 1975.

[Nat] I. Natanson, *Theory of Functions of a Real Variable, Vols. 1–2*, F. Ungar, New York, 1955.

[Nel] E. Nelson, *Feynman integrals and the Schrödinger equation*, J. Math. Phys. **5** (1964), 332–343.

[PR] H. O. Peitgen and P. Richter, *The Beauty of Fractals*, Springer-Verlag, New York, 1986.

[Pet] K. Petersen, *Ergodic Theory*, Cambridge Univ. Press, 1983.

[Pi] J. Pierpont, *The Theory of Functions of a Real Variable, Vols. 1–2*, Dover, New York, 1959.

[Ray] D. Ray, *Sojourn times and the exact Hausdorff measure of the sample path for planar Brownian motion*, Trans. AMS **106** (1963), 436–444.

[RS] M. Reed and B. Simon, *Methods of Mathematical Physics*, Academic Press, New York, Vols. 1–2, 1975; Vols. 3–4, 1978..

[RN] F. Riesz and B. Sz. Nagy, *Functional Analysis*, Ungar, New York, 1955.

[Rog] C. Rogers, *Hausdorff Measures*, Cambridge Univ. Press, 1970.

[Roh] V. Rohlin, *On the fundamental ideas of measure theory*, Functional Analysis and Measure Theory, AMS Transl., Vol. 10, Amer. Math. Soc., Providence, RI, 1962, pp. 1–54.

[RST] J. Rosenblatt, D. Stroock, and M. Taylor, *Group actions and singular martingales*, Ergodic Theory and Dyn. Sys. **23** (2003), 293–305.

[Roy] H. Royden, *Real Analysis*, Macmillan, New York, 1963.

[Ru] W. Rudin, *Real and Complex Analysis*, McGraw-Hill, New York, 1976.

[Sak] S. Saks, *Theory of the Integral*, Warsaw, 1937.

[Sat] K. Sato, *Lévy Processes and Infinitely Divisible Distributions*, Cambridge Univ. Press, Cambridge, 1999.

[Sch] L. Schwartz, *Théorie des Distributions*, Herman, Paris, 1950.

[Si] B. Simon, *Functional Integration and Quantum Physics*, Academic Press, New York, 1979.

[Spa] E. Spanier, *Algebraic Topology*, McGraw-Hill, New York, 1966.

[Spi] M. Spivak, *A Comprehensive Introduction to Differential Geometry, Vols. 1–4*, Publish or Perish Press, Berkeley, CA, 1979.

[St] E. Stein, *Singular Integrals and Differentiability Properties of Functions*, Princeton Univ. Press, Princeton, NJ, 1970.

[StS] E. Stein and R. Shakarchi, *Real Analysis*, Princeton Univ. Press, Princeton, NJ, 2005.

[StW] E. Stein and G. Weiss, *Introduction to Fourier Analysis on Euclidean Spaces*, Princeton Univ. Press, Princeton, NJ, 1971.

[Stb] S. Sternberg, *Lectures on Differential Geometry*, Prentice Hall, New Jersey, 1964.

[T1] M. Taylor, *Partial Differential Equations, Vols. 1–3*, Springer-Verlag, New York, 1996.

[T2] M. Taylor, *Differential forms and the change of variable formula for multiple integrals*, J. Math. Anal. Appl. **268** (2002), 378–383.

[TS] S. J. Taylor, *The exact Hausdorff measure of the sample path for planar Brownian motion*, Proc. Cambridge Phil. Soc. **60** (1964), 253–258.

[Wh] H. Whitney, *Geometric Integration Theory*, Princeton Univ. Press, Princeton, NJ, 1957.

[Yo] K. Yosida, *Functional Analysis*, Springer-Verlag, New York, 1965.

[Zie] W. Ziemer, *Weakly Differentiable Functions*, Springer-Verlag, New York, 1989.

[Zim] R. Zimmer, *Ergodic Theory and Semisimple Groups*, Birkhauser, Boston, 1984.

Symbol Index

Subjext Index

Titles in This Series

76 **Michael E. Taylor,** Measure theory and integration, 2006
75 **Peter D. Miller,** Applied asymptotic analysis, 2006
74 **V. V. Prasolov,** Elements of combinatorial and differential topology, 2006
73 **Louis Halle Rowen,** Graduate algebra: Commutative view, 2006
72 **R. J. Williams,** Introduction the the mathematics of finance, 2006
70 **Seán Dineen,** Probability theory in finance, 2005
69 **Sebastián Montiel and Antonio Ros,** Curves and surfaces, 2005
68 **Luis Caffarelli and Sandro Salsa,** A geometric approach to free boundary problems, 2005
67 **T.Y. Lam,** Introduction to quadratic forms over fields, 2004
66 **Yuli Eidelman, Vitali Milman, and Antonis Tsolomitis,** Functional analysis, An introduction, 2004
65 **S. Ramanan,** Global calculus, 2004
64 **A. A. Kirillov,** Lectures on the orbit method, 2004
63 **Steven Dale Cutkosky,** Resolution of singularities, 2004
62 **T. W. Körner,** A companion to analysis: A second first and first second course in analysis, 2004
61 **Thomas A. Ivey and J. M. Landsberg,** Cartan for beginners: Differential geometry via moving frames and exterior differential systems, 2003
60 **Alberto Candel and Lawrence Conlon,** Foliations II, 2003
59 **Steven H. Weintraub,** Representation theory of finite groups: algebra and arithmetic, 2003
58 **Cédric Villani,** Topics in optimal transportation, 2003
57 **Robert Plato,** Concise numerical mathematics, 2003
56 **E. B. Vinberg,** A course in algebra, 2003
55 **C. Herbert Clemens,** A scrapbook of complex curve theory, second edition, 2003
54 **Alexander Barvinok,** A course in convexity, 2002
53 **Henryk Iwaniec,** Spectral methods of automorphic forms, 2002
52 **Ilka Agricola and Thomas Friedrich,** Global analysis: Differential forms in analysis, geometry and physics, 2002
51 **Y. A. Abramovich and C. D. Aliprantis,** Problems in operator theory, 2002
50 **Y. A. Abramovich and C. D. Aliprantis,** An invitation to operator theory, 2002
49 **John R. Harper,** Secondary cohomology operations, 2002
48 **Y. Eliashberg and N. Mishachev,** Introduction to the h-principle, 2002
47 **A. Yu. Kitaev, A. H. Shen, and M. N. Vyalyi,** Classical and quantum computation, 2002
46 **Joseph L. Taylor,** Several complex variables with connections to algebraic geometry and Lie groups, 2002
45 **Inder K. Rana,** An introduction to measure and integration, second edition, 2002
44 **Jim Agler and John E. McCarthy,** Pick interpolation and Hilbert function spaces, 2002
43 **N. V. Krylov,** Introduction to the theory of random processes, 2002
42 **Jin Hong and Seok-Jin Kang,** Introduction to quantum groups and crystal bases, 2002
41 **Georgi V. Smirnov,** Introduction to the theory of differential inclusions, 2002
40 **Robert E. Greene and Steven G. Krantz,** Function theory of one complex variable, third edition, 2006
39 **Larry C. Grove,** Classical groups and geometric algebra, 2002
38 **Elton P. Hsu,** Stochastic analysis on manifolds, 2002

TITLES IN THIS SERIES

37 **Hershel M. Farkas and Irwin Kra,** Theta constants, Riemann surfaces and the modular group, 2001

36 **Martin Schechter,** Principles of functional analysis, second edition, 2002

35 **James F. Davis and Paul Kirk,** Lecture notes in algebraic topology, 2001

34 **Sigurdur Helgason,** Differential geometry, Lie groups, and symmetric spaces, 2001

33 **Dmitri Burago, Yuri Burago, and Sergei Ivanov,** A course in metric geometry, 2001

32 **Robert G. Bartle,** A modern theory of integration, 2001

31 **Ralf Korn and Elke Korn,** Option pricing and portfolio optimization: Modern methods of financial mathematics, 2001

30 **J. C. McConnell and J. C. Robson,** Noncommutative Noetherian rings, 2001

29 **Javier Duoandikoetxea,** Fourier analysis, 2001

28 **Liviu I. Nicolaescu,** Notes on Seiberg-Witten theory, 2000

27 **Thierry Aubin,** A course in differential geometry, 2001

26 **Rolf Berndt,** An introduction to symplectic geometry, 2001

25 **Thomas Friedrich,** Dirac operators in Riemannian geometry, 2000

24 **Helmut Koch,** Number theory: Algebraic numbers and functions, 2000

23 **Alberto Candel and Lawrence Conlon,** Foliations I, 2000

22 **Günter R. Krause and Thomas H. Lenagan,** Growth of algebras and Gelfand-Kirillov dimension, 2000

21 **John B. Conway,** A course in operator theory, 2000

20 **Robert E. Gompf and András I. Stipsicz,** 4-manifolds and Kirby calculus, 1999

19 **Lawrence C. Evans,** Partial differential equations, 1998

18 **Winfried Just and Martin Weese,** Discovering modern set theory. II: Set-theoretic tools for every mathematician, 1997

17 **Henryk Iwaniec,** Topics in classical automorphic forms, 1997

16 **Richard V. Kadison and John R. Ringrose,** Fundamentals of the theory of operator algebras. Volume II: Advanced theory, 1997

15 **Richard V. Kadison and John R. Ringrose,** Fundamentals of the theory of operator algebras. Volume I: Elementary theory, 1997

14 **Elliott H. Lieb and Michael Loss,** Analysis, 1997

13 **Paul C. Shields,** The ergodic theory of discrete sample paths, 1996

12 **N. V. Krylov,** Lectures on elliptic and parabolic equations in Hölder spaces, 1996

11 **Jacques Dixmier,** Enveloping algebras, 1996 Printing

10 **Barry Simon,** Representations of finite and compact groups, 1996

9 **Dino Lorenzini,** An invitation to arithmetic geometry, 1996

8 **Winfried Just and Martin Weese,** Discovering modern set theory. I: The basics, 1996

7 **Gerald J. Janusz,** Algebraic number fields, second edition, 1996

6 **Jens Carsten Jantzen,** Lectures on quantum groups, 1996

5 **Rick Miranda,** Algebraic curves and Riemann surfaces, 1995

4 **Russell A. Gordon,** The integrals of Lebesgue, Denjoy, Perron, and Henstock, 1994

3 **William W. Adams and Philippe Loustaunau,** An introduction to Gröbner bases, 1994

2 **Jack Graver, Brigitte Servatius, and Herman Servatius,** Combinatorial rigidity, 1993

1 **Ethan Akin,** The general topology of dynamical systems, 1993